RECONCEPTUALIZING STEM EDUCATION

Reconceptualizing STEM Education explores and maps out research and development ideas and issues around five central practice themes: Systems Thinking; Model-Based Reasoning; Quantitative Reasoning; Equity, Epistemic, and Ethical Outcomes; and STEM Communication and Outreach. These themes are aligned with the comprehensive agenda for the reform of science and engineering education set out by the 2015 PISA Framework, the US Next Generation Science Standards and the US National Research Council's A Framework for K-12 Science Education. The new practice-focused agenda has implications for the redesign of preK-12 education for alignment of curriculum-instruction-assessment; STEM teacher education and professional development; postsecondary, further, and graduate studies; and out-of-school informal education. In each section, experts set out powerful ideas followed by two eminent discussant responses that both respond to and provoke additional ideas from the lead papers. In the associated website (http://waterbury.psu.edu/summit) highly distinguished, nationally recognized STEM education scholars and policymakers engage in deep conversations and considerations addressing core practices that guide STEM education.

Richard A. Duschl is the Kenneth B. Waterbury Chaired Professor in Secondary Education, Department of Curriculum and Instruction, College of Education, The Pennsylvania State University, USA.

Amber S. Bismack is a Ph.D. student, Department of Educational Studies (Science Education), School of Education, The University of Michigan, USA.

Teaching and Learning in Science Series
Norman G. Lederman, Series Editor

Berry/Friedrichsen/Loughran
Re-examining Pedagogical Content Knowledge in Science Education

Taber
Student Thinking and Learning in Science: Perspectives on the Nature and Development of Learners' Ideas

Rennie/Venville/Wallace (Eds.)
Integrating Science, Technology, Engineering, and Mathematics: Issues, Reflections, and Ways Forward

Rosenblatt
Rethinking the Way We Teach Science: The Interplay of Content, Pedagogy, and the Nature of Science

Linder/Östman/Roberts/Wickman/Erickson/MacKinnon (Eds.)
Exploring the Landscape of Scientific Literacy

Abell/Appleton/Hanuscin
Designing and Teaching the Elementary Science Methods Course

Akerson (Ed.)
Interdisciplinary Language Arts and Science Instruction in Elementary Classrooms: Applying Research to Practice

Wickman
Aesthetic Experience in Science Education: Learning and Meaning-Making as Situated Talk and Action

Duschl & Bismack (editors)
Reconceptualizing STEM Education

Visit **www.routledge.com/education** for additional information on titles in the Teaching and Learning in Science Series

RECONCEPTUALIZING STEM EDUCATION

The Central Role of Practices

Edited by Richard A. Duschl
and Amber S. Bismack

Routledge
Taylor & Francis Group
NEW YORK AND LONDON

First published 2016
by Routledge
711 Third Avenue, New York, NY 10017

and by Routledge
2 Park Square, Milton Park, Abingdon, Oxon OX14 4RN

Routledge is an imprint of the Taylor & Francis Group, an informa business

Library of Congress Cataloging-in-Publication Data
Waterbury Summit (Conference) (2013 : Pennsylvania State University)
Reconceptualizing STEM education : the central role of practices / edited by Richard A. Duschl and Amber S. Bismack.
pages cm.— (Teaching and learning in science)
Papers originally presented at the Waterbury Summit held August 7–9, 2013 at the Pennsylvania State University.
Includes bibliographical references.
1. Science—Study and teaching—United States—Congresses. 2. Technology—Study and teaching—United States—Congresses. 3. Engineering—Study and teaching—United States—Congresses. 4. Mathematics—Study and teaching—United States—Congresses. I. Duschl, Richard A. (Richard Alan), 1951– editor of compilation. II. Bismack, Amber S., editor of compilation. III. Title.
Q181.W279 2016
507.1'073—dc23
2015015854

ISBN: 978-1-138-90103-2 (hbk)
ISBN: 978-1-138-90104-9 (pbk)
ISBN: 978-1-315-70032-8 (ebk)

Typeset in Bembo
by Apex CoVantage, LLC

Richard A. Duschl
To Stephanie, we are there, and I am so pleased to have you at my side and as my guide.

Amber S. Bismack
To Brian, thank you for your support and encouragement, and I am so happy to spend life hand in hand with you.

CONTENTS

PREFACE

The edited volume reports on the proceedings of the Waterbury Summit held during August 2013. Approximately 50 scholars gathered for three days to explore thematic practices associated with learning and teaching STEM disciplines—Science, Technology, Engineering, and Mathematics. The participants included Penn State faculty and graduate students and nationally recognized scholars brought together to examine K-16 STEM education. Names, bios, and the Summit sessions appear immediate after the Preface. The broad goal was to articulate research and development goals with respect to the reform agendas that advocate a shift to practices in STEM learning and teaching. The shift to practices is about getting students to use knowledge, to reason, to model, to explain, to critique, and to communicate. The Summit was a conversational conference interspersed with panels, discussions, and several social events providing ample time for exploring five thematic practices judged to be viable 'common ground' among the STEM disciplines.

The US *Next Generation Science Standards (NGSS)* and the US National Research Council's *A Framework for K-12 Science Education* herald "a new vision of science in education." The *Framework* and *NGSS* set out a comprehensive agenda for the reform of science and engineering education coordinated around three dimensions: Science and Engineering Practices, Crosscutting Concepts, and Disciplinary Core Ideas. The new agenda has implications for 1) the redesign of preK-12 education for alignment of curriculum-instruction-assessment, 2) STEM teacher education and professional development, 3) post-secondary STEM education, 4) informal STEM education, and 5) graduate studies in STEM education. Hence, the audience for the book is preK-16 education stakeholders nationally and internationally. Additionally, and importantly, there are poorly understood implications for how policy decisions will be impacted and how higher education, informal science education, and public media will contribute to the public's engagement in science.

Advocates for integrating STEM disciplines must attend to the dual problems of bridging across and within the S, T, E, and M. This bridging involves more than just highlighting the S, T, E, and M components of a problem or lesson, but more intentionally emphasizing the aspects—or in this case *practices*—that cross all four fields. When adopting a practices orientation, important and distinctive epistemic structures and cognitive demands occur when building, refining, critiquing, and/or communicating STEM knowledge claims and methods. Is there a common ground to be had? I conducted an unscientific poll of colleagues from the STEM disciplines to test the idea that across and within the STEM diciplines "*Systems Thinking*," "*Model-based Reasoning*," and "*Quantiative Reasoning*" could be considered common practices. Person after person agreed! These are by no means the only practices that cross the four STEM fields, but they are unique in that they represent an understanding of each field's content and structure necessary to engage with the interdisciplinary practices. This compares to other practices, such as communicating information, which does occur in all four STEM fields, but in ways that are unique to those distinct fields and not necessarily interdisciplinary. To these, the Summit Advisory Board—Penn State colleagues Greg Kelly, Tom Keiter, Nancy Tuana, and Carla Zembal-Saul—suggested two additional practice themes: "*Equity, Epistemic, and Ethical Outcomes*" and "*STEM Communication and Outreach.*"

The volume explores and maps out research and development ideas and issues around the five themes for the reconceptualization of STEM education. All writers, respondents, panelists, and participants received the plenary paper "STEM Education Standards and Reforms: An Overview" that sets out recent policy and research agendas and appears in a modified form as Chapter 1. Six months prior to the Summit, I met up with Jim Greeno and Drew Gitomer in Pittsburgh to flesh out a broad based research agenda predicated on the issues and perspectives raised in the plenary paper. Jim and Drew's sage comments were also shared with Summit attendees and appear as a coda in Chapter 1. Each of the five themes of the book includes two commissioned papers that set out poignant ideas from experts followed by two eminent discussant responses that provoked additional ideas and audience responses to the lead papers. The four Summit commissioned Poster Presentations appear at the end of Themes I, II, III, and IV.

- Theme I, *Systems Thinking* examines the importance of understanding a system as an interrelated and interdependent web of components and highlights the relevance of developing systems thinking skills among K-16 students and taking a systems approach in STEM education interventions. **Poster Presentation I**: Virtual and Authentic Web-based Ecological Inquiries.
- Theme II, *Model-based Reasoning*, a central important form of reasoning in STEM disciplines, highlights the need for science curriculum/course redesigns and for new designs for both initial STEM teacher education and

ongoing professional development. **Poster Presentation II**: Cultivating Systems Thinking and Engagement through an Intercollege Minor in Sustainability Leadership.

- Theme III, *Quantitative Reasoning*, is taken up as an important way of thinking about the core practices of data modeling, measurement, analyzing data, and representing data that exist across STEM disciplines. **Poster Presentation III**: Fundamentals of Climate Science.
- Theme IV, *Equity, Epistemic, & Ethical Outcomes* are scrutinized to argue for a greater emphasis on and place for scientific ethics and values in STEM programs to inform STEM decision making and facilitate understandings of knowledge about science. **Poster Presentation IV**: Fossil Mammals and Biotic Response to Climate Change.
- Theme V, *STEM Communication & Outreach*, draws on the importance of using technology as a tool to support teachers' learning of STEM fields, to engage the public in productive science discussions, and to influence policy decisions regarding STEM education.

In the associated Summit website (waterbury.psu.edu), edited video presentions can be viewed of the highly distinguished, nationally recognized STEM education scholars and policymakers engaging in deep conversations addressing core practices that guide STEM education. Shirley Malcolm, American Association for the Advancement of Science (AAAS), gave the opening Keynote Address and Richard Duschl the closing Keynote Address. Leona Schauble, Vanderbilt University, and Paul LeMahieu, Carnegie Institute for the Advancement of Teaching, provided summation statements of the conference. The panel presentations and subsequent Q&A sessions represent some of the most insightful deliberations of the Summit.

The Waterbury Summit was a year in the making and involved a large contingent of Pennsylvania State University faculty, staff, and students. Thanks go out to the Summit Advisory Board, Greg, Nancy, Carla, and Tom. Yann Shio Ong and Armend Tahirsylaj provided critical support with Summit arrangements, as well as making major contribution to the Summit Summary Chapter. Logistics were masterfully handled by Chriss Schultz and Karen Dinunzio from the Penn State Conferences & Institutes Office. Thanks go out to Kevin Conaway and his Penn State Public Broadcasting crew for the expert video recording of the Summit and to Dean David Monk, College of Education, for his support and guidance. And special acknowledgements go to Amber Bismack who started at Penn State as my Research Assistant for the Summit, and after significant contributions to the plenary paper, the coordination of the Summit, and leading the writing of the summary chapter it was my honor to have Amber join me in editing the Summit volume and securing the book contract with Routledge. Amber and I are fortunate to have had Naomi Silverman as Commissioning Editor and Norm Lederman as Series Editor.

Richard A. Duschl

INTRODUCTION

Coordinating PreK-16 STEM Education Research and Practices for Advancing and Refining Reform Agendas

Richard A. Duschl, Amber S. Bismack, James Greeno, and Drew H. Gitomer

Among the many challenges facing the transformation of PreK-16 science teaching and learning is that of coordinating conceptual, epistemological, and social competencies and learning goals in disciplinary domains. From very early ages, we confront phenomenon that do not present themselves in tangible concrete meaningful ways. Consider the movement of the sun through the sky and the changing images of the moon. Many of these events are hard to comprehend because of the spatial or temporal scales and the interacting mechanisms involved both within and across Earth systems. There are numerous challenges for understanding and communicating evidence and explanations of such events.

When we consider STEM disciplines and computing challenges in terms of our PreK-16 education system, there are two agendas. First, the education system must sustain a STEM workforce of students to serve as the next generation of technicians, scientists, engineers, and mathematicians who will help research the grand challenges. Second, and just as important, the education system must also develop a public who is engaged in scientific issues and can make informed policy decisions regarding the grand challenges that are increasingly STEM and computationally driven. These two desired outcomes will require a new vision of STEM education for PreK-16 students: one that will provide deeper and broader learning experiences. Deeper in the sense that understanding the complexity of scientific models and engineering design solutions necessitates acquiring a range of core disciplinary ideas and a suite of science and engineering practices. Broader in the sense that these ideas and practices are inherently interdisciplinary, as well as cognitively and epistemically challenging when working between and among the S "science," T "technology," E "engineering," and M "mathematics." The consideration of science and engineering systems as a means for coordinating and forging STEM engagements represents a robust context for the design of curriculum, instruction, and assessment models that can address the two educational challenges.

We begin by presenting our framing of the relationships between systems thinking, model-based reasoning, and quantitative reasoning. Implications for current STEM and science education are examined. We then examine recent developments regarding the run up to and the writing and dissemination of the United States *Next Generation Science Standards*. Next, an overview of the now dominant 'learning sciences' perspective for teaching and learning is presented. Here we focus on the new perspectives of what constitutes effective science learning and teaching environments. The review of the 'learning sciences' sets up a discussion of what represents a new frontier in STEM education research—learning progressions (LP). The "Learning Progressions and Model-Based Reasoning" section follows and it examines LP frameworks being developed, particularly around theory/model-building. The intent is to demonstrate and argue that PreK-16 science education projects should adopt a LP perspective on developing curriculum, instruction, and assessment models.

Framing of Quantitative Reasoning and Model-Based Reasoning in Science Education

The interdisciplinary nature of science, as demonstrated by experts (e.g., Nersessian, 2002, 2008), requires socially constructing knowledge about the interactions among various STEM disciplines, in order to explain the physical, human, and created worlds. This means understanding the relationships of natural and human systems, their ever-changing nature, and how they influence and are influenced by other systems. To internalize this understanding means to internalize a systems thinking approach toward viewing and analyzing phenomena and processes.

Engaging in systems thinking involves being able to explain the system or components of a system under study and then predict the impacts the system will have on surrounding systems (Assaraf & Orion, 2005). To do this, a model is developed as an explanatory and predictive tool. It can be used to explain phenomena, collect data, predict outcomes (NRC, 2012a), or persuade others (Lehrer & Schauble, 2006). We recognize that there are many different types of models and distinguishing between them is beyond the scope of this chapter. However, we embrace the idea that there is a "common element" of all models, which is their use as "external aids to reasoning—they are cognitive prostheses" (Duschl & Grandy, 2008, p. 21).

When scientists participate in the theory building practices of science, they engage in model-based reasoning, which involves the development and use of varying forms of representations and the subsequent feedback and re-design of the model (Lehrer & Schauble, 2002, 2006). We characterize model-based reasoning as the cognitive processes involved with developing model(s)—often varying mathematical entities based on data—to build or re-define a theory. Duschl and Grandy (2008) argue that models "stand between data and theory . . . and . . . models can be influenced by both data revisions and changes in theory commitments" (p. 23). Here the notion is that of a theory as "a family of models" and a model as "an abstract mathematical entity" (Duschl & Grandy, 2008; Giere, 1988, 1999). When engaging in model-based reasoning, scientists are engaging in the cognitive processes found within the intermediary realm between data and theory that involves the iterative process of analyzing data to then represent it in a way that either supports current theories or informs the development of new theories in order to explain aspects of a system. This type of reasoning is critical to the "doing" of science, since it incorporates analyses, explanations, and communication about systems.

Modeling involves the use of different representations, some of which are quantitative in nature (NRC, 2007), "Mathematics in all its forms is a symbol system that is fundamental to both expressing and understanding science. Often, expressing an idea mathematically results in noticing new patterns or relationships that otherwise would not be grasped" (p. 153). Both model-based reasoning and quantitative reasoning involve an iterative process of analyzing, modeling, communicating, evaluating, and redesigning models to explain scientific phenomena or processes.

Mayes, Peterson, and Bonilla (2013) define quantitative reasoning in science as "mathematics and statistics applied in real-life, authentic situations that impact an individual's life as a constructive, concerned, and reflective citizen. [Quantitative reasoning] problems are context-dependent, interdisciplinary, and open-ended tasks that require critical thinking and the capacity to communicate a course of action" (p. 6). The authors outline a framework for quantitative reasoning that includes 1) the quantification act, 2) quantitative literacy, 3) quantitative interpretation, and 4) quantitative modeling. When engaging in quantitative reasoning, a scientist would engage in a process that involves conceptualizing data quantitatively (e.g., quantification act), applying various mathematical skills to analyze and evaluate the data, interpreting the results from the mathematical analysis of the data (e.g., quantitative literacy), and developing a quantitative model to represent the data (i.e. graph). This characterization of quantitative reasoning in science represents how intertwined this practice is with model-based reasoning and how it is used to characterize components of a system.

Figure 1.1 is a representation that characterizes the relationship between model-based reasoning and quantitative reasoning, within the context of systems thinking.

Quantitative reasoning and model-based reasoning are related, in that they involve analyzing, structuring, explaining, and representing data. When engaging in quantitative reasoning, one uses mathematical skills and concepts to analyze data of a natural or humanistic system and represent it as a model. The model is then used to make predictions, which leads to the collection and quantitative

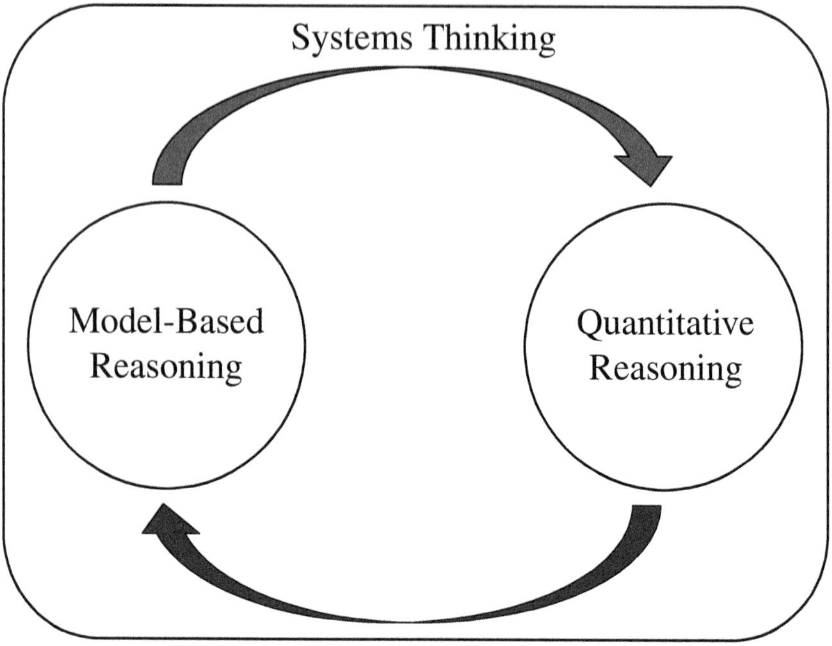

FIGURE 1.1 Relationship between systems thinking, model-based reasoning, and quantitative reasoning

analysis of new data. This informs the redesign of the model, which then influences how that model is used to reason about the systems. Both practices involve an iterative process in the development of models. Quantitative reasoning and model-based reasoning support a systems thinking habit of mind by explaining the mechanisms and interactions of systems. For this reason, model-based reasoning and quantitative reasoning occur within systems thinking.

Advances in technology, especially computing power for simulations, visualizations, and sensors, and in scientific understanding have shifted both the focus and the methods of the sciences. The focus, for several decades, has shifted toward a systems-based approach, while the methods have shifted toward model-based science. These two trends have strong implications for the inclusion of engineering and engineered systems as a context for STEM education. Models are increasingly being used for exploring and explaining complex, dynamic systems.

In a global society where future large-scale, scientifically based decisions occur, it is important to develop a systems thinking orientation toward the world. To do so, there is a need for understanding the explanatory and predictive functions of models. Developing models that represent quantitative data requires applying mathematical tools to explain quantitative data and often leads to a quantitative model of the phenomena (Mayes et al., 2013). In current K-12 classrooms, science does not regularly connect to mathematics. It is important to remember, "science is a quantitative discipline, which means it is important for educators to ensure that students' learning in science coheres well with their learning in mathematics" (NGSS Lead States, 2013, p. 10).

There is a need to move away from teaching isolated facts to a more integrated education system that values the need to support students' understanding of how science is quantitative. Including and developing quantitative reasoning and model-based reasoning in the science classroom is an important step toward developing students' abilities to analyze and model quantitative data to support an understanding of systems thinking.

US National Science Education Reports

The systems perspective has implications for educational practices and policy and is represented in recent National Research Council (NRC) reports—*Taking Science to School* (NRC, 2007), *A Framework for K-12 Science Education: Practices, Crosscutting Concepts and Core Ideas* (NRC, 2012a), and *Education for Life and Work: Developing Transferable Knowledge and Skills in the 21st Century* (NRC, 2012b)—through their emphasis on deeper learning that connects the "what" of science with the "how" and "why." Though the audiences differ with each report, all share in the push toward an integration of conceptual, epistemic, and social competencies within science education and beyond. Here we present a brief overview of the major recommendations and conclusions from these three recent NRC reports.

A Framework for K-12 Education and the Next Generation Science Standards

The NRC (2007, 2012a) reports have set a new course for US science education. The recommendations from the *Framework* have three implications for science education, and the integrative STEM approach: (1) science education should be coordinated around three dimensions—crosscutting concepts, core ideas, and practices (see Figure 1.2), (2) the practices should represent both science and engineering, and (3) the alignment of curriculum, instruction, and assessment should be implemented through the development of learning progressions that function across grade bands.

Model-based reasoning and quantitative reasoning are particularly salient in the *Framework* with regard to the science and engineering practices. To engage in model-based reasoning, one needs to engage in all eight of the above-mentioned science and engineering practices, which is often an iterative process involving the design and redesign of the model to best explain all components in question. It involves ongoing communication with others and evaluation of the model (Lehrer & Schauble, 2006). Engaging in model-based reasoning is not only a

BOX S-1

THE THREE DIMENSIONS OF THE FRAMEWORK

1 Scientific and Engineering Practices
1. Asking questions (for science) and defining problems (for engineering)
2. Developing and using models
3. Planning and carrying out investigations
4. Analyzing and interpreting data
5. Using mathematics and computational thinking
6. Constructing explanations (for science) and designing solutions (for engineering)
7. Engaging in argument from evidence
8. Obtaining, evaluating, and communicating information

2 Crosscutting Concepts
1. Patterns
2. Cause and effect: Mechanism and explanation
3. Scale, proportion, and quantity
4. Systems and system models
5. Energy and matter: Flows, cycles, and conservation
6. Structure and function
7. Stability and change

3 Disciplinary Core Ideas
Physical Sciences
PS1: Matter and its interactions
PS2: Motion and stability: Forces and interactions
PS3: Energy
PS4: Waves and their applications in technologies for information transfer

Life Sciences
LS1: From molecules to organisms: Structures and processes
LS2: Ecosystems: Interactions, energy, and dynamics
LS3: Heredity: Inheritance and variation of traits
LS4: Biological evolution: Unity and diversity

Earth and Space Sciences
ESS1: Earth's place in the universe
ESS2: Earth's systems
ESS3: Earth and human activity

Engineering, Technology, and Applications of Science
ETS1: Engineering Design
ETS2: Links among engineering, technology, science, and society

FIGURE 1.2 Three dimensions of *The Framework* (NRC, 2012a, p. ES3)

process engaged in by scientists, but is advocated in the *Framework* to be foundational in science instruction for all students, whether interested in a STEM career or not.

Since science is a quantitative field grounded in measurements, it is imperative that quantitative reasoning be given a prominent place in the science curriculum, as emphasized in the *Framework* (see practices 2–5 in Figure 1.2); e.g., using mathematics to identify patterns, analyze and interpret data, model physical phenomena, and quantitatively communicate the structure and function of the phenomena (Mayes et al., 2013). The push in the *Framework* for the integration of the science and engineering practices, crosscutting concepts, and core ideas emphasizes that in order for students to engage in science, the students must participate in the "discourses by which such ideas are developed and refined," which only occurs in the context of science (NRC, 2012a, p. 218). The important shift is away from an emphasis on merely knowing to an emphasis on using knowledge. This calls for science instruction to give priority to depth over breadth:

> Understanding the core ideas and engaging in the scientific and engineering practices helps to prepare students for broader understanding, and deeper levels of scientific and engineering investigation, later on—in high school, college, and beyond. One rationale for organizing content around core ideas comes from studies comparing experts and novices in any field. Experts understand the core principles and theoretical constructs of their field, and they use them to make sense of new information or tackle novel problems. Novices, in contrast, tend to hold disconnected and even contradictory bits of knowledge as isolated facts and struggle to find a way to organize and integrate them (NRC, 1999). The assumption, then, is that helping students learn the core ideas through engaging in scientific and engineering practices will enable them to become less like novices and more like experts (NRC, 2012a, p. 25).

Education for Life and Work

The emphasis on deep learning also occurs in the NRC (2012b) report that addresses learning of 21st century competencies (often referred to as 21st century skills) and the important roles they have for deep learning. Deep learning involves fostering an "interconnected network" of five knowledge types:

- Facts, statements about the characteristics or relationships of elements in the universe;
- Concepts, which are categories, schemas, models, or principals;
- Procedures, or step-by-step processes;
- Strategies, general methods; and
- Beliefs about one's own learning (NRC, 2012b, p. 84; modified from Mayer, 2011).

This integration of the "types of knowledge" is similar to the integration emphasized in the *Framework* in that it incorporates all three dimensions, as well as the beliefs about one's own learning.

To develop this "deeper learning," the report recommends the inclusion of "21st century competencies"—such as problem solving, critical thinking, collaboration—in instruction. The 21st century competencies include the integration of domain-specific knowledge and skills while supporting and extending the science and engineering practices of the *Framework*. The competencies are divided into three categories: cognitive, intrapersonal, and interpersonal, but those most closely intertwined with quantitative reasoning and model-based reasoning are cognitive competencies, with some interpersonal competencies (see Figure 1.3).

As identified in Figure 1.3, many of the competencies identified as having the strongest overlap between the three dimensions of the *Framework* and the competencies from *Education for Life and Work* support the development of model-based reasoning and quantitative reasoning. Most are cognitive competencies (e.g., critical thinking, information literacy, reasoning and argumentation, innovation). For example, nonroutine problem solving is similar to the focus of quantitative reasoning on ill-defined problems. This overlap highlights the emphasis placed on students engaging in the *doing* of science. The importance of doing is not just emphasized in the science education community, but in the informal education community, as well.

The other consideration is how this report extends science learning. The other two 21st century competencies (intrapersonal and interpersonal) are not heavily represented in the overlap in Figure 1.3. Intrapersonal competencies are related to managing one's behavior and emotions (e.g., flexibility initiative, appreciation for diversity, metacognition) and interpersonal competencies are related to expressing one's ideas and interpreting and responding to others (e.g., communication skills, collaboration, responsibility, conflict resolution) (NRC, 2012b).

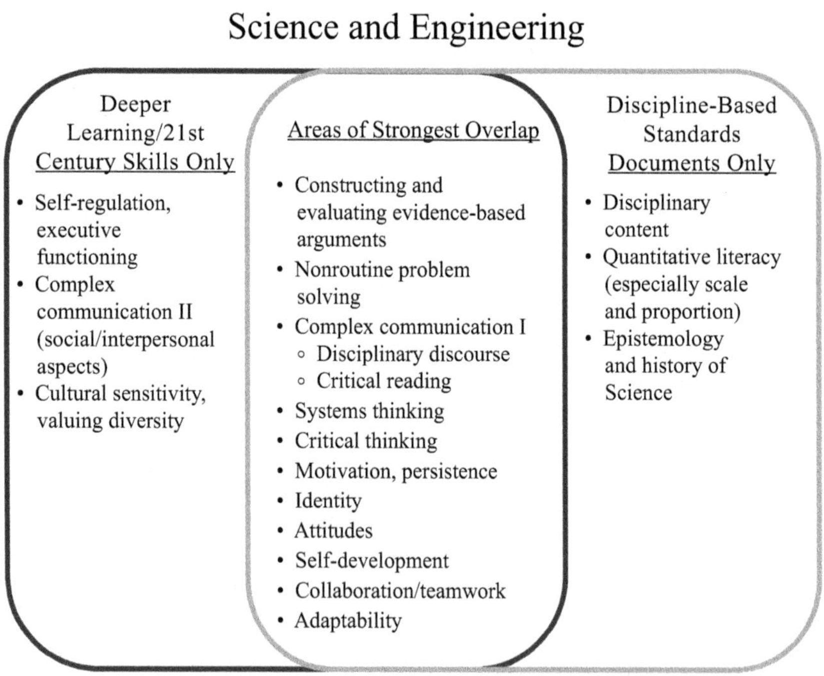

FIGURE 1.3 Overlap between selected competencies from *Education for Life and Work* and the three dimensions from *The Framework* (NRC, 2012b, p. 138).

Some intrapersonal competencies, such as metacognition, are known to be important to the learning of science (e.g., Magnusson & Palincsar, 2005) and should be included in the new science reform agenda. There should also be clarification of the skills within the interpersonal competencies and how these relate to their application in science education.

Learning Sciences and Science Learning: An Overview

This section presents a selected review of the literature on science learning and teaching that is guided by the concomitant and ongoing developments in cognitive/learning sciences and science studies. The overview focuses on a few salient topics that capture the vibrant debates and current challenges among researchers that have emerged when the study of science learning, science discourse, and scientific inquiry is examined in contexts (e.g., conceptual, epistemological, and social), at different ages (e.g., preschool, K-8, secondary, college, adult) and in various learning environments (e.g., formal and informal).

STEM learning, when viewed generally as the growth of knowledge, has many parallels with scientific and engineering inquiry as a set of knowledge building and refining activities and practices. These activities and practices progress from experiments to models and explanatory theories. Models are seen as cognitive tools that sit between experiments and theory (Giere, 1988, 2002; Nersessian, 2002, 2008). What has come to gain traction is the view of science and science learning as fundamentally a model building and refining enterprise. *Taking Science to School (TSTS)* (NRC, 2007) takes the position that the teaching and learning of science should be based on an image of science that sees the growth of knowledge as involving the following epistemic and social practices:

1. Building theories and models,
2. Constructing arguments, and
3. Using specialized ways of talking, writing, and representing phenomena.

This tripartite perspective on school science reflects a synthesis of ideas about the growth of knowledge and the nature of scientific reasoning taken from the learning sciences community and from the science studies community. Model-based reasoning becomes a critical component of this tripartite perspective, as it involves all three practices to characterize natural and human systems.

The learning sciences emerged from the earlier constructivist theories of learning and from the pioneering research in the cognitive sciences. Our deeper understanding of how children's thinking is fundamentally different from that of adults coupled with richer understandings of expertise, representation, reflection, problem solving, and thinking provided a foundation for a major tenet of the learning sciences; "students learn deeper knowledge when they engage in activities that are similar to the everyday activities of professionals who work in a discipline" (Sawyer, 2006, p. 4). Similarly, philosophers started to realize that any attempts to account for the growth of scientific knowledge or theory change needed to view inquiry through the natural human mental processes and modes of acquiring knowledge. This philosophical perspective aligns somewhat with research on informal learning that reveals the importance of participation structures and the development of practices in culturally valued activities (Cole, 1996; NRC, 2009). Focusing on scaffolding, apprenticeship, legitimate peripheral participation, and guided participation, informal learning researchers provided "broader units of analysis . . . these views move beyond the study of individuals alone to consider how learning occurs within enduring social groups such as families and communities" (Bransford et al., 2006, p. 24).

Advances in our understandings about learning have occurred in tandem with our richer understandings about the growth of knowledge within STEM disciplines. Essentially, we are learning how to learn with respect to the natural and designed world and about learning itself. Ideas from the interdisciplinary research communities labeled *learning sciences* and *science studies* are extending our understandings of science learning, science practices, scientific knowledge, and scientific discourse (Duschl, 2008; Duschl & Grandy, 2008). Cognitive, historical, sociological, and anthropological studies of individuals working in knowledge building contexts reveals the importance of practices to the professional activities in these knowledge growth communities. With respect to the scientific disciplines, cognitive models of science (Giere, 1988; Goldman, 1986; Kitcher, 1993; Thagard, 1992) coupled with sociocultural models of science (Knorr-Cetina, 1999; Kuhn, 1996; Longino, 1990, 2002) have established the importance that models and modeling, visual representations, knowledge exchange mechanisms, and peer interactions have in the advancement and refinement of knowledge. In brief, doing science takes place in complex settings of cognitive, epistemic, and social practices.

Research on learning is moving away from a focus on general principles of learning to a focus on developing domain specific knowledge (e.g., the epistemic, cognitive, social, and cultural factors that influence the growth of knowledge in STEM domains). New images of science coupled with new images of learning have, in rapid succession, led to new perspectives about the foundations of science and thus of STEM education. *TSTS* (NRC, 2007) and the *Framework* (NRC, 2012a) recommend that science learning be organized around select conceptual knowledge frameworks and practices that, in turn, are coordinated around core content and learning progressions. Several reports (e.g., NRC, 2002, 2005, 2006, 2007) include information about the development of scientific knowledge and practices across grade levels in K-12 and between and within courses in post-secondary STEM education. Such sequencing and coherence of teaching and learning is a major new orientation for STEM education research and development and comes under the headings learning progressions and teaching sequences.[1] The study of science and engineering systems involves complex thinking concerning interacting systems. Understanding how systems work requires a great deal more than knowing the essential principles.

What the current research in cognitive development and philosophy of mind suggests is that very young children have a surprising capacity for reasoning and prior knowledge in select domains (Keil, 1989; Subrahmanyam, Gelman, & Lafosse, 2002). The current research on cognitive development and reasoning in science also demonstrates that context matters in terms of content, learning environment, and learning goals (Atran, 2002; Koslowski & Thompson, 2002; Siegal, 2002). That is, learning is linked to the domain within which learning takes place and is dependent on the acquisition of select practices and ways of representing and communicating science ideas and critiques. Consequently, *core knowledge* learning and *learning progression* designs for the alignment of curriculum, instruction, and assessment are seen as robust areas for future science learning research.

Embedding research on science learning within specific contexts (e.g., core ideas and crosscutting concepts) has produced valuable insights into pathways or trajectories of learning in the disciplines (Catley, Lehrer, & Reiser, 2005; Smith et al., 2006). However, the research on learning in contexts challenges many of the *received views* of child and adolescent science learning. These domain-general views assume that development involves broad mental structures that facilitate mastery of a variety of tasks. Examining the research on children's learning and capacities for representation provides insights on how domain-specific learning frameworks can serve as a foundation for model building and systems thinking in science.

Children's engagements in pretend play, in which one object stands in for another, is a beginning notion of symbolism. Early understandings of words as representing objects or actions are also indicative of emerging symbolic capacities. Engagement with measurement and data representation can be introduced early on as the PrePS curriculum (Gelman & Brenneman, 2004) demonstrates. Preschool children can sort objects based on size, color, shape, or other features and then be guided to display this information in the form of lists, tables, and simple graphs. Children can compare measurements, for example shoe size and height of children in different classes, as well as chart growth in these quantities over time (Gelman & Brenneman, 2004). Understandings about counting, measuring, and illustrating patterns provide a necessary foundation for developing more sophisticated notions of descriptive statistics and data modeling, as well as quantitative reasoning.

Research on elementary students' abilities to measure and represent data suggest that young children can engage in productive discussions about aspects of an object to measure and how these data should be graphically represented (Lehrer, Kim, & Schauble, 2007; Lehrer, Jaslow, & Curtis, 2003; Lehrer & Schauble, 2000a, 2000b, 2002). Lehrer and Schauble (2004) employed a design study approach to investigating students' understanding of natural variation through learning and reasoning about distribution in a data-modeling context. Students studied the growth of native plants and the influences of different growth conditions. Over a two-month period, students' reasoning related to "distribution" and "natural variation" significantly improved. This depth of understanding developed out of students' experiences in generating, evaluating, and revising models of data. The students' invented and teacher-guided representations of data served as a focus for discussions about simple statistical qualities of data and different forms of representations for illustrating different features of data patterns (Lehrer & Schauble, 2004).

The extensive research on infants and young children's cognitive development underscores the multitude of knowledge resources and reasoning capabilities children bring to formal schooling. Young learners are anything but empty minds. Whether or not we choose to capitalize on learner's emerging scientific reasoning abilities depends on how we construe the goals of science learning and how they can be achieved. A focus on understanding the doing of science and how scientific knowledge is developed and evaluated will entail building on students' emerging capacities for representation, model-building, quantification, casual reasoning, and the like. Three critical aspects of the nature of contexts and situations that are embedded in most views and research on learning within domain specific contexts are issues of *authenticity, collaboration,* and *inquiry* (Blumenfeld et al., 2006). Authenticity, within the context of STEM learning, focuses on embedding the learning within the learners' everyday world and the practices of the discipline. Collaboration encourages the sharing and contrasting of ideas with other individuals within a community who are engaged in similar tasks and who have similar aims. Finally, inquiry motivates STEM learners to engage in problem stating and solving activities, which require planning, synthesis, and evaluation skills using relevant domain specific content knowledge.

If the focus of science education is on the accumulation of scientific facts and essential principles devoid of using that information to propose explanations and predictions, then it is not clear how one might capitalize on the emerging understandings. Thus, the NRC research and policy documents argue for a STEM education that focuses on the investigative and discourse practices, as well as practices for 21st century skills, which are embedded in model/theory building/refining. Research informs us that implementing such a building and refining learning environment allows students to bring significant conceptual resources that can, and should be, used as leverage for developing more sophisticated understandings of the scientific and engineering enterprises throughout schooling.

Grounded strongly in perspectives from philosophy of science, philosophy of mind, and developmental psychology, the interdisciplinary approach to understanding STEM learning, knowing, and doing has established in no uncertain terms that learning, cognition, and reasoning are contingent on context and content. The strong recommendation from *TSTS* and the *Framework* is that the teaching of conceptual knowledge should not be independent of science and design practices. In short, our understandings of the growth of scientific knowledge and reasoning are philosophically and psychologically grounded (Carruthers et al., 2002). Each domain has contributed to our understandings about *learning how to learn*. The emerging consensus is that learning and teaching ought to be grounded in and informed by conceptual, epistemological, and social structures and practices.

Learning Progressions and Developmental Pathways

The preceding learning science research overview makes very clear that sequence and coherence are paramount. There are developmental pathways children need to follow to enhance both conceptual understandings and participation in essential discourse practices.

Taking Science To School (NRC, 2007) and the *Framework* make the recommendation that science learning needs to be strongly grounded in model-based approaches that focus on the use and consideration of evidence for posing, building, and refining models. Science and engineering practices are an important component wherein the critique and communication discourse practices reside. The science and engineering practices and practices for 21st century skills require time to develop.

Thus, the recommendation in *TSTS* (NRC, 2007), the *Framework* (NRC, 2012a), the *Next Generation Science Standards* (NGSS Lead States, 2013), and *Education for Life and Work* (NRC, 2012b) is that science learning be organized into longer sequences—learning progressions (LPs)—that serve as vertical pathways of learning across grade levels and as teaching sequences horizontally within any instructional year. The rationale is that facilitating the learning of core knowledge and practices that are critical for development of scientific knowledge and reasoning is complex, takes time, and requires instruction-assisted development grounded in sound assessment practices. Thus, the content of LPs includes the core knowledge, epistemic practices (e.g., science talk and argumentation) and social practices (e.g., critique, communication and representation) that characterize a domain of science and/or engineering. The recommendation from *TSTS* is that:

> The core concepts used in this practice [learning progressions] would be dramatically fewer in number than those currently focused on or included in standards and curriculum documents . . . a grade-level teacher would need to be concerned not only with the relevant "slice" of a given core idea in her particular grade, but also with the longer continuum of learning that K-8 students experience. Thus, teachers and science teacher educators . . . would need to build structures and social processes to support the exchange of knowledge and information related to core concepts across grade levels. (NRC, 2007, p. 61)

To support the "exchange of knowledge" or transfer, deep understanding of a topic is necessary. LPs have the potential to characterize how that deeper understanding might develop across grade levels, for particular core ideas or science and engineering practices. The agenda for supporting extended deeper learning of 21st century competencies also appears in *Education for Life and Work.* This emphasis on LPs, whether explicitly mentioned or not, is now filtering into many other aspects of education. There is a need to consider how these different aspects of education relate to one another and can be aligned to better support students' development across the PreK-12 years.

Corcoran, Mosher, and Rogat (2009) convened several workshops of experts exploring LPs to look at two questions:

1. What promise might LPs have for improving instruction in schools?
2. What further might be required to make the promise real?

LPs are empirically grounded and testable hypotheses about how students' understandings of and abilities to use core ideas become more sophisticated over time. Four features characterizing LPs (Corcoran, et al, 2009) include:

1. Targeting core and generative disciplinary understandings and practices that merge science content with science practices;
2. Lower and upper boundaries that describe entry assumptions and exiting expectations for knowing and doing;
3. Descriptions of levels or steps of progressing achievement; and
4. Purposeful curriculum and instruction that mediates targeted student outcomes.

The consideration of LPs represents a shift in emphasis away from teaching that focuses on what we know to teaching that focuses on how we came to know and develop scientific knowledge and why we believe what we know over alternatives. Within the how and why approaches reside "assessment for learning" opportunities to make thinking visible as students engage in talk and argumentation and in modeling and representation. Corcoran et al. (2009) state that "progressions can play a central role in supporting the needed shift toward adaptive instruction" (p. 9) and that the following are seen as possible learning outcome benefits of establishing LPs:

- Providing a basis for setting standards that are tighter and more clearly tied to instruction;
- Providing reference points for assessment to report on levels of progress and thereby facilitate teacher interventions and instruction-assisted development; and
- Informing the design of curricula that are aligned with progressing students forward (e.g., assessments for learning).

An examination of the growth of scientific knowledge as provided by longitudinal studies around LPs (Corcoran et al., 2009) and by science studies scholars (Nersessian, 2008) can provide some helpful insights on how to proceed with the discipline-based education curriculum redesign agenda. Corcoran and Silander (2009) conducted a review that examined the effects that different instructional strategies had on high school students' learning. The strategies included *interdisciplinary teaching, cooperative learning, problem-based learning, adaptive instruction, inquiry, and dialogic teaching*. The results found that well-designed student grouping strategies supported student learning by offering challenging tasks and allowing students to express their ideas and questions. In addition, adaptive instruction supports student learning by using formative assessments to monitor student variation and adapt instruction in response to students' progress and needs.

A key component of LPs is instruction-assisted development that, like adaptive instruction, is grounded in robust learning performances (Wilson, 2009), which serve as "assessments for learning" (Black & William, 1998). The extant alternative is the selection of topics and sequences based on a logical analysis of content domains and personal experiences with teaching (e.g., American Association for the Advancement of Science [2001] *Atlas of Science Literacy*; scope and sequence curriculum frameworks common in school districts).

In a review and analysis of learning progression and teaching sequence research, Duschl, Maeng and Sezen (2011) draw a distinction between "Validation LPs" and "Evolutionary LPs" (See Table 1.1). Only Evolutionary LPs are conducting research that attends to the development of foundational knowledge and to thorough descriptions of LP instruction-assisted pathways. The Validation forms (e.g., developing and testing assessment models; testing discourse strategies or instructional interventions), while valuable, are only components or constituents of science learning and thus better labeled as teaching sequences that can be precursors to, but are not necessarily, learning progressions.

LP designs and research, whether evolutionary or validation, need to be longitudinal: following learners and learning across several grades or, in the case of postsecondary education, across courses. This eliminates the teaching sequence investigations that examine learning within single units that entail short durations of instruction as LP research. Teaching sequences that focus on the conceptual demands of core idea domains but eschew the science practices are problematic. Typically such hypothetical learning progressions begin from reviews of strand maps, curriculum guides, or standards frameworks.

Canonical knowledge is not the same as learners' foundational knowledge. What drops out in many teaching-sequence formatted trajectories is the consideration of the inherently diverse learners' perspectives where the foundational knowledge resides. If the research studies conceptual development without also examining how learners use knowledge when engaging in a science practice(s), then the research should not be considered LP research. Considerations of knowledge use and its coupling with science practices are important criteria for LP research.

TABLE 1.1 Validation LPs and Evolutionary LPs

Validation LPs	*Evolutionary LPs*
LP based on validating a standards-based progression: instruction as intervention	LP based on sequencing of teaching experiments across multi-grades: instruction as refining progression
Theory-driven top/down approach	Evidence-driven bottom/up approach
Upper anchors as college readiness	Upper anchors as targeted literacy
Uses assessments to confirm learning models	Uses assessments to explore learning models
Progress variable steps and targets are fixed	Progress variable steps and targets are flexible
Adopts a misconception-based 'Fix It' view of conceptual change instruction	Adopts an intuition-based 'Work with It' view of conceptual change instruction
Theory building as conceptual change	Model building as conceptual change
Domain general orientation to topic selection	Domain specific orientation to topic selection

The content of LPs—core ideas and practices—can also be informed by science studies research that examines the practices researchers and designers employ. For example, Nersessian (2002; 2008) studies the cognitive basis of model-based reasoning in science. Her recent research studies the cognitive practices of biomolecular scientists and biomedical engineers working together on interdisciplinary problems concerning cultivating and engineering tissues. The work is guided by the premise that "studying inquiry practices in research laboratories could lead to development of effective pedagogical strategies for improving the instructional laboratory" (2008, p. 72). Nersessian maintains, everyone is a learner—undergraduates, Ph.D. candidates, post-doctoral researchers, and lab directors. Nersessian refers to such contexts as "agentive learning environments" and found several significant features:

- With conceptual and methodological knowledge and skills distributed, everyone makes contributions;
- The organization is non-hierarchical—no one person is the expert, neophyte members can contribute and achieve legitimacy and identity;
- Interactional structures allow for membership routes into the laboratory that motivate learning; and
- Multiple social support systems bolster resiliency in a research context that has frequent failures.

Commenting on the potential bridges from science labs to science classrooms and recognizing the differences, she writes, these contexts have "their own unique constraints and affordances that need to be figured into the development of strategies for learning and using model-based reasoning . . . the point is that the *kinds of reasoning processes* should aim to approximate those of a scientist." (2008, p. 78).

Brickman, Gormally, Armstrong, and Hallar (2009) studied the effects of inquiry-based laboratory settings on undergraduate students' science literacy skills. They found that students who received an inquiry-oriented laboratory instruction demonstrated increased science process skills, including quantitative reasoning. With the new three dimensional learning science agenda aiming to increase integration and coherence across science instruction in PreK-16 classrooms, the current and future PreK-12 students will be receiving a very different educational experience than previous undergraduate students. The questions then become: How will the postsecondary science community adjust to meet the needs of these future undergraduate students? What does this mean for the current lecture-driven postsecondary science community? How can future undergraduate STEM research begin changing the undergraduate learning environments to be more coherent and cross-curricular?

Learning Progressions and Model-Based Reasoning

Descriptions of learning progressions involve establishing a beginning point and an ending point that can span months, semesters, or years. *TSTS* (NRC, 2007) refers to the beginning point as the "lower anchor," which represents the knowledge children bring with them to school. This beginning knowledge is often grounded in sensory-based observations of commonly occurring natural events. In this way the lower anchor disciplinary concepts of LPs are said to be accessible to learners since they have some awareness of the phenomenon. The "upper anchor" represents the expectations we have of students learning at the end of the LP. That is, what students should know and be able to do.

The lower anchors of LPs often consist of macroscopic events, which are easily visible or related to students' everyday experiences. This characteristic of LPs ensures the target concepts of LPs are accessible to learners. For example, Mohan et al.'s (2009) LP on the carbon cycle was based on five focused macroscopic events that are familiar to students. The lower anchor of this LP focuses on intuitive accounts that macroscopic events are the result of natural tendencies by differing agents and enablers. In other words, the growth of plants is a natural process enabled by food, water, or sunlight. Mohan et al. labeled this kind of reasoning and accounts as *force-dynamic,* which is closely related to children's informal everyday experiences—making this LP appear to be accessible to early years learners.

Lehrer and Schauble (2012), in a component LP, recognize that the theory of evolution has several everyday knowledge entailments that could be productive resources for developing scientific explanations. These are observable and measurable differences (1) between or within species, (2) with changes overtime in individuals' growth or population fluctuations, and (3) between organisms' structural features and habitat. The lowest levels of the three entailments involve describing qualitative differences, observing and describing the current state of an organism or group of organisms, and posing questions about where an organism lives. Thus, the lower anchor is accessible in that it is related to basic practices for understanding the evolutionary concept through variability, change, and ecology.

The upper anchor represents the learning goals of the LP and emphasizes using knowledge and practices. *TSTS* (NRC, 2007) represents upper anchors as the successive adoption of more accurate scientific understanding and increasing sophisticated science practices that together establish societal expectations for science literacy. The upper anchor goals and performance expectations will obviously vary depending on the targeted ending grade. Not unlike the accessibility issue for the lower anchor, the upper anchor should also attend to issues of appropriately targeted learning goals.

While this upper anchor may appear to be highly abstract, the learning pathway, across multiple grades, begins with accessible macroscopic characteristics in the lower anchor. Through a process of iterative design based research implementations and refinements of the curriculum sequence, researchers are working on re-organizing the domain-specific concepts and doing so in the context of scientific decision-making.

Between the lower and the upper anchor there are the intermediate steps. Wiser et al. (2009) and Smith et al. (2010) adopt a conceptual framework that stresses the importance of the intermediary levels when developing learning progressions. Using terminology such as anchor points, stepping-stones, lever concepts, and linchpins, they describe instruction-assisted conceptual development that is based on learners' extant knowledge. Taking a narrower perspective (grades 3–5) on the Learning Progression for Matter (LPM), the goal is to help learners bridge from lower to upper anchors using stepping stones that are intermediate steps in the bridging processes of the knowledge network. Lever concepts are core placeholder concepts present in the lower anchor (e.g., weight) that bridge to important upper anchor concepts (e.g., mass, density). Lever concepts are salient in students' everyday thinking and intimately related to other ideas. Linchpins are organizers that express the structural aspects and/or relations among concepts in the upper anchor. Linchpins then are tools that make it possible to re-conceptualize the lower anchor lever concepts. Thus, the stepping-stones, lever concepts, and linchpins were applied to the LPM as interventional instruction strategies, to support re-conceptualizations that progress students from the lower anchors to the upper anchors.

LPs should also include progress levels related to science practices. Gunckel, Covitt, Salinas, and Anderson (2012) developed a learning progression about water in socio-ecological systems that incorporated various elements of the three dimensions from the *Framework*. Each level was defined by the extent to which understanding of each factor was elaborated: 1) structures and systems, 2) scale, 3) scientific principles, 4) representations, and 5) dependency and human agency. A student was considered to represent a level 4 understanding and strong model-based reasoning when they demonstrated cohesive scientific understanding and applicability of each of the five factors. This required students to represent understanding of how to "[develop] and use models, [interpret] data, [construct] explanations, and [engage] in arguments from science" (p. 860).

Many of the current K-8 science programs have lean low-level reasoning demands required of students (Ford, 2005; Hapgood et al., 2004; Metz, 1995; NRC, 2007). Curriculum that addresses domain-general reasoning skills and surface level knowledge dominate over curricula addressing core knowledge and domain-specific reasoning opportunities that meaningfully integrate knowledge. The reasoning-lean curriculum approaches (a) tend to separate reasoning and learning into discrete lessons thus glossing over the salient themes and big ideas of science thereby making American curricula "a mile wide and an inch deep" (Schmidt, McKnight, & Raizen, 1997) and (b) in the case of middle school textbooks, tends to present science topics as unrelated items with little or no regard to relations among them (Keisdou & Roseman, 2002).

Ohlsson (1992) recognized that the focus on teaching scientific theories did not include using the theories; missing were cognitive processes involved with theory articulation and refinement. Ford (2005) examined 3rd grade students' engagement with a kit-based unit on *Rocks and Minerals* and found that the principal learning goals for the lessons was classification reasoning. Descriptive observational features of rocks and minerals were used to assign rocks to types and kinds, while the curriculum learning goals lacked any expectation for using information from rocks and minerals to tell a story about the rocks. Ford concluded that the lessons in the kit were impoverished and underestimated the known capabilities of children to engage in science.

Research on young children's learning demonstrates that children entering school are well equipped cognitively and socially to engage in theory and model building. The role of modeling natural phenomenon and then reasoning from those models has led Ford (2008), Herrenkhol and Guerra (1998), Lehrer and Schauble (2004, 2006), Smith (2007), among others, to investigate ways to design classroom learning environments that promote students' theory and model building reasoning. Lehrer and Schauble (2006) report on a 10-year longitudinal research project that examines planned instructional sequences across grades K-5, focusing on model-based reasoning and instruction in science and mathematics. Critical to the design of these learning environments is engagement in analogical mapping of students' representational systems and emergent models to the natural world. Important instructional supports are coordinated around three forms of collective activity: (a) finding ways to help students understand and appropriate the process of scientific inquiry, (b) emphasizing the development and use of varying forms of representations and inscriptions, and, (c) capitalizing on the cyclical nature of modeling (p. 381).

When designing learning environments, students need support in engaging in argumentation using model-based reasoning. Herrenkohl, Palincsar, DeWater, and Kawasaki (1999) studied elementary students' abilities to engage in structured discussions about predictions and theory building in the context of sinking and floating objects. The students developed conceptual models of why some objects floated and others did not and then shared those conceptual models (i.e., theories) with their peers. The structured discussions allowed students to engage in scientific argumentation that involved critiquing their conceptual models and refining them to better align with and predict data. They gained a deeper understanding of problem solving in science, which is an important cognitive competency that supports quantitative reasoning and model-based reasoning.

Sandoval (2005) argues that having a better understanding of how scientific knowledge is constructed makes one better at doing and learning science. The goal is to engage students in a set of practices that build models from patterns of evidence and that examine how what comes to count as evidence depends on careful observations and building arguments. Schauble, Glaser, Duschl, Shultz, and Johns (1995) found that students participating in sequenced inquiry lessons with explicit epistemic goals (e.g., evaluating causal explanations for the carrying capacity performance of designed boats) showed improved learning over students who simply enacted the investigations. They found that students' understanding of the purposes of experimentation made a difference. Other researchers (Kelly & Crawford, 1997; Sandoval & Reiser, 2004; Songer & Linn, 1991; Toth, Suthers, & Lesgold, 2002) have found positive learning effects of students working with and from evidence and seeing discourse and argumentation as a key feature of doing science.

Additional insights for the design of reflective classroom discourse environments comes from research by Rosebery, Warren, and Conant (1992), Smith, Maclin, Houghton, and Hennessey (2000), van Zee and Minstrell (1997), and Herrenkohl and Guerra (1998). Rosebery et al.'s (1992) study spanned an entire school year while that of Smith et al. (2000) followed a cohort of students for several years with the same teacher. Both studies used classroom practices that place a heavy emphasis on (a) requiring evidence for claims, (b) evaluating the fit of new ideas to data, (c) justifications for specific claims and (d) examining methods for generating data. Engle and Conant (2002) refer to such classroom discourse as "productive disciplinary engagement" when it is grounded in the disciplinary norms for both social and cognitive activity.

The research by van Zee and Minstrell (1997) shows the positive gains in learning that come about when the authority for classroom conversation shifts from the teacher to the students. Employing a technique they call the *reflective toss*, van Zee and Minstrell found that students become more active in the classroom discourse with the positive consequence of making student thinking more visible to both the teacher and the students themselves. Herrenkohl and Guerra (1998) examined the effect on student engagement of guidelines for students who constituted the audience, that is, the scaffolding was on listening to others. The intellectual goals for students were predicting and theorizing, summarizing results and relating predictions, theories, and results. By adopting audience role assignments, students were directed to develop a *question chart* that would support them in their assigned roles (e.g., What questions could we ask when it is our job to check summaries of results?). Following the framework developed by Hatano and Inagaki (1991), Herrenhkohl and Guerra used the audience role procedures to engage students in (a) asking clarification questions, (b) challenging others' claims, and, (c) coordinating bits of knowledge. The focus on listening skills and audience roles helps to foster productive community discourse around students *thinking in science*.

Summary and Future Directions

Researchers studying science learning and STEM education are learning that with proper supports (e.g., instruction-assisted development, assessment for learning) and sequencing (e.g., immersion units and learning progressions) young children and adolescents are capable of complex reasoning and engaging in sophisticated scientific critique and communication practices. Theory-building, modeling, and other forms of scientific reasoning are possible when learners are provided with multiple opportunities that sustain engagement with select scientific practices over time (e.g., predicting, observing, testing, measuring, counting, recording, collaborating, and communicating). When sustained engagement and instruction-assisted development occurs, learners develop images of systems thinking, of the nature of science, and of scientific inquiry as an enterprise that is fundamentally a model-based reasoning and refining process. Viewing formal and informal learning environments as a scientific community where learners participate in science practices and discourse processes akin to professional communities is under studied and more research is needed.

Research on developmental trajectories/progressions that examine learning and reasoning is important and, while it is informed by lab studies and cross-age interview data, it must go further in order to establish a stronger empirical base. One aspect of going further is to study the pathways, trajectories, or progressions where learning occurs—in the study of learning environments where student learning is taking place. This includes studying the progression of learning into undergraduate STEM environments.

As we focus on a systems thinking approach to science with model-based reasoning and quantitative reasoning as tools to characterize components and processes within systems, we must consider how this new approach impacts the development of learning progressions. As advocated in the recent NRC reports, learning progressions should incorporate the three dimensions (core ideas, crosscutting concepts, and science and engineering practices), as well as the cognitive, intrapersonal, and interpersonal competencies and do so across the span of K-16 education.

The *Next Generation Science Standards* emphasize this deeper learning by using an evidence centered design approach toward the integration of the three dimensions in the new performance expectations. Students are expected to demonstrate understanding of the core ideas by applying that understanding through the use of various science practices. This application represents the need for evidence to support understanding of the core science concepts. By applying their understanding, students are making their thinking visible for both themselves and their teachers. Teachers can then use this information to inform their instruction of how to support students in continued depth of science learning and how to apply that understanding to new ideas and the practices of science.

Another aspect of going further is to study how teachers are engaging in such new "pathway" sequences of instructional units. That is, how do teachers come to understand the alignments and coherence among curriculum-instruction-assessment that, in turn, inform and frame instruction-assisted development. There needs to be more research about the design of tasks and assessments that make thinking visible and thus inform and guide instruction and learning. Here is where teacher feedback and teacher/peer mediation guides learners to increasingly higher levels of sophistication in understanding concepts and using concepts in the implementation of practices. The research agendas highlighted in the various NRC reports will be complex given the new images we have of science, of capable young learners, of science and engineering participatory practices, of 21st century competencies, and of the important role context plays in motivating the understanding and evaluation of science knowledge and engineered systems. But the rewards will be many as we develop richer understandings about the cultivation and motivation of PreK-16 STEM learning and teaching.

Coda: Research Priorities and Challenges for the Waterbury Summit Goals

Prior to the Waterbury Summit, Drew Gitomer, James Greeno, and Richard Duschl met for two days to engage in discussions about the broader challenges the field faced with regard to research and implementation of curriculum, instruction, and assessment, as well as the implications the forthcoming Summit held for K-16 STEM education. Based on those discussions, Gitomore and Greeno elaborated on the research challenges facing the field with regard to particular aspects of STEM education, which are expanded upon in the following sections. Gitomer's focus was on the issues of improving and modifying measures and assessments to accommodate new learning and performance goals. Greeno's focus was on the cognitive and sociocultural issues surrounding learning and reasoning within the complex multidimensional knowledge contexts and practices posed by the *Framework's* reform agendas.
Drew H. Gitomer

Observation Protocols

Teaching observation protocols have received significant attention in recent years because of their prominent place in teacher evaluation systems, but they have also been used to support professional development and as research tools that characterize classroom interactions.

Every observation protocol provides a lens through which participants, both the observer and the observed, can view instruction. Each protocol provides a language that can help to support a common understanding of what is valued in instruction and what constitutes evidence of different aspects, or dimensions, of instruction. Therefore, the design and/or selection and use of protocols that are coherent with the values described in recent science reform efforts are important considerations.

Observation protocols vary on a number of relevant dimensions. The most widely used protocols are generic with respect to grade levels and subject matter. However, there are some protocols that are specific to subject areas and provide much greater specificity about interactions around the teaching and learning of content and reasoning within specific domains. Protocols also vary in the extent to which they attend to students and their interactions with the teacher and with other students. Traditional observation protocols (and some in use today) primarily make use of a lens that focuses only on the teacher—the questions she asks, the behavior she corrects, and the explanations she provides. More recent protocols use the lens of interaction—in considering questioning, for example, it is not just that the teacher asks a question, for it matters whether students pick up on the question and whether meaningful interchanges occur.

Interestingly, studies of protocols have shown us that observers have a difficult time making consistent judgments about certain aspects of classrooms, particularly those things associated with instruction. While psychometricians talk about this in terms of poor rater reliability, what this really represents is a lack of common understanding of what it means to teach and learn content, to engage in meaningful questioning and feedback, and to support students' reasoning and problem solving. Thus, the use of these protocols not only can tell us something about teaching, they can reveal the broader challenge of establishing common understanding and language about concepts such as those described in the *Framework*.

The vision of science education expressed in the *Framework* pushes us to think about the lens we want to use to observe science classrooms. Observation protocols could be very useful in describing the characteristics of instruction that are associated with particular valued practices. Observation protocols can also give guidance to how the teaching of core ideas looks in classroom teaching. Each of these protocols could be supplemented with rich descriptions and video examples that help to give shape and meaning to attempted enactments of the *Framework*.

Expanding Common Expectations and Indicators

The vast majority of assessments in science focus on content of a particular domain and in some cases, reasoning about that content. But the *Framework* takes a very deliberate stance that science education should not simply be a series of shallow forays into a broad range of content areas. In fact, by explicitly focusing on crosscutting themes and valuing deep progressions over limited exposure to broad content, the *Framework* takes the stance that such an approach builds a set of skills, interests, dispositions, and conceptual orientations that will support sustained engagement of science beyond the specific content studied in school.

And so, how do we articulate and monitor student progression on these more abstract constructs? Before addressing that question, we should clarify why we would want to add these things to the assessment menu? Simply tracking one more set of competencies as part of an accountability paradigm does not seem particularly compelling or desirable. But, if these indicators can be used as part of a system to inform and cultivate these crosscutting competencies, then they may have some significant utility, for the indicators should be markers of what it looks like if progress is being made on these critical objectives.

Assessments can be indicators, but not all indicators need to be assessments. What would be the markers that students are developing interest in science and ways of engaging those interests? Are they increasingly capable of asking productive questions as they engage with new science problems and ideas? Are they more disposed to look at problems that can be addressed through scientific investigation or an engineering solution? Is there evidence that students have more developed understandings of the nature of science and scientific inquiry as they move through the curriculum? Could some of these indicators be created by observing students in different situations as they engage with science? Could indicators such as participation in non-school science activities or advanced and optional science coursework be used as indicators of greater interest in science?

In doing this kind of exercise, it is critical to first identify the valued outcomes that we would hope for this new vision of science education. Then, we need to think through what would be interesting and useful evidence to provide insight into progress toward these valued outcomes at student, classroom and institutional (e.g., school, district) levels.

Other Problem Contexts

How might we think about assessing student learning with respect to three STEM themes: systems thinking, model-based reasoning, and quantitative reasoning? Certainly, this has been a long-standing challenge for the assessment community for reasons that extend into curriculum and curricular policy.

For example, large-scale assessments such as the National Assessment of Educational Progress (NAEP) (National Assessment Governing Board, 2008) do not make strong assumptions about the particular curriculum that students have encountered. Therefore, in attempting to gather assessment evidence on a crosscutting theme (e.g., systems thinking), the solution has been to create artificial problems that address the theme, but in ways that do not assume mastery of any particular content. Thus, the tasks become abstracted reasoning processes that are divorced from the critical ability of engaging those reasoning processes around and together with rich content.

We might think about how the *Framework* and resulting curricular efforts can allow for the design of assessments and problem contexts that are grounded in content around core concepts and ask for students to engage in deep reasoning around that content. To the extent that particular assumptions about curriculum can be made, the potential for developing assessments that inquire about these STEM themes in ways that reflect legitimate participation in science reasoning becomes much more tractable.

This then forces us to consider the nature of assumptions that can be made under the *Framework* and how that then frees us up to develop new kinds of assessments that do not seem as abstracted from core content as those that have been used in national and international assessments.
James Greeno

Coherence

The authors of the *Framework* (NRC, 2012a) wrote, "The framework is motivated in part by a growing national consensus around the need for greater coherence—that is, a sense of unity—in K-12 science education. Too often, standards are long lists of detailed and disconnected facts, reinforcing the criticism that science curricula in the United States tend to be "a mile wide and an inch deep" (p. 10). This urge for greater coherence is well grounded in research. We now understand from research on expertise that well-developed knowledge in a domain is integrated, and when material is presented lacking coherence, it is difficult or impossible for it to be learned successfully (NRC, 1999; Tulving & Donaldson, 1972).

At the same time, this urge for greater coherence presents one side of a serious dilemma. The other horn can be expressed in the form of a question: Is scientific knowledge coherent? My impression is that it is not, and to the extent that it is (with a couple of outstanding exceptions) the price of apparent coherence is superficiality. Consider the *Framework*'s crosscutting concepts. The first three—(1) patterns; (2) cause and effect: mechanism and explanation; and (3) scale, proportion, and quantity—seem to be abstractions that don't find their way explicitly into actual scientific work. This characterization could be wrong, and under "needed research," developing and studying a curriculum around the crosscutting ideas proposed in the *Framework* could be a productive project. But if my conjecture is right, there is a reason to be dubious about this version of the crosscutting ideas. The worry about shallowness needs to be addressed by something with depth, and almost all of these "crosscutting ideas" are more like a view from 30,000 feet—inclusivity without depth.

Of course, scientists do have deep understanding and knowledge of something, but for most scientists, their knowledge is not about science in general, it is in regard to a problem they are working on. Quoting the *National Science Education Standards* (NRC, 1996), the *Framework* authors wrote "The natural and designed world is complex; it is too large and complicated to investigate and comprehend all at once. Scientists and students learn to define small portions for the convenience of investigation" (NRC, 2012a, pp. 91–2). A scientist, or a group of scientists, making a claim are accountable for relating it, and the associated evidence, to other work on the same problem. Scientists are not usually held accountable for relating their project to science in general.

The coherence dilemma is reinforced when we consider the practices dimension of the *Framework*. The inclusion of scientific and engineering practices in the *Framework*'s recommended program is admirable, but the discussion of them is too weakly connected with the contents of scientific and engineering knowledge. Consider the first practice—"asking questions (for science) and defining problems (for engineering)"—as an example. Scientific questions and engineering problems do not come out of thin air. Progress on a meaningful question or problem has the prospect of advancing the field. Scientists and engineers work on problems and questions whose solutions and answers are going to move their field beyond where it is at the time. Perhaps science classes could keep track of what they've been able to explain and solve so far, and formulate questions and problems that seem like promising next issues to tackle. (There are examples in math education of doing this for geometry [e.g., Fawcett, 1938]). It may be a mistake to treat a practice of defining a scientific question or engineering problem in a way that can make it seem that the sources of these are just idle curiosities.

Assessment

The idea of assessing students based on their accomplishments and contributions has been in the air for many years (e.g., Meier, 2002). The field should engage in research and development that would support doing that. It may be necessary to give up the idea of a single, standardized version of assessment. Developing the capabilities of scientific inquiry, thinking, and discourse may need to go on in ways that allow for significant differences in the contents of activity in which different students participate. For example, some teachers and students could become adept in scientific practices and core concepts in biology, while others could become adept at a topic in chemistry. It might not make sense to try to assess all of them with a single test.

Research On Learning

The *Framework* and other documents set an ambitious agenda for improving science education. There is much about the learning of science that would be required by this agenda that we do not know or understand. An agenda of designing, developing, and studying curricula, teaching, learning progressions, learning activities, and assessments could be taken on by the field. This would be a very ambitious undertaking and would sternly test the viability of design-based research.

For example, it could be productive to undertake a design-based research study built around the crosscutting concepts that are set out in the *Framework*. I expect the results would be disappointing, but the effort could be productive for understanding aspects of learning and teaching about which we currently do not have sufficient knowledge.

It will be important for the process of developing the next set of science education standards to take an experimental stance. There is much we need to learn about learning and teaching science. An agenda of improvement of science education curricula, teaching, learning progressions, learning activities, and assessment should include provisions to advance our scientific understanding of these processes.

Note

1. Research reviews on learning progressions in science, learning trajectories in mathematics, and teaching sequences can be found in Corcoran, Mosher and Rogat (2009); Daro, Mosher and Corcoran (2011); Duschl, Maeng and Sezen (2011); and Alonzo and Gotwals (2012).

References

American Association for the Advancement of Science. (2001). *Atlas of Science Literacy: Mapping K-12 Learning and Goals*. Washington, DC: AAAS.

Assaraf, O.B., & Orion, N. (2005). Development of system thinking skills in the context of earth system education. *Journal of Research in Science Teaching, 42*(5), 518–60.

Atran, S. (2002). Modular and cultural factors in biological understanding: An experimental approach to the cognitive basis of science. In P. Carruthers, S.P. Stich, & M. Siegal (Eds.), *The Cognitive Basis of Science* (pp. 41–72). Cambridge, UK: Cambridge University Press.

Black, P. & William, D. (1998). Assessment and classroom learning. *Assessment in Education, 5*(1), 7–74.

Blumenfeld, P.C., Kempler, T.M., & Krajcik, J.S. (2006). Motivation and cognitive engagement in learning environments. In Sawyer, R.K. (Ed.), *The Cambridge handbook of the learning sciences* (pp. 475–88). New York: Cambridge University Press.

Bransford, J., Barron, B., Pea, R., Meltzoff, A., Kuhl, P., Bell, P., Stevens, R., et al. (2006). Foundations and opportunities for an interdisciplinary science of learning. In K. Sawyer (Ed.), *The Cambridge Handbook of the Learning Sciences* (pp. 19–34). New York: Cambridge University Press.

Brickman, P., Gormally, C., Armstrong, N., & Hallar, B. (2009). Effects of inquiry-based learning on students' science literacy skills and confidence. *International Journal for the Scholarship of Teaching and Learning, 3*(2), 1–22.

Carruthers, P., Stich, S., & Siegal, M. (Eds.) (2002). *The Cognitive Basis of Science.* Cambridge, UK: Cambridge University Press.

Catley, K., Lehrer, R., & Reiser, B. (2005). *Tracing a prospective learning progression for developing understanding of evolution.* Paper Commissioned by the National Research Council Board of Testing and Assessment, Committee on Test Design for K-12 Science Achievement. Washington, DC: National Research Council.

Cole, M. (1996). *Cultural Psychology: A Once and Future Discipline.* Cambridge, MA: Belknap Press.

Corcoran, T., & Silander, M. (2009). Instruction in high schools: The evidence and the challenge. *The Future of Children: America's High Schools, 19*(1), 157–83. (www.futureofchildren.org).

Corcoran, T., Mosher, R., & Rogat, A. (2009). *Learning Progressions in Science: An Evidence-based Approach to Reform.* (CPRE Research Report #RR-63). Philadelphia, PA: Consortium for Policy Research in Education.

Duschl, R. (2008). Science education in 3 part harmony: Balancing conceptual, epistemic and social learning goals. *Review of Research in Education, 32,* 68–291.

Duschl, R., & Grandy, R. (2008). Reconsidering the Role of Inquiry in School Science: Framing the Debates. In R. Duschl & R. Grandy (Eds.), *Teaching Scientific Inquiry: Recommendations for Research and Implementation* (pp. 1–37). Rotterdam, Netherlands: Sense Publishers.

Duschl, R, Maeng, S., & Sezen, A. (2011). Learning progressions and teaching sequences: A review and analysis. *Studies in Science Education, 47* (2), 119–77.

Engle, R., & Conant, F. (2002). Guiding principles for fostering productive disciplinary engagement: Explaining an emergent argument in a community of learners classroom. *Cognition and Instruction, 20*(4), 399–483.

Fawcett, H.P. (1938). The nature of proof; a description and evaluation of certain procedures used in a senior high school to develop an understanding of the nature of proof.

Ford, D. (2005). The challenges of observing geologically: Third grades descriptions of rock and mineral properties. *Science Education, 89,* 276–95.

Ford, M. (2008). 'Grasp of Practice' as a reasoning resource for inquiry and nature of science understanding. *Science and Education, 17,* 147–77.

Gelman, R., & Brenneman, K. (2004). Science pathways for young children. *Early Childhood Research Quarterly, 19*(1), 150–58.

Giere, R. (1988). *Explaining Science: A Cognitive Approach*. Chicago: University of Chicago Press.

Giere, R. (1999). *Science Without Laws*. Chicago: University of Chicago Press.

Giere, R. (2002). Scientific cognition as distributed cognition. In P. Carruthers, S. Stich, & M. Siegal (Eds.), *The Cognitive Basis of Science* (pp. 285–99). Cambridge, UK: Cambridge University Press.

Goldman, A. (1986). *Epistemology and Cognition*. Cambridge, MA: Harvard University Press.

Gunckel, K., Covitt, B., Salinas, I., & Anderson, C. (2012). A learning progression for water in socio-ecological systems. *Journal of Research in Science Teaching, 49*(7), 843–68.

Hapgood, S., Magnusson, S., & Palincsar, A.S. (2004). Teacher, text, and experience: A case of young children's scientific inquiry. *The Journal Of The Learning Sciences, 13*(4), 455–505.

Hatano, G., & Inagaki, K. (1991). Sharing cognition through collective comprehension activity. In L.B. Resnick, J.M. Levine, & S.D. Teasley (Eds.), *Perspectives on Socially Shared Cognition* (pp. 331–48). Washington DC: American Psychological Association.

Herrenkohl, L., & Guerra, M. (1998). Participant structures, scientific discourse, and student engagement in fourth grade. *Cognition and Instruction, 16*(4), 431–73.

Herrenkohl, L., Palincsar, A., DeWater, L., & Kawasaki, K. (1999). Developing scientific communities in classrooms: A sociocognitive approach. *Journal of the Learning Sciences, 8*(3–4), 451–493.

Keil, F. (1989). *Concepts, Kinds and Cognitive Development*. Cambridge, MA: MIT Press.

Keil, F. (2003). That's life: Coming to understand biology. *Journal of Human Development, 46*, 369–377.

Kelly, G., & Crawford, T. (1997). An ethnographic investigation of the discourse processes of school science. *Science Education, 81*(5), 533–59.

Kesidou, S., & Roseman, J. (2002). How well do middle school science programs measure up? Findings from Project 2061's curriculum review. *Journal of Research in Science Teaching, 39*(6), 522–49.

Kitcher, P. (1993). *The Advancement of Science: Science Without Legend, Objectivity Without Illusions*. New York: Oxford University Press.

Knorr-Cetina, K. (1999). *Epistemic Cultures: How Science Makes Knowledge*. Cambridge, MA: Harvard University Press.

Koslowski, B., & Thompson, S. (2002). Theorizing is important, and collateral information constrains how well it is done. In P. Carruthers, S.P. Stich, & M. Siegal (Eds.), *The Cognitive Basis of Science* (pp. 171–92). Cambridge, UK: Cambridge University Press.

Kuhn, T. (1996). *The Structure of Scientific Revolutions, 4th Edition*. Chicago: University of Chicago Press.

Lehrer, R., & Schauble, L. (2000a). Inventing data structures for representational purposes: Elementary grade students' classification models. *Mathematical Thinking and Learning, 2*, 49–72.

Lehrer, R., & Schauble, L. (2000b). The development of model-based reasoning. *Journal of Applied Developmental Psychology, 21*(1), 39–48.

Lehrer, R., & Schauble, L., Eds. (2002). *Investigating Real Data in the Classroom: Expanding Children's Understanding of Math and Science*. New York: Teachers College Press.

Lehrer, R., & Schauble, L., (2004). Modeling natural variation through distribution. *American Educational Research Journal, 41*(3), 635–79.

Lehrer, R., & Schauble, L. (2006). Cultivating model-based reasoning in science education. In K. Sawyer (Ed.), *The Cambridge Handbook of the Learning Sciences* (pp. 371–87). New York: Cambridge University Press.

Lehrer, R., & Schauble, L. (2012). Seeding evolutionary thinking by engaging children in modeling its foundations. *Science Education, 96*(4), 701–24.

Lehrer, R., Jaslow, L., & Curtis, C. (2003). Developing understanding of measurement in the elementary grades. In D.H. Clements & G. Bright (Eds.), *Learning and Teaching Measurement.* 2003 Yearbook (pp. 100–21). Reston, VA: National Council of Teachers of Mathematics.

Lehrer, R., Kim, M., & Schauble, L. (2007). Supporting the development of conceptions of statistics by engaging students in measuring and modeling variability. *International Journal of Computers for Mathematics Learning, 12*, 195–216.

Longino, H. (1990). *Science as Social Knowledge.* Princeton, NJ: Princeton University Press.

Longino, H. (2002). *The Fate of Knowledge.* Princeton, NJ: Princeton University Press.

Magnusson, S.J., & Palincsar, A.S. (2005). Teaching to promote the development of scientific knowledge and reasoning about light at the elementary school level. In M.S. Donovan & J.D. Bransford (Eds.), *How Students Learn: History, Mathematics, and Science in the Classroom* (pp. 421–74). Washington, DC: National Academies Press.

Mayer, R. (2011). Instruction based on visualizations. In R.E. Mayer and P.A. Alexander (Eds.), *Handbook of Research on Learning and Instruction* (pp. 427–45). New York: Routledge.

Mayes, R.L., Peterson, F., & Bonilla, R. (2013). Quantitative reasoning learning progressions for environmental science: Developing a framework. *Numeracy, 6*(1), Article 4.

Meier, D. (2002). The Power of Their Ideas: Lessons for America from a Small School in Harlem. Boston: Beacon Press.

Metz, K. (1995). Reassessment of developmental constraints on children's science instruction. *Review of Educational Research, 65*, 93–127.

Mohan, L., Chen, J., & Anderson, C.W. (2009). Developing a multi-year learning progression for carbon cycling in socio-ecological systems. *Journal of Research in Science Teaching, 46*, 675–98.

National Assessment Governing Board (2008). NAEP 2009 science framework development: Issues and recommendations. Washington, DC: US Department of Education.

National Research Council (1996). *National Science Education Standards.* Washington, DC: National Academy Press. http://www.nap.gov

National Research Council (1999). *How People Learn.* Washington, DC: National Academy Press. http://www.nap.gov

National Research Council (2002). *Scientific Research in Education.* Washington, DC: National Academy Press. http://www.nap.gov

National Research Council (2005). *Systems for State Science Assessment.* Washington, DC: National Academy Press. http://www.nap.gov

National Research Council (2006). *Rising Above the Gathering Storm.* Washington, DC: National Academy Press. http://www.nap.gov

National Research Council (2007). *Taking Science to School: Learning and Teaching Science Kindergarten to Eighth Grade.* Washington, DC: National Academy Press. http://www.nap.gov

National Research Council (2009). *Engineering in K-12 Education: Understanding the Status and Improving the Prospects.* Washington, DC: National Academy Press. http://www.nap.gov

National Research Council (2012a). *A Framework for K-12 Science Education: Practices, Crosscutting Concepts, and Core Ideas.* Washington, DC: National Academy Press. http://www.nap.gov

National Research Council (2012b). *Education for Life and Work: Developing Transferable Knowledge and Skills in the 21st Century.* Washington, DC: National Academy Press. http://www.nap.gov

Nersessian, N. (2002). The cognitive basis of model-based reasoning in science. In P. Carruthers, S. Stich, & M. Siegal (Eds.), *The Cognitive Basis of Science* (pp. 133–53). Cambridge, UK: Cambridge University Press.

Nersessian, N. (2008). Inquiry: How science works; Model-based reasoning in scientific practice. In R. Duschl & R. Grandy (Eds.), *Teaching Scientific Inquiry: Recommendations for Research and Implementation* (pp. 57–79). Rotterdam, Netherlands: Sense Publishers.

NGSS Lead States (2013). *Next Generation Science Standards: For States, By States.* Retrieved from www.nap.edu/catalog/18290/next-generation-science-standards-for-states-by-states.

Ohlsson, S. (1992). The cognitive skill of theory articulation: A neglected aspect of science education. *Science & Education, 1*(2), 181–92.

Rosebery, A., Warren, B., & Conant, F. (1992). Appropriating scientific discourse: Findings from language minority classrooms. *The Journal of the Learning Sciences, 2*(1), 61–94.

Sandoval, W. A. (2005). Understanding students' practical epistemologies and their influence on learning through inquiry. *Science Education, 89*(4).

Sandoval, W. A. & Reiser, B. J. (2004). Explanation-driven inquiry: Integrating conceptual and epistemic scaffolds for scientific inquiry. *Science Education, 88*, 345–72.

Sawyer, R. K., (Ed.) (2006). *The Cambridge Handbook of the Learning Sciences.* New York: Cambridge University Press.

Schauble, L., Glaser, R., Duschl, R., Schultz, S., & John, J. (1995). Students' understanding of the objectives and procedures of experimentation in the science classroom. *The Journal of the Learning Sciences, 4*(2), 131–66.

Schmidt, W. H., McKnight, C. C., & Raizen, S. A. (1997). *A Splintered Vision: An Investigation of US Science and Mathematics Education.* Boston: Kluwer Academic Publishers.

Siegal, M. (2002). The science of childhood. In P. Carruthers, S. P. Stich, & M. Siegal. (Eds.), *The Cognitive Basis of Science,* (pp. 300–15). Cambridge, UK: Cambridge University Press.

Smith, C. (2007). Bootstrapping processes in the development of students' possessive common sense matter theories: Using analogical mappings, thought experiments, and learning to promote conceptual restructuring. *Cognition and Instruction, 25*(4), 337–98.

Smith, C., Maclin, D., Houghton, C., & Hennessey, M. G. (2000). Sixth-grade students' epistemologies of science: The impact of school science experience on epistemological development. *Cognition and Instruction, 18*(3), 285–316.

Smith, C., Wiser, M., Anderson, C., & Krajcik, J. (2006). Implications of research on children's learning for assessment: Matter and atomic molecular theory. *Measurement: Interdisciplinary Research and Perspectives, 4,* 11–98.

Smith, C. L., Wiser, M., & Carraher, D.W. (2010). *Using a comparative, longitudinal study with upper elementary school students to test some assumptions of a learning progression for matter.* Paper presented at the annual meeting of the National Association for Research on Science Teaching, Philadelphia, PA.

Songer, N. & Linn, M. (1991). How do students' views of science influence knowledge integration? *Journal of Research in Science Teaching, 28*(9), 761–84.

Subrahmanyam, K., Gelman, R., & Lafosse, A. (2002). Animates and other separable moveable objects. In E. Fordes & G. Humphreys (Eds.), *Category Specificity in Brain and Mind* (pp. 341–373). London: Psychology Press.

Thagard, P. (1992). *Conceptual Revolutions*. Princeton, NJ: Princeton University Press.

Toth, E., Suthers, D., & Lesgold, A. (2002). "Mapping to know": The effects of representational guidance and reflective assessment on scientific inquiry. *Science Education, 86*(2), 264–86.

Tulving, E., & Donaldson, W. (Eds.) (1972). *Organization of Memory*. New York: Academic Press.

van Zee, E., & Minstrell, J. (1997). Using questioning to guide student thinking. *The Journal of the Learning Sciences, 6*, 227–69.

Wilson, M. (2009). Measuring progressions: Assessment structures underlying a learning progression. *Journal of Research in Science Teaching, 46*(6), 716–30.

Wiser, M., Smith, C., Doubler, S., & Asbell-Clark, J. (2009). *Learning progressions as a tool for curriculum development: Lessons from the inquiry project*. Paper presented at the Learning Progressions in Science (LeaPS) Conference, Iowa City, IA, USA.

THEME I

Systems Thinking

2

THINKING ABOUT A SYSTEM AND SYSTEMS THINKING IN ENGINEERING

Thomas A. Litzinger

Introduction

"System" is a concept that is central to learning and teaching engineering. In engineering analysis courses, students' thinking is focused on a system that they wish to model mathematically in order to understand its behavior (i.e., students are thinking about a system). A major focus of analysis courses is having students learn to use well-established procedural knowledge.

In engineering design courses, on the other hand, students engage in "design thinking." In a traditional design process, one of the first steps is engaging clients or customers to determine their expectations. If "systems thinking" is folded into the design process, students explore a much broader design space than the one represented by the needs of the customer. Thus, integrating systems thinking into design clarifies the rich context in which engineering design occurs.

The next section of this chapter discusses the importance of systems thinking in engineering and its place in engineering undergraduate curricula. The concept of a system and definitions of systems thinking are then discussed along with the general characteristics of analytical, design and systems thinking. Example problems for analytical thinking, design thinking, and systems thinking in mechanical engineering are presented. Next, two basic approaches to the teaching of systems thinking are described and the broad goal statements related to systems from *A Framework for K-12 Science Education* (National Research Council [NRC], 2012) (hereafter referred to as the *Framework*) are mapped to a set of learning objectives. The chapter concludes with a discussion of opportunities and challenges in teaching students to apply systems thinking in an engineering context.

Systems Thinking in Engineering Undergraduate Curricula

Addressing the issues of the 21st century will require a new type of engineer and a new approach to the practice of engineering—requiring engineers who can integrate systems thinking fully and seamlessly into their work. Some writers have referred to this new approach as "Holistic Engineering." In her chapter within the book *Holistic Engineering Education*, Charla Wise asserts that "(t)he days of engineers creating designs, structures, and platforms as an independent product without an understanding of the end product, the intended use, and the unintended consequences are over" (Wise, 2010, p. 228). Within the same book, Priscilla Guthrie states that "Engineering endeavors should be holistic, with context taking precedence, setting the stage for, and playing a prominent role in the collaborative development of solutions" (Guthrie, 2010, p. 94). According to de Weck, Roos and Magee (2011), engineers working on today's complex technological systems must

> interact with a host of socioeconomic complexities and "externalities "—impacts, either positive or negative, that are not a direct part of the artifact or even a self-contained system or process under consideration. It used to be that engineers, even those who were beginning to understand that these externalities might matter, did not worry about them in their designs. Today, these externalities must be factored into the design process. It is all about broadening the boundaries of the system, because the system has, in some ways, swamped us (p. 16).

Current accreditation standards for US engineering programs do not explicitly include systems thinking. However, the criterion related to design implies that students should engage in such thinking:

> Students should have an ability to design a system, component, or process to meet desired needs within realistic constraints such as economic, environmental, social, political, ethical, health and safety, manufacturability, and sustainability (ABET, 2013).

The inclusion of sustainability in the definition provides a strong connection to systems thinking because it provides an excellent context in which to have students practice systems thinking. The definition of design in the accreditation criteria is:

> Engineering design is the process of devising a system, component, or process to meet desired needs. It is a decision-making process (often iterative), in which the basic sciences, mathematics, and the engineering sciences are applied to convert resources optimally to meet these stated needs (ABET, 2013, p. 4).

All US accredited engineering programs must have a "curriculum culminating in a major design experience" (p. 4), which provides an excellent opportunity for students to engage in systems thinking. When "systems thinking" is integrated into the design process, students explore environmental, social, societal and political factors, and both long and short timescale behavior of the system. Having students apply systems thinking during their design process creates opportunities to discuss the purpose and importance of engineering in modern society and the level of creativity that is required to address complex global issues.

Systems and Systems Thinking in Engineering

So what is systems thinking? This is a difficult question to answer in a general way because systems thinking is applied in so many contexts from organizational management to ecology to engineering. Further complicating the definition of systems thinking is the fact that it has a number of intellectual traditions including cybernetics, general systems theory, system dynamics, soft systems, and complexity theory (Ramage & Shipp, 2009). However, one need not adopt any particular theoretical perspective to discuss the key elements of systems thinking that are appropriate for engineering education, especially if the focus is on K-16 education.

Before describing systems thinking in an engineering context, we need to first define an engineering system. Meadows offers a general definition of a system as "an interconnected set of elements that is coherently organized in a way that achieves something" (Meadows & Wright, 2008, p. 11). de Weck, Roos and Magee (2011) define an engineering system as a "class of systems characterized by a high degree of technical complexity, social intricacy, and elaborate processes, aimed at fulfilling important functions in society" (p. 31). Merging these two definitions leads to the following: an engineering system is an interconnected set of technical elements characterized by a high degree of complexity, elaborate processes, and social intricacy aimed at fulfilling an important function in society.

There are five key constructs in this definition: function, technical elements, interconnectedness, complexity, and social intricacy. Function is the purpose of the system (e.g., the function of the commercial air transportation system is to move people and goods around the US and the world). Technical elements are human-made artifacts of sufficient technical complexity that engineers must be involved in their design and manufacture (e.g., cellphones, cars, aircraft and computers). Complexity and interconnectedness may best be explained by example. The commercial air transportation system, which consists of thousands of aircraft operating across hundreds of airports, has a very high level of complexity and interconnectedness. The interconnectedness exists on multiple levels. Flights from one airport to another form one level of interconnectedness as evidenced when delays at a major hub ripple though the entire system.

Other interconnectedness exists between the air transportation system and the US and global economies. The last construct, social intricacy, refers to the fact that complex engineering systems "rely on people and their organizations for the design, manufacturing, and operation of the system, and are influenced by the societal . . . context as well" (de Weck et al. 2011, p. 185). Complex engineering systems do not exist inside a bubble independent of the people who design, build, and use them, nor are engineering systems free from societal and political influences.

Analytical, Design, and Systems Thinking

A significant portion of engineering courses in undergraduate curricula focus on the development of analysis skills (Sheppard, Macatangay, Colby, & Sullivan, 2008). A major focus of such courses is having students learn to apply well-established procedural knowledge. Much of the teaching and learning in these courses is based on a "reductionist" approach (i.e., reducing a complex device or engineering system down to its simplest components and then studying those simplest components).

In analysis courses, students may study groupings of components such as sets of devices that make up an electric circuit, a simple mechanical system, or a refrigerator. However, even in these cases, the approach is to study the smallest set of components possible to address the problem of interest. Typically, the first step in the analysis process is to identify the system and to separate it from its surroundings so that interactions between the system and its surroundings can be identified as a prelude to mathematical modeling. This step is followed by making assumptions and approximations about the system so that the appropriate set of equations to model the system can be written. These equations are then solved, and in some cases, students reflect upon the results of the analysis and justify why they are reasonable. The cognition involved in the analysis process is predominantly convergent.

In engineering design courses, students engage in "design thinking" which requires both divergent and convergent thinking. The goal of the design process is to find an optimal solution for customers while minimizing costs so that profit is maximized, thereby addressing the needs of the owners of the company and/or shareholders. In a traditional design process, one of the first steps is engaging clients or customers to determine their expectations. Using methods of engineering design, students define concrete specifications based on the needs of customers or clients; these specifications then guide the design process and become part of the criteria against which designs are evaluated. The students use divergent thinking processes in generating multiple designs that satisfy the design criteria. Following the divergent phase, students select preferred solutions, which then go through further refinement and additional evaluations against criteria until the final design is selected.

TABLE 2.1 Comparison of analytical, design, and systems thinking in engineering

	Analytical Thinking	*Design Thinking*	*Systems Thinking*
Goal of process	Characterize a device, component, or set of components through mathematical modeling	"Create optimal solutions for select stakeholders." (Coughlan and Ponto, 2011) "Designing things right (technically correct, efficient, emotionally meaningful or successful in the marketplace)" (Cardenas et al. 2010, p. 287)	"Optimize performance of overall system (Coughlan and Ponto, 2011) "Designing the right things (i.e. aligned with the needs of the present without compromising future generations)" (Cardenas et al. 2010, p. 287)
Focus of cognitive processes	Behavior of device, component or set of components	Performance of product/system against the needs of stakeholders	Interactions within the system; performance of complete system
Convergent/Divergent Thinking	Convergent thinking dominant	Both	Both, with even greater use of divergent thinking than in design
Time scale/temporal behavior	Often focuses on an instant in time and short time scales	May focus on different time scales, both short and long term	Considers short and long time scales, and evolution over time; seeks patterns of change
Nature of solution process	Often linear due to level of simplification used in analysis	Iterative/recursive	Iterative/recursive
Types of solutions considered	Technical solution	Technical solution	Explores all possible solutions, not just technical ones (Coughlan and Ponto, 2011)

If "systems thinking" is integrated into the design process, students explore a much broader design space during the divergent thinking phase than the design space defined by the needs of the customer and financial considerations. Systems thinking has been characterized as holistic and "expansionist" (Pourdehnad, Wexler, & Wilson, 2011). Whereas engineering analysis focuses on the components or small sets of components, engineering systems thinking focuses on the much larger, more complex systems, sometimes referred to as a system of systems, and their interactions with natural and sociopolitical systems. Through systems thinking, students consider much more than the needs of the customers and financial criteria. In exploring potential designs, they also consider interactions of the designed system on people, organizations, society, government and the environment. They also consider both long and short time scales including effects of the design on future generations.

Table 2.1 summarizes characteristics of analytical thinking, design thinking, and systems thinking in engineering. It is important to note that all three types of thinking are critical for the successful execution of real-world engineering projects. The relative importance of each type of thinking depends on the complexity of the project.

Before leaving this section, it is important to explain "systems engineering," which is not synonymous with systems thinking. According to Guthrie (2010, p. 94), systems engineering focuses on "the design and manufacture of complex products" and was built on "processes developed in the last quarter of the 20th century to support government space and defense programs." The tools of systems engineering provide systematic approaches to identifying, tracking, and achieving the performance goals of a complex technological product such as a fighter aircraft, which is composed of many sub-systems. In many ways, systems engineering is an example of a highly structured design process. Much like traditional design thinking, systems engineering ignores many externalities that are considered when systems thinking is applied to design. Here, systems engineering will be considered to be a special type of design process. It will not be explicitly discussed in order to avoid confusion between the terms systems thinking and systems engineering.

Examples of Analytical, Design, and Systems Thinking in an Engineering Context

To illustrate the distinctions among analytical, design, and systems thinking, examples in a mechanical engineering context are presented in Table 2.2. Each example is related to wind energy. Problem statements and an illustrative set of learning objectives for the three different types of thinking are provided.

For the engineering analysis example, the system being considered is the wind turbine itself (Figure 2.1). At the most abstract level of analysis, the students can completely ignore the details of the blades and treat them as a disk as illustrated in Figure 2.2. The gray lines in the figure illustrate the boundary between the surroundings and the system being considered, which is the air flowing through the wind turbine. Having identified and graphically represented the appropriate system, the students will proceed to simplify the equations of conservation of mass and momentum, or they will write the appropriate forms from memory. They will then do algebraic manipulation to combine the equations until they arrive at an expression for mechanical power produced by the turbine. Finally, they will determine the maximum power using methods of differential calculus. This type of problem represents the simpler end of the spectrum of problems that are solved in introductory engineering analysis courses. At the other end of the spectrum, students in a senior course on the design of wind turbines could spend weeks developing and applying a computer code to simulate the flow of air over the blades of the wind turbine in order to determine the optimal geometry.

FIGURE 2.1 Image of modern wind turbine

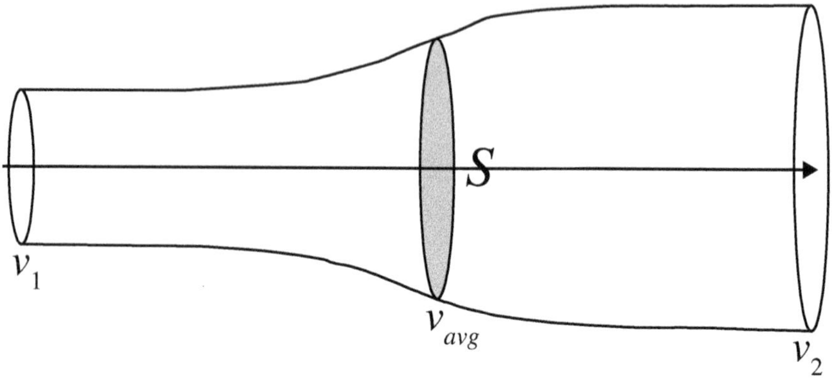

FIGURE 2.2 System diagram for analysis of wind turbine

The design thinking project summarized in Table 2.2 includes phases of the design process that involve establishing the need for a new product, determining customer needs, translating customer needs into product specifications, generating multiple design concepts, selecting the best design based on customer needs, performing system level design and analysis, building and testing a prototype, doing detailed design and analysis of production design, and planning for manufacturing. Complementing the overall design process are methodologies that are intended to enhance overall product quality while minimizing costs of production and impact on the environment such as "Design for Assembly" and "Design for Environment." Typically, students will write reports and make presentations on their designs. They will also apply skills and knowledge from their analysis courses.

To complete the systems thinking example presented in Table 2.2, students would have to engage in a wide range of learning activities including writing goal statements, defining the system, and exploring the complex behavior of the system across technological, social, political, environmental, and financial dimensions. In order to ensure that the students are making reasonable progress, it would be necessary to have intermediate deliverables. The first intermediate deliverable might be their goal statements along with verbal and graphical definitions of the system that they intend to study to ensure that students are considering a system of sufficient complexity to address their goals. Other intermediate deliverables could include a description of key interactions among components of the system and a summary of major sources of information that will be used. In addition to requiring the use of systems thinking skills, this project would require students to apply their skills in locating, critically evaluating, and synthesizing information from a variety of sources.

TABLE 2.2 Examples of analytical, design, and systems thinking appropriate for junior or senior mechanical engineering students

Analytical thinking	*Design Thinking*	*Systems Thinking*
Apply conservation of mass and momentum to determine the maximum possible power that can be generated by an ideal wind turbine. Model the wind as one-dimension, uniform flow and treat air as an incompressible fluid. State and justify any additional approximations and assumptions involved in arriving at your solution.	The new framework for science education includes engineering design, which creates a new market for educational products. Teams will design and fabricate a model wind turbine that 3rd and 4th grade students can use to experience the design process. For this product development process, the customers will include students and faculty members from Science Education.	In order to reduce emissions of greenhouse gases, many governments around the world are implementing legislation to encourage the construction of wind farms. The Commonwealth of Pennsylvania has hired your consulting firm to recommend guidelines for new wind farms in PA that optimize environmental and societal impacts. Your firm was hired because of its expertise in applying systems thinking to complex technological systems. Your deliverable for this project is a detailed report summarizing your recommended guidelines and the information on which they are based.
Learning objectives: Students will • Sketch the system and its boundary, then indicate all mass, momentum and energy flows • Identify assumptions and approximations • Combine conservation of mass, momentum, and energy to determine the power from the turbine • Apply differential calculus to determine maximum power	Learning Objectives: Student teams will • Execute a systematic design process, which begins with determining customer needs, including safety • Apply their analysis skills to predict the power output of their wind turbine • Conduct experiments to measure the performance of their wind turbines • Demonstrate effective team skills	Learning Objectives: Student teams will • Define the system that includes technological components as well as natural, social, and political components • Identify key interactions among the components of the system • Consider all phases of the life cycle of wind farms from materials' extraction to end of life disposal/recycling • Consider the integration of wind power into the existing electrical power production systems serving PA on the overall effectiveness of wind energy in reducing green house gas emissions. • Locate and critically evaluate sources of information necessary to complete the project

Teaching Systems Thinking

Córdoba-Pachón (2011) describes two approaches to teaching systems thinking, "abstracting" and "engaging," which he links to Abbott's conceptions of professional knowledge (Abbott, 1988). In the abstracting mode of teaching, the instructor teaches by "*abstracting* and disseminating existing knowledge of systems thinking" (p. 44). Use of the "abstracting" mode of teaching will help students learn the concepts of systems thinking and its basic tools. However, students who are taught in this way are not likely to learn how to *do* systems thinking. In the engaging mode of teaching, the focus shifts to teaching students about systems thinking as a "strategy for learning about relevant areas . . . sustainability, project management, or corporate social responsibility" (p. 48). Students are asked to apply the methods of systems thinking to real problems of interest to them such as the sustainability of a city or a business relocation to another country (p. 49).

Regardless of the approach used to teach systems thinking, teachers need a concrete set of learning objectives describing the key knowledge, skills, and attitudes that are central to systems thinking. The literature provides a few examples of learning objectives for systems thinking. Stave and Hopper (2007) propose a taxonomy of learning objectives for systems thinking based on a survey of the literature and experts on systems thinking. The Waters Foundation, a non-profit educator's group focused on teaching systems thinking, has defined a set of learning objectives that are described as "habits of systems thinkers." Froyd et al. (2007) derived a set of 20 learning objectives for systems thinking in engineering based upon four broad characteristics of systems thinking and six levels of Bloom's taxonomy. In addition, Zhang and Vanasupa (2011) offer a set of learning objectives for systems thinking in engineering that include both cognitive and affective objectives.

Building upon these four sets of learning objectives, a set of cognitive learning objectives for systems thinking in engineering education was created and mapped to systems and system modeling statements within the *Framework* (NRC, 2012). This set of learning objectives is presented in Table 2.3. In addition to cognitive outcomes, research on systems thinkers in engineering (Davidz & Nightingale, 2008) suggests that teachers should also help students develop curiosity, open-mindedness, and tolerance for uncertainty. In addition, Davidz and Nightengale's research identified a creative thinking ability as central to successful systems thinking as well as strong communication skills, because complex systemic problems can only be addressed by teams.

The set of learning outcomes in Table 2.3 can be mapped to a progression of sophistication in the types of assignments that students undertake. At the lowest cognitive level, students could be asked to learn and apply key terminology related to systems thinking. At the next level students could be given a context and goal (e.g., designing new policies or technology to reduce food waste in their school), and then be asked to define and describe the system they need to study for the design process. They could be asked to provide representations of their

TABLE 2.3 Possible cognitive learning objectives related to systems thinking

	Related statements from crosscutting themes found in the Framework	*Possible Learning Objectives*
1. Apply basic terminology and concepts	Define relevant terms/concepts including system, component, boundary, resources, flow, feedback, and emergent properties. (p. 92)	Define key terms/concepts Apply key terms/concepts appropriately in describing a system
2. Define the system	"Specify the boundary of the system being modeled" (p. 93)	Define system boundary verbally and in graphical representations Justify the selected boundary
3. Identify and characterize interactions	Identify "what interactions are occurring . . . and to recognize that they all involve transfers of energy, matter, and (in some cases) information among parts of the system" (p. 93)	Identify interactions within the system and between the system and its surroundings; classify the type of interaction as mass, energy, or information Identify processes that involve feedback Identify long and short term behavior of the system Consider systems from different perspectives to enhance completeness of the model Create a *verbal* description of interactions within a system and between a system and its surroundings Create a *graphical* representation of interactions within a system and between a system and its surroundings Create a *mathematical* description of the interactions within a system and between a system and its surroundings

(*Continued*)

TABLE 2.3 Continued

	Related statements from crosscutting themes found in the Framework	*Possible Learning Objectives*
4. Create models of the system	Create *models* of systems at different levels from "lists and simple sketches to detailed computer simulations or functioning prototypes" (p. 92)	Create a *verbal* description of a system Create *graphical* representations of a system Create and solve a *mathematical* model of a system; identify and justify all approximations and assumptions used in developing the mathematical model
5. Solution process		Develop a goal statement for the problem (e.g. "optimize the impact of wind farms") Define the appropriate system and describe it as fully as possible including all interactions within the system and between the system and its surroundings Generate possible approaches Evaluate approaches Repeat processes to refine, as needed, the goal statement, system description, interactions and models, possible solutions, etc.

system in graphical form. Moving up in progression of challenge, students could be asked to identify interrelationships within the system and between the systems and it surroundings. Students could also be asked to quantify characteristics of the system (e.g., how much food is used annually at their school and how much is wasted). Depending on the level of the students and the tools available, they could be asked to create some form of mathematical representation of the system. Applying systems thinking to problems related to food would bring many different topics into the discussion such as hunger in the US and other countries, the competition to use corn for food versus biofuel production, and energy use in agriculture.

Whatever methods are chosen to teach systems thinking, a fundamental skill to be developed must be the ability to comprehend the scope and scale of a complex technological system intertwined with natural and sociopolitical systems. At the very least, we must strive to help students expand their awareness of the interactions that occur among components of such a system so that they will consider a larger space for exploring solutions. In developing this ability, however, we much take care not to allow the complexity to overwhelm our students (or ourselves!). As pointed out by de Weck, Roos and Magee (2011), "There must be some limits and structure to what is considered to constitute 'holistic,' or we would have to study every issue from the perspective of the entire universe—a debilitating approach to any progress" (p. 46). So educators must not only help students to see the whole system, but also help them to define the boundary around the system to include those elements that are necessary for achieving an optimal solution to the problem under consideration for the complete system.

Opportunities and Challenges

Teaching systems thinking and engineering design has great potential to enhance the education of K-12 students and engineering undergraduates. Systems thinking makes clear the rich context in which engineering design takes place. It affords the opportunity to discuss the purpose and importance of engineering in modern society and the level of creativity that is required to address complex, global issues. A systems context for teaching engineering would also provide a perfect opportunity to address many of the topics listed in engineering accreditation criteria for design including environmental, social, political, and ethical topics.

In spite of the potential of systems thinking to enhance the teaching and learning in engineering programs, systems thinking is not yet part of many US engineering programs. Fortunately, examples of the successful integration of systems thinking do exist that can serve as models for integrating systems thinking and engineering. Sustainability is a common context through which systems thinking is introduced into engineering programs. The relatively new engineering program at James Madison University, founded in 2005, was created with the goal of integrating systems thinking, sustainability, and design (Nagel, Pierrakos, Pappas, & Ogundipe, 2011). Carnegie Mellon (Davidson, Hendrickson, & Matthews, 2007) and Michigan Technological University (Hokanson, Phillips, & Mihelcic, 2007) also have successful programs built around sustainability.

Preparing teachers, whether K-12 or engineering faculty members, to teach and guide the learning of systems thinking will present significant challenges. Preparing teachers for addressing lower level learning objectives in Table 2.3 should be straightforward; this teaching would likely be done in the "abstracting" mode described by Córdoba-Pachón (2011). However, preparing teachers to guide students as they undertake projects that involve systems thinking (i.e., Córdoba-Pachón's "engaging" mode) will be much more challenging. Such project-based learning is a much less ordered form of teaching that often places unforeseen demands on the teacher. As such, only teachers with a high degree of confidence in their knowledge and ability to apply methods of systems thinking and to scaffold learning are likely to undertake such an approach. Systems thinking projects also have potential for students to become frustrated and therefore require careful planning and management from the teacher.

Preparing K-12 teachers to teach systems thinking in an engineering design context brings the additional challenge of preparing them to teach engineering design. So, K-12 teachers are faced not only with learning about systems thinking, but they must also learn about engineering design. Furthermore, they do not have the technical expertise to address the types of problems that could be addressed by an engineering instructor. Therefore, creating professional development experiences that would prepare K-12 teachers to teach systems thinking in an engineering design context presents significant challenges.

Designing such a professional development experience for teachers is an opportunity to apply systems thinking approaches in an instructional design context. Indeed, lessons from the history of change in US educational processes clearly indicates the need for a systems approach to avoid repeating past failures, which ignored the needs of individual teachers and the demands placed on them by the complex system in which they do their work (Litzinger & Lattuca, 2014). Developing appropriate instructional materials for systems thinking and design, and professional development workshops for teachers to learn about and adapt the materials, would seem to be one path forward. As a minimum, the team to design such materials, and the workshops, should involve teachers who will use the materials, engineering faculty members with experience in teaching systems thinking and design, and instructional designers. The involvement of school administrators would also enrich the design process. Ideally, the design process would be a "learning process for participants; structuring ongoing interactions among those who seek to enact change with those who are being asked to implement that change." (Litzinger & Lattuca, 2014, p. 389).

3

DIAGNOSTIC INSTRUCTION

Toward an Integrated System for Classroom Assessment

Jim Minstrell, Ruth Anderson, and Min Li

Introduction

The Next Generation Science Standards (NGSS Lead States, 2013), which highlight a blending of knowledge and practice, provide an exciting opportunity to re-examine the very nature of assessment and its role throughout our educational system (Heritage, 2010; Pellegrino, 2013). As we consider how to measure deep learning and seek to foster it within the classroom, we must also think about how the entire system of assessment—from the classroom to large scale testing—can contribute rich information about what and how students are learning. Efforts are currently underway at the national level to create a more coherent and coordinated technology assisted system of assessment. However, the plan seems to be predominantly "top down" and does not incorporate classroom level assessment. Nevertheless, if our long term intention is (as we hope) to become a "learning" educational system (i.e., adaptable, flexible, responsive to changing needs and constraints), then all levels of the system need to be actively involved.

In this chapter, we focus on the subsystem of classroom assessment and propose one vision for revising its structure to make it more systematic and better positioned to participate in a larger system of assessment. Within the context of science education, we address three important challenges to effective classroom assessment. The first is that of integrating or "blending" assessment within instruction and learning by design into a single productive cycle. Assessment is often viewed as an "add on" by teachers and students alike—an event that manages to "take time away," "interrupt," or otherwise disturb the learning process. In response, some researchers have suggested conceptualizing assessment *for* learning (i.e., formative assessment) and even assessment *as* learning not within a measurement paradigm but within a larger paradigm of learning (Heritage, 2010; Shepard, 2005). This sizeable challenge to focus on deep learning requires teachers and students to adopt a vision of the classroom that gives primacy to student expression and interactions rather than to teacher moves and to envision each learning activity as an assessment opportunity that in turn informs a subsequent design or choice of learning experience.

A second challenge is that of providing high quality assessment data to which educators, and larger education systems, can respond productively. Assessments intended to inform instruction need to support a nuanced interpretation of student learning from which to design actions that target specific learning needs in order to move forward (Wiliam, 2008). Currently, when teachers do engage in the interpretation of student work and responses, their evaluations rarely go beyond "got it/didn't get it" (Otero, 2006). Then subsequent instruction is mostly limited to "re-teaching" of the same materials or topically related actions (e.g., another problem about forces and motion) rather than one that is targeted to a specific learning need (e.g., the need to associate *accelerated motion* with a *constant net force*, etc.) (Minstrell, 1984). Teachers need rich assessment data for making sense of and talking about student expressions of learning as they plan for subsequent instruction.

A third related challenge involves the constraints of time and data management. Pellegrino has recently observed: "For teachers to effectively implement assessment as part of their pedagogy, they need tools for presenting tasks and collecting and scoring student performance. They also need smart systems that provide actionable information about the meaning and implications of student performance relative to instruction and student learning . . ." (Pellegrino, 2013, p. 323). As we consider "smart" systems, however, we need to be wary of quests to create the equivalent of "teacher proof curricula" in the realm of assessment. Whatever the technology, teachers should be familiar with the structure and fluent in the language behind the technology that will be used to represent student learning. Moreover, such technologies must be well aligned with the other instructional elements of the classroom so that it does not become yet another "add on."

In response to the above challenges, we propose an integrated system of three complementary parts: First, we present a theoretical framework that provides a vision or "blueprint" for integrating and blending assessment and instruction as teachers design and implement lessons that monitor and build on students' thinking and practices. Next we explain "Facets," a framework for interpreting and describing learner conceptions and practices in relation to specific learning goals. The Facets framework is the instructional "engine" that drives the blended instruction; it provides the basis for a small grain analysis of student understanding and thereby tightens the cycle of assessment and instruction. Finally, we present one example of a technology-supported "tool box" to facilitate teachers in their implementation of such a blended instruction—one which is compatible with both the theoretical and interpretive frameworks.

Each of the three parts is essential to a smooth and coherent blending of assessment and instruction. Remove the support tools and the process becomes burdensome and unfeasible; remove the interpretive framework and the instructional decision-making becomes less targeted; remove the theoretical or instructional framework and a perspective and coherent vision are more difficult to maintain.

We recognize that the individual pieces we present are not unique in the field and that others have been or are in the process of developing tools and frameworks with similar characteristics and purposes. However, the compatibility of the pieces we are presenting is unique. All three parts were developed in concert over several years and together provide a coherence that is not easily achieved when assembling independent tools and instruments. Moreover, each of the three has been favorably vetted in classrooms. The interpretive framework, for example, which constitutes the backbone of the system, is grounded in dozens of years of classroom research as well as practitioner wisdom and experience and has been used by teachers and researchers for over twenty years. This framework and a growing body of supporting empirical research encouraged us to present this as a promising system of blended assessment, instruction, and learning. Even though we present each individual part of the system in the pages that follow, we emphasize their interdependence and the need for installing and testing the system as a whole rather than as individual, isolated interventions.

Theoretical Framework for Blending Assessment and Instruction

At the heart of effective classroom assessment lies the need to engage with student thinking as it develops. As Margaret Heritage has observed, "Only by keeping a very close eye on emerging learning through formative assessment can teachers be prospective, determining what is within the students' reach, and providing them experiences to support and extend learning through which students can then incorporate new learning into their developing schema" (Heritage, 2010, p. 8).

Figure 3.1 represents Building on Learner Thinking (BOLT): a framework for guiding design and implementation of learning environments that blends assessment and instruction into an *assessment as learning* process (Black & Wiliam, 2009). In order to accentuate the focus on learners and learning, this framework emphasizes primacy of students' ideas and experiences and teacher responsive actions over teacher-directed moves that students follow.[1] BOLT is not tied to any particular curriculum or teaching strategy. Rather, it is intended to guide teachers in coherently fostering, supporting, and monitoring student thinking as it develops from naïve to more sophisticated through a process that emphasizes model development and validation through observation, rational argument, community (i.e., class) consensus, and knowledge generation.

The squares from left to right represent a progression of ideas from students' initial ideas (A) on the left to whole-class consensus ideas (E) on the right, which

BUILDING ON LEARNER THINKING

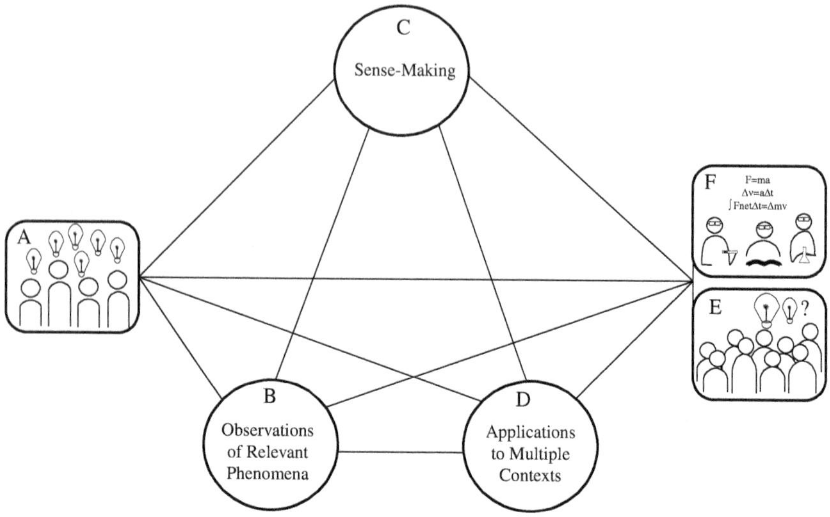

FIGURE 3.1 BOLT framework

are built from the experiences and initial ideas and arrived at by collaboratively generating, testing and evaluating ideas and making collective sense. By the end of an instructional unit those consensus ideas represent foundational knowledge (Duschl, Bismack, Greeno, & Gitomer, this volume) that can be aligned with the formal learning goals (F), in the form of the canonical ideas and representations of science. Planning for BOLT entails anticipating productive resources and problematic ideas, reasoning and actions (A), as well as the learning goals (F). The latter are subsumed within the NGSS or some other set of standards.

The circles and connections in the middle of the Framework represent the "dance" of scientific inquiry between observable or problematic situations (B and D) and inferences to describe and make sense of the observations (C). Simple phenomenological situations that elicit students' thinking related to critical issues (B) often lead learners to observations inconsistent with their initial ideas (A). Other phenomenological situations (D) present students with opportunities to extend contexts or test limits of and/or transfer newly derived class consensus ideas (E). Opportunities to make sense of, reason, and relate claims students generate (C) about phenomena (B and D) complete the "dance" between observations and inferences that lead from initial ideas (A) to ideas more consistent with phenomena (E and F).

The framework of BOLT provides a window through which to map the flow of instruction and the extent to which it addresses student thinking. For example, typically K-12 science instruction presents a collection of strategies to help students become aware of and know *what* is already known (i.e., the science canon), but with little attention to *how* it came to be known (Osborne, this volume). Students are presented with formal, canonical products of science, and then asked to engage with text, problems, or verification lab activities to see how those abstract formal ideas and representations can be used. This approach shortcuts the goal of helping students experience and understand *how* they know, i.e., how one comes to know and believe ideas through students participating in and experiencing the practices of science.

A BOLT approach ideally fosters the development of the ideas from the onset: learners participate in actions consistent with scientific practices to build ownership and understanding of how knowledge gets developed. Individual and collective sensemaking and coming to class consensus underscore the central role of student and teacher dialogue in the process of coming to know and understand within the BOLT framework. Classroom discourse and curricular activities, represented by the segments in the middle of the BOLT framework stimulate students' reflections on their thinking and actions in connecting observations and inferences. Again, integral to all of these instructional moves is their connectivity to students' developing ideas and representations. Such connectivity is made explicit through questions and activities comprising the instructional moves constantly seeking to elicit, elaborate on, and respond to students' initial and evolving ideas (Schoenfeld, 2002).

Pressing for Integration of Assessment, Instruction, and Learning

BOLT is a synthesis framework that reifies the vision of the NRC framework, integrating science content and practices. The framework intentionally focuses on the learner experiences and teacher responsiveness, guiding teachers' attention as they plan and implement activities to address and build on learner thinking. In this sense, BOLT supports the blending of assessment, learning, and instruction. Assessment is subsumed within the BOLT framework and therein takes on a very diagnostic nature, pursuing depth of understanding through cycles of eliciting student thinking, interpreting learner responses, and contingent actions based on identified learning needs of students (Bell, 2002).

Through its emphasis on constantly attending to student thinking, a BOLT approach presses for a tight and explicit loop between what will be assessed, what is being assessed, and what needs to be taught/learned for teachers and students, respectively. BOLT teachers need to systematically follow and build responsively on learner thinking. Each instructional move in the BOLT framework presumes taking account of the learners' thinking, such as initial ideas as well as evolving ideas when students have collected and analyzed experiment data, so that the instruction can be effectively designed and/or modified to address students' learning needs and move toward the desirable learning goals. Such a strong focus on attending to student thinking requires some way of monitoring and describing student ideas. The next section of this chapter presents a framework for doing just that.

Interpretive Framework to Describe Student Learning

To effectively integrate assessment with instruction and learning in the classroom there needs to be an ongoing flow of high quality [formative] assessment data that can be interpreted and taken up by both teacher *and* students to shape the process of learning (e.g., Bell, 2002; Brookhart, 2009; Ramaprasad, 1983; Sadler, 1989). For this to happen, at least two things need to be in place. First, the interpretation of the assessment data needs to be sufficiently nuanced (going beyond "got it" vs. "didn't get it") so that the needs and next steps in learning and teaching are clear and specific. Second, and more importantly, the language used to describe the status of learning must be understandable to teachers and students alike. In this section, we describe such language, which we call Facets of Learner Thinking, or "Facets."

Facets and Facet Clusters constitute one promising scheme as a "language of learning" for describing and interpreting student thinking (Penuel et al. 2013; Scalise & Gifford, 2006). A Facet-based approach to instruction presumes a "knowledge in pieces" perspective of learning (diSessa, 1988; diSessa, 1993). That is to say, when learners are confronted with a puzzling phenomenon, situation, or problem, they draw on pieces of knowledge and reasoning from past experiences to construct an explanation or solution to the problem. The perspective assumes that learners' naive ideas are not exclusively problematic, but contain productive pieces or elements to be built upon or revised and improved upon during subsequent learning opportunities (Minstrell & van Zee, 2003).

Facets refer to the particular individual constructions students make and which represent a student's conceptual or mental model applied tentatively to the present situation. A Facet Cluster, on the other hand, is a collection of Facets grouped around a domain-specific set of phenomena (e.g., forces on interacting objects) or a particular concept (e.g., average velocity). Facet Clusters seek to represent the constellation of ideas, reasoning, and skills that students bring to new learning. Each Facet Cluster contains Goal Facets, which are analogous to the targeted NGSS disciplinary core content ideas, as well as Problematic Facets, which include the alternative, intuitive ideas that students exhibit when learning science[2]. The Clusters and corresponding Facets are derived from research on students' misconceptions and conceptual development (Camp, Clement, & Brown, 1994; Driver, Squires, Rushworth, & Wood-Robinson, 1994; McCloskey, 1984; Minstrell, 1982, 1984, 1992, 2001) or from practitioner knowledge and experience (Minstrell & Kraus, 2005). Figure 3.2 provides a Facet Cluster example for Newton's Third Law, within a strand focused on Nature of Forces.

Facet Cluster

00 The student understands that all forces arise out of an interaction between two objects and that these forces are equal in magnitude and opposite in direction.

01 All forces arise out of an interaction between two objects.

02 The force pairs are equal in magnitude.

03 The force pairs are opposite in direction.

---

40 The student identifies equal force pairs, but indicates that both forces act on the same object. (For the example of a book at rest on a table, the gravitational force down on the book and the normal force up by the table on the book are identified as an action-reaction pair.)

50 The student uses the effects of a force as an indication of the relative magnitudes of the forces in an interaction.

51 More damage indicates one of the interacting objects exerted a larger force.

52 If an object is at rest, the interaction forces must be balanced.

53 If an object moves, the interaction forces must be unbalanced.

54 If an object accelerates, the interaction forces must be unbalanced.

60 The student indicates that the forces in a force pair do not have equal magnitude because the objects are dissimilar in some property (e.g., bigger, stronger, faster).

61 The 'stronger' object exerts a greater force.

62 The moving object or a faster moving object exerts a greater force.

63 The more active or energetic object exerts more force.

64 The bigger or heavier object exerts more force.

90 The student believes that inanimate/passive objects cannot exert a force.

FIGURE 3.2 Facet cluster example

As Figure 3.2 illustrates, Goal Facets are situated at the top of the Facet Cluster (above the line) while Problematic Facets are organized in ascending order of urgency at the bottom (below the line).[3] The two-digit codes roughly rank problematic Facets from least urgent (20's–40's) to most urgent to address (80's–90's). In other words, the larger the number, the more problematic the Facet. Facets at the bottom of the cluster may not be the worst idea, but often identify a fundamental idea that will be very critical to address if the student is to understand the goal Facet. For example, in the case of this facet 90, the student will need prerequisite experiences that will allow her to include passive actions (e.g., table supporting a book) as well as active actions (hand supporting a book) in her fundamental understanding of what a force means and can do. Thus, this prerequisite (seemingly unrelated) idea is included in the Facet Cluster about the Third Law.

The wording of individual Facets paraphrase what students actually say or do. Consequently, Facets can be identified and used by teachers and students to work with students' ideas or may serve as common points of reference in conversations among learning specialists across disciplines (e.g., teachers, researchers, test/measurement experts, etc). In this way, Facets could become a common language for describing student learning pathways or steps to the learning goals.

Facet Clusters are related to, but should not be confused with, learning progressions. On the one hand, the goal Facets within a cluster represent an unpacking of standards statements and anticipated progressions of learning for a unit.[4] The problematic Facets, on the other hand, represent the multiple and variable ideas or learning constructs that students actually demonstrate on their way to understanding the learning goal. For this reason, their ordering within the cluster is a rough (not fixed) ranking rather than a clear progression. Since students bring different experiences and knowledge to new learning situations, the obstacles they encounter on their way to more explanatory models may differ as well.

Because each Facet represents a small piece of a larger conceptual construct, assessments based on Facets are necessarily diagnostic. A Facet "diagnosis" reveals the specific obstacles that students are encountering, helping to pinpoint *how* a student seems to be thinking and what is conceptually problematic with respect to making sense of phenomena rather than simply *whether or not* the student got it right. This obviously supports more streamlined and responsive instruction.[5] Rather than casting a "large net" and hoping to help the majority of learners, lessons can be designed to target the specific learning obstacles (i.e., Facets) that emerge through assessment.

Within the classroom, a language of learning such as Facets is useful for streamlining and blending assessment and instruction into one seamless cycle. However, an ongoing stream of rich and nuanced data means a greater level of data collection and management. The acknowledgment of this dilemma has (in part) prompted learning and assessment experts to call for a new generation of classroom assessment technologies to help ease the burden (Pellegrino, 2013). What good is a rich language of learning such as Facets unless it is feasible for use by teachers in their work day to day? In the next section, we present a suite of learning and teaching tools that were specifically designed to make the Facets language of learning usable by teachers with their students.

Online Support Tools Based on Facets and BOLT

It is clear that the systematic use of any interpretive framework in the classroom requires tools and strategies that can be easily integrated into classroom routines. Here we describe Diagnoser (Diagnoser.com), a comprehensive and coordinated set of online tools developed specifically to assist teachers in building on learner thinking in the classroom. Built around the language of Facets and based on a BOLT approach to instruction, Diagnoser has been providing teachers and their students with a BOLT-compatible toolbox for more than ten years. The specific tools are briefly described below.

> **Elicitation questions** (EQs, Figure 3.3), together with a follow-up whole class discussion, open up students' individual thinking early in a learning unit in order to reveal the range of student ideas. EQs provide students with key content issues and other students' ideas, and provide the teacher with the "lay of the land" assessment for instructional needs. The questions and subsequent discussion also stimulate student curiosity and motivation to learn.

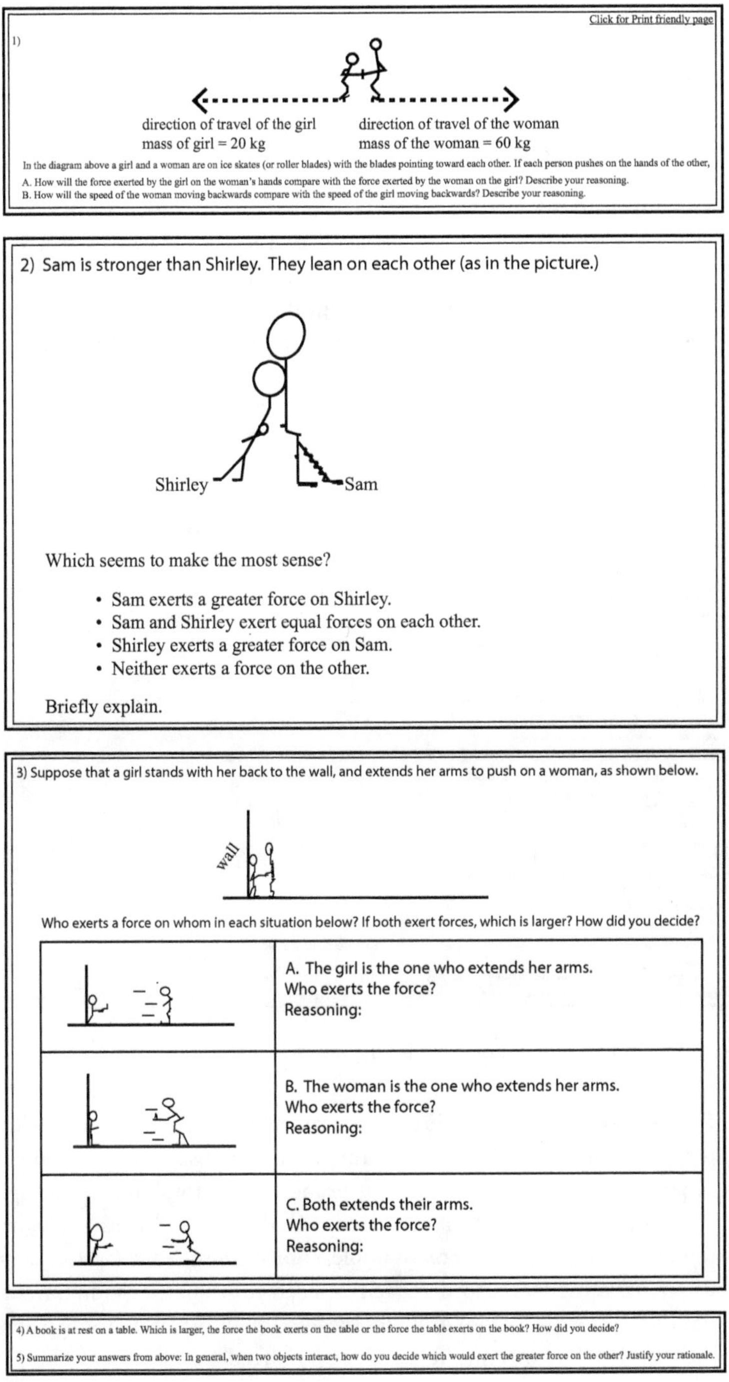

Click for Print friendly page

1)

direction of travel of the girl
mass of girl = 20 kg

direction of travel of the woman
mass of the woman = 60 kg

In the diagram above a girl and a woman are on ice skates (or roller blades) with the blades pointing toward each other. If each person pushes on the hands of the other,
A. How will the force exerted by the girl on the woman's hands compare with the force exerted by the woman on the girl? Describe your reasoning.
B. How will the speed of the woman moving backwards compare with the speed of the girl moving backwards? Describe your reasoning.

2) Sam is stronger than Shirley. They lean on each other (as in the picture.)

Shirley

Sam

Which seems to make the most sense?

- Sam exerts a greater force on Shirley.
- Sam and Shirley exert equal forces on each other.
- Shirley exerts a greater force on Sam.
- Neither exerts a force on the other.

Briefly explain.

3) Suppose that a girl stands with her back to the wall, and extends her arms to push on a woman, as shown below.

wall

Who exerts a force on whom in each situation below? If both exert forces, which is larger? How did you decide?

	A. The girl is the one who extends her arms. Who exerts the force? Reasoning:
	B. The woman is the one who extends her arms. Who exerts the force? Reasoning:
	C. Both extends their arms. Who exerts the force? Reasoning:

4) A book is at rest on a table. Which is larger, the force the book exerts on the table or the force the table exerts on the book? How did you decide?

5) Summarize your answers from above: In general, when two objects interact, how do you decide which would exert the greater force on the other? Justify your rationale.

FIGURE 3.3 Elicitation questions from Diagnoser

At first glance, the above may seem like any number of pre-tests or entry tasks. However, it is important to note that each item of this Elicitation activity is linked to a specific Facet within the Facet Cluster so that each student's response will provide the teacher with a specific "diagnosis" of student understanding (strengths as well as weaknesses).

> **Developmental lessons** (DLs, Figure 3.4) are designed to help students explore experiences and ideas related to the learning goals that emerge during the elicitation activity. Based on student responses to elicitation questions, teachers can choose or adapt developmental lessons to build on strengths and/ or to challenge problematic thinking related to key ideas.

Print Friendly Page

Forces as Interactions Developmental Lesson

Activity A: Testing whether a stronger object exerts a larger force than a weaker object when the two objects interact.

Attach a spring scale to a strong magnet and another spring scale to a weak magnet and compare the forces. Try a variety of experiments like pulling on each spring scale, pulling until the magnets separate, holding one spring scale stationary aganist a desk (maybe taping it down) or the wall and pulling on the other.

- What does your force data (from the spring scales) indicate about the size of the forces in each case?

Now try using a magnet attached to a spring scale and a paper clip attached to another spring scale. Repeat the experiments. Think about similarties/differences with your previous experiments.

- What general conclusions can you make?

Activity B: Bigger, stronger, heavier person pulling on smaller, lighter, weaker person.

Chose the biggest and smallest people in your group to try a gentle "tug of war" using spring scales. Have each person hold a spring scale and attach the scales together with a piece of string (or attach the hook of one scale to the hook of the other.)

- How do the forces compare?
- What happens to the forces if one person quits pulling as hard?

Now hold one scale in place (or attach it to a solid, immovable object) and pull on the other scale.

- How do the forces compare?

Now try using a spring or a rubber band in between the spring scales. Compare forces.
Think about the sizes of the forces in each case and the similarties/differences between the experiments.

- What general conclusions can you make?

Activity C: Active, energetic objects pulling on other objects.

Attach a spring scale to a box and, using another scale attached to the first scale, pull the box across the table or floor at a constant speed. Compare the readings on each scale.

Now, repeat the experiment with a heavier box and compare the force readings. Repeat a third time with a different weight box.

- Think about how the force readings compared in each case.
- What general conclusions can you make?

Activity D: Pushing forces with spring dynamics carts

Use two equal massed dynamics carts with spring-loaded plungers. Put the carts end to end so that the plungers are touching each other. Try experiments such as relesing one plunger and observing the motion of each cart. Now, reverse the process using the other cart.

- How do the 'pushing forces" of each cart compare in your experiments?
- what is your evidence?

Now try releasing both plungers at the same time (this may be challenging). Observe the motion of the carts in this case.

- what does this indicate about the "pushing forces"?

Hold one cart in place while the other cart pushes. Now hold one cart in place and have it push. Think about the forces involved.

- What is your evidence?
- What inferences can you make about relative sizes of the forces based on your experiments?

Follow-up activity for all groups: Problems involving drawing force diagrams dealing with moving objects and stationary objects. Have students identify the forces acting on each object.

FIGURE 3.4 Developmental lesson from Diagnoser

> **Formative assessment question sets** (QSets) consist of approximately ten items in multiple-choice, numerical response, and/or constructed response forms and are designed to probe student thinking around specific subtopics of science content associated with the Facet Cluster (see Figure 3.5 for one example item). For multiple-choice items, each option is coded to one or more facets so that each "incorrect" response (sometimes along with student explanation) provides meaningful information for teachers (via the Teacher Report) and for students (via feedback screens). The pattern of each student's responses determines the branched path she/he will take through each question set.

Jennifer and Katie stand and lean on each other.

Jennifer weighs 150 pounds and katie weighs 120 pounds.

Which one pushes harder on the other?

[a] ○ Katie must push harder because she weighs less and has to compensate for having less weight.

[b] ○ Jennifer and Katie push on each other with the same size force because force pairs are always equal.

[c] ○ Jennifer pushes harder because she weighs more.

[d] ○ It depends on whether Jennifer or Katie moves.

FIGURE 3.5 An example Forces as Interactions multiple-choice question from Diagnoser

The Teacher Report (TR, Figure 3.6) displays a "diagnostic profile" of specific problematic and goal facets exhibited by students in the class (individually and collectively) based on their responses to a question set. The Teacher Report includes a summary for each class and for individuals. Links to explanations of specific Facets and to prescriptive activities are also provided.

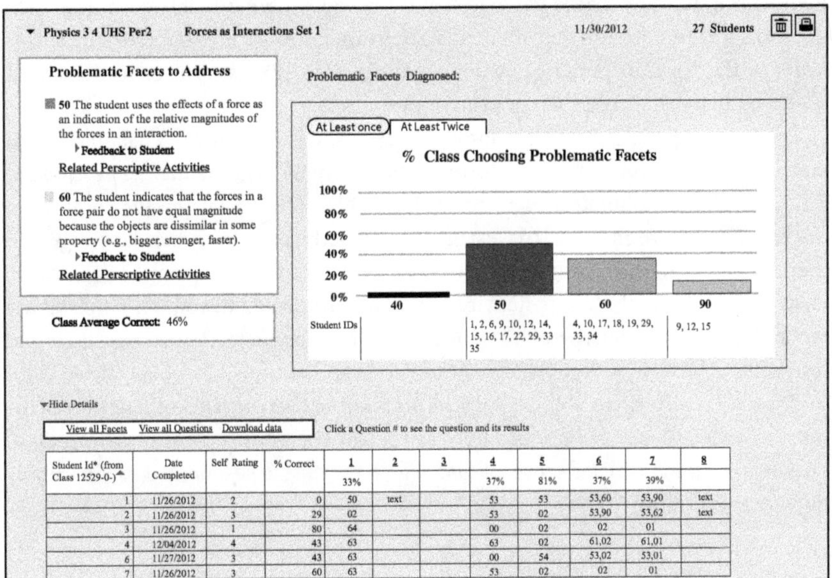

Student Id* (from Class 12529-0-)	Date Completed	Self Rating	% Correct	1	2	3	4	5	6	7	8
				33%			37%	81%	37%	39%	
1	11/26/2012	2	0	50	text		53	53	53,60	53,90	text
2	11/26/2012	3	29	02			53	02	53,90	53,62	text
3	11/26/2012	1	80	64			00	02	02	01	
4	12/04/2012	4	43	63			63	02	61,02	61,01	
6	11/27/2012	3	43	63			00	54	53,02	53,01	
7	11/26/2012	3	60	63			53	02	02	01	

FIGURE 3.6 A screen shot showing a teacher report for a class of 27 students

Prescriptive activities (PAs, Figure 3.7) are designed to target the specific problematic ideas diagnosed through the student question sets and summarized in the teacher report. Often an extension of the developmental lesson, these may be given to individuals or groups of students in order to tailor instruction to meet or extend specific student needs.

Activity to address Facet 50:

Print Friendly Page

1. Consider the results of the following set up:

Materials needed:

- Two identical force scales, e.g. two big laboratory spring scales for pulling (100n scales work well) OR two bathroom scales for pushing adjusted to zero when held oriented up and down rather than horizontally. (In all situations when spring scales are used, calibrate the scales first to make sure they give identical readings when pulled against each other or when lifting the same object.)
- One cart, wagon or skateboard that is sturdy and stable enough for a person to ride safely on it. Use common sense to do this experiment safely

Procedure:

One person sits on the wagon, holding one scale. The other person connects the hook of his scale onto the hook of the other person's scale. Predict which scale will read the larger value and why. Then the standing person (and his spring scale) pulls on the seated person (and her spring scale).

Compare the two scale readings when the cart is not moving, when it is moving at a constant velocity and when the cart is accelerating.

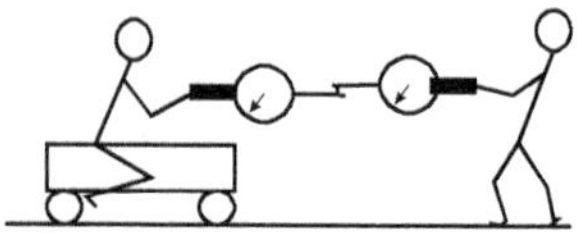

For pushing, the two hold their bathroom type scales against each other and the standing person pushes while each reads his or her scale. Sometimes it is easier to have someone looking over the shoulder of the two people pushing the scales read the scale.

FIGURE 3.7 Prescriptive activity to address Facet 50 in Forces as Interactions

Diagnoser tools mediate the interpretive framework (Facets) for teacher use in the classroom. As students complete "question sets" related to a specific content, their responses to the multiple choice and numerical response items are each linked to specific Facets. These are then recorded and summarized in the Teacher Report (Figure 3.6) that is generated automatically after students complete each question set. In addition, subsequent instructional moves are suggested by Diagnoser in the form of prescriptive activities based on the patterns of student ideas observed from their responses to the question set.

Interpreting and acting on student responses is extremely challenging for teachers to effectively engage in formative assessment (e.g., Schneider & Gowan, 2011; Stecher, Frohbieter, & Greenwald, 2011). All the Diagnoser tools were designed to ensure that student learning is assessed using the Facets language and that the subsequent instructional activities are attentive to students' problematic ideas described in the assessment report. Thus the tools become a time-saving technology for teachers, which include carefully constructed assessment tasks and instructional activities organized around Facets and clusters of student ideas. With such tools, teachers are able to streamline their assessment and instruction planning in order to not only adequately assess students' conceptual understanding, but also accurately make sense of and effectively use assessment data for subsequence instruction.

Putting It All Together: Using the System in the Classroom

What would it look like to blend assessment with instruction in the classroom using the three-part system we have been describing? In this section, we share a classroom vignette based on teaching experiences (our own and that of teacher colleagues) to illustrate.

Mr. Murphy's 8th grade physics class has been investigating forces. His students can identify forces acting on an object and have come to believe that the several forces that might be acting on an object need to balance each other to keep an object "at rest," and if forces are unbalanced (a net force), the object will move in some way. The students have also learned how to use a "Free Body Diagram" (FBD) to represent forces, using the direction and relative lengths of arrows to represent forces, and labeling each force with what was exerting the force.

For the next few days they will extend their idea of forces beyond a simple push or pull of an object to noting that forces always come in pairs (i.e., that forces always involve an interaction between two objects). The formal learning target for this sub-unit is for students to understand and be able to use Newton's Third Law, which in the NGSS language (MS-PS2-1) states, "for any pair of interacting objects, the force exerted by the first object on the second object is equal in strength to the force that the second object exerts on the first, but in the opposite direction."

Planning

Mr. Murphy wants the students to generate their own understanding. He uses the BOLT framework for planning the flow of subunit activities, including where he anticipates needing to assess. Mr. Murphy also consults the relevant Facet Cluster to familiarize himself with the typical (and problematic) learner ideas he might encounter from his students. (This will prime him to hear and see those ideas in his students' written and spoken responses.) Mr. Murphy then chooses a Diagnoser Elicitation activity that represents various situations in which two objects interact with each other.

Generating New Knowledge, Starting with Sharing Initial Ideas

The first day of the unit, Mr. Murphy has students respond to the elicitation questions individually in writing. He walks around the room to get a sense of what they have written. (Based on the Facet themes he observes, Mr. Murphy can begin to anticipate the discussions he will need to facilitate.) After writing their responses, students share their responses in conversation—first with tablemates and then with the larger group. There is no evaluation or critique (by students or teacher) as students share their ideas. Mr. Murphy simply records the ideas on the board, asking for clarification when needed. Students articulate their ideas as clearly as possible—for themselves and others. (Some students will have slightly revised their responses in light of others' ideas.) After it appears that the students have shared the major ideas, he asks if they have any knowledge or experience that seems contrary to any of the ideas expressed by other students.

What is recorded stands as a public record of the "lay of the land" of ideas. The dominant student thinking (in this case, focused on relative properties of the interacting object or relative effects obtained by the objects) is evident.

After so much sharing without evaluation, the students are motivated to know "What are the right answers?" In response, Mr. Murphy has them reduce the many ideas recorded on the board to the key issues and questions that have emerged: e.g., the bigger, stronger, faster object in the pair will exert the greater force, and/or whether the object that makes the other one move more or creates more damage will exert the greater force. Then Murphy asks them "How could we find out?"

Testing Initial Ideas

The issues, questions and ideas on the board are the seeds for proto-hypotheses to investigate. Students set about designing experiments to test their ideas, grouped with others who share the same idea. Mr. Murphy was able to use the Facet Cluster to anticipate the ideas students would likely express. By consulting the related Developmental Lessons in Diagnoser, he was also able to have the necessary materials on hand for students to use in setting up their investigations. Murphy has chosen situations that will allow testing of student ideas that represent problematic Facets.

While students are experimenting, Mr. Murphy has a chance to assess students' skills at conducting experiments and to help them make adjustments. He notices that one group that seems to be confirming their initial (problematic) idea that the bigger/heavier/stronger object will exert the greater force. He sees the "big" student and the "small" student are each pushing bathroom spring scales against the wall rather than against the other person's scale. He questions them about the meaning of two objects *interacting* with each other. Other students suggest it means the objects have to be pushing on each other, not on a third thing like the wall. He wants students to have data from which they can come to evidence-based conclusions.

Sensemaking and Coming to Consensus

When students have finished their experiments, Mr. Murphy facilitates a sensemaking discussion of the data directed at coming to a class consensus of what they know so far and how they know it. The results from their experiments suggested that there was no difference between the sizes of the two forces. Mr. Murphy takes notes on the board regarding what the students thought at the beginning and how they tested their ideas and predicted outcomes. The class summarizes the results of the several experiments and comes to the consensus that the forces during interaction always are "equal and opposite."

After a brief discussion of how the results were surprising, Mr. Murphy then reads out-loud the formal statement of Newton's Third Law and congratulates the class on their experimental work to test their initial ideas and on their willingness to reject their original idea, hold tentatively the new idea of the Third Law, and see if it helps to explain other situations.

Applying and Challenging New Knowledge

Mr. Murphy assigns a reading about Newton's Third Law and a related online question set from Diagnoser. The questions ask students to identify and compare forces in several new situations, some in which the Third Law applies and some in which it doesn't. These allow students to identify the Third Law forces in various contexts and to revisit new learning from the earlier sub-unit on identifying forces. The students receive immediate feedback from the online system regarding their reasoning around the key concepts. Meanwhile, Mr. Murphy reviews the Teacher Report to assess how individual students, and the class as a whole, are able to make sense of their new understanding in new contexts. The results suggest that in some situations the class still believes that the object, which causes the greater effect must be exerting the greater force. Murphy chooses a Prescriptive Activity related to that high frequency Facet to conduct as a class experiment the next day. The Teacher Report also identifies three students who are still having difficulty with the Facet that passive objects do not exert forces. Mr. Murphy decides to work with that small group separately to review experiences that led the class as a whole to reject that Facet. Finally, for homework, he tells the class to add their new ideas, evidence and reasoning to their ongoing essay on what they believe about the Nature of Forces.

Discussion of the Narrative about Mr. Murphy

In the Mr. Murphy narrative above, we note the centrality of dialogue (student-student(s) as well as student-teacher) to mediate engagement with ideas and reasoning about the phenomena depicted in the activities. The teacher balanced the press for individual reasoning with the press to come to collective understanding as a class. Meanwhile, multiple rich phenomenological contexts and multiple representations provided opportunities for students to express and negotiate their developing ideas and for the teacher to assess their application. Most importantly, the various components and connections suggested in the BOLT framework were instantiated through the teacher's use of Facets and the tools of Diagnoser. Facets of student thinking helped the teacher to anticipate, interpret, and address those ideas. The teacher's familiarity with the relevant Facets effectively "primed" him to hear and remain focused on the problematic ideas related to the learning goal. At the same time, Diagnoser tools significantly streamlined the work of gathering and interpreting assessment information, and later that of prioritizing and focusing subsequent instruction. The tools in Diagnoser were used by the teacher to assist in selecting activities with the most diagnostic values of uncovering student thinking, interpreting what students say and do, and in identifying and choosing subsequent actions to address problematic ideas, as well as developing goal knowledge and skills.

With the BOLT framework, Facets language, and Diagnoser tools, the teacher has a coordinated set of resources for planning and implementing curricular activities and dialogue that will aid in building on learner thinking toward deep understanding of worthy learning goals. Although he has taken on responsibility of attending to student thinking, the tools make that job doable and more effective in promoting learning.

A Note on Research Related to the BOLT System

The Murphy vignette provides more than a hypothetical glimpse of what the use of the three-component system would look like in a classroom. The resources of the three-component system described in this chapter currently exist and are being used by teachers across the US and abroad. Each of the tools has been extensively vetted by middle and high school teachers. There is also a growing body that includes studies of classrooms in which teachers skillfully use Facets as an interpretive framework for assessment and instruction and have shown evidence of highly responsive teaching, as well as student learning gains (Hunt & Minstrell, 1996; Minstrell, 1984; Minstrell & Matteson, 1994; Penuel et al., 2013). A recent study of teachers' use of BOLT for planning and implementing classroom activities that attend to students' thinking showed significant student gains over students in a comparison group (DeBarger, Minstrell, Ayala, Freed, & Vasquez, 2013). Similarly, research on the use of Diagnoser tools has found that the more question sets students complete, the better they perform on assessments of student performance (Minstrell & Kraus, 2005), with the greatest gains exhibited by students of teachers who not only used multiple tools, but also worked with a "blended" perspective of assessment and learning and actively attending to student thinking (i.e., a BOLT perspective).[6]

Each of the above studies has focused on one or two of the three components (BOLT, Facets, and Diagnoser). In order to understand the challenges and affordances of the complete system (as well as its impact on teaching and learning), studies of the complete system still need to be conducted.

Conclusion

In this chapter, we presented an example classroom system of blended assessment and instruction based on three existing resources (i.e., BOLT, Facets of learner understanding, and Diagnoser tools). Our aim was to present a vision of more meaningful and streamlined science learning that would also result in more productive assessment results to contribute to the larger assessment system and especially to student learning. This included addressing the challenges of

(1) integrating assessment with learning and instruction, (2) increasing the instructional responsiveness of classroom assessment, and (3) efficiently and effectively using the instructional time with the support of diagnostic tools. BOLT provides the "vision" of blended instruction; a framework for planning and implementation that emphasizes creating learner and learning-driven experiences that afford opportunities for rich and nuanced assessment of thinking and practice. Facets of learner thinking present the "language" to interpret, characterize, and respond to students' actual learning as it develops. Finally, the tools and resources of Diagnoser serve to instantiate the first two components, emphasizing the need for tight alignment with the other two components in order to optimize instruction.

The three components were designed to work together as a system for planning and implementing a learning environment that builds on learning and thinking. The omission of any one component severely limits the system's effectiveness.[7] The BOLT framework helps teachers focus and connect activities to the learning process and thus support students' learning and sensemaking and avoid simply choosing topically related activities.

In this chapter, we have focused on a classroom system. At the same time, however, we recognize its potential for participating in a larger system of assessment and instruction. Imagine, for example, if the language of Facets were carried through to even a portion of assessment items within a state or district assessment. Aggregated summative data exhibiting high frequency of specific problematic Facets could be used to tune curricula and activities to address those problematic Facets the next year. The results of such an assessment would not only be more nuanced (and perhaps more actionable at the classroom level), but potentially would also produce improved summative assessment results. In this way, an interpretive framework such as Facets has the potential of "funneling up" useful information from the classroom to other levels of the assessment system (e.g., district and state).

Assessment information collected in the classroom—even for formative purposes—can and should be considered as a critical source of evidence for system accountability purposes (Mandinach, Gummer, & Muller, 2011). Moreover, the better the quality of that data, the greater its potential use for program decision-making and other purposes throughout the system (cf. Broadfoot & Black, 2004; Frederiksen & White, 2004; Herman, Osmundson, Ayala, Schneider, & Timms, 2005; Wilson, 2004).

In presenting BOLT, Facets, and Diagnoser, we have demonstrated one set of learning system components, including a theoretical framework for relating evolving ideas and activities and dialogue to promote learning, a backbone language for interpreting learner words and actions, and an instantiation of tools to help automate the assessment, instruction, and learning. We have shown how the components might fit together to provide a single coherent system that is usable and helpful for teachers in planning and implementing a learning environment.

This systematic approach obviously requires a research agenda to examine and evaluate the complex dynamics of the entire system—not just its individual parts. This will require both immediate and long-term work among measurement specialists, science educators, teachers, and scientists. It will be important for researchers and practitioners to bring together some of the work that has been done in the three realms of learning, instruction, and assessment and to organize those efforts into one single coherent system (where possible) that has some level of automation and is accessible to teachers and school leaders (Penuel, this volume.)

While the proposal of a single system may seem ambitious, we have only to recall how far we have come in the last decade with the standards movement to envision the possibility of realizing this vision. To what extent can we engage in the sort of systemic thinking, consistent effort, and perseverance that we are calling upon from teachers and students to demonstrate in the classroom? As a community of researchers, curriculum and assessment developers, and professional development experts, we should ask ourselves this question and determine to what extent we can hold ourselves accountable for conducting responsive research that serves the great systemic needs of educators and learners.

Finally, as Penuel (this volume) advocates, research is needed on the connectivity and sustainability of innovation among the various levels of the education system as a whole. "With potentially seismic changes in the mechanisms, nature, and organization of education must also come changes in educational assessment" (Bennett, 2013, p. 2). However, failure to move forward with a learner-driven theory for assessment and to properly implement the assessments will continue to blunt efforts to improve education, as past experience with performance assessment and portfolios, and more recent struggles with formative assessment have shown (Shepard, 2005). Continued commitment to assessment practices that are separated from instructional validity and utility "may well result in the appearance of gains at the tremendous cost of suppressing important aspects of learning, attending to the wrong things in instruction, and conveying to students a distorted view of science" (Coffey & Alberts, 2013, p. 489). As the field re-examines the larger system of educational assessment with expectations for cohesion with curricula, alignment with NGSS, and formative values to improve learning and instructional decisions, such an assessment system must consider (and probably begin) in the classroom. Only by taking account of classroom assessment that is fully immersed in the flow of teaching and learning for teachers and students to engage with all the time can the promises of current standards-based educational reforms be fully realized.[8] [9]

Notes

1. For a more elaborate explanation of the BOLT framework, see Minstrell, Anderson, & Li (2011).
2. For detailed descriptions of Facets and Facet Clusters, see Minstrell (1992; 2001).

3. Facet Clusters vary in their number of associated goal and problematic facets, reflecting the different sets of facts, reasoning, and skills learners apply in understanding different science concepts.
4. Facet Clusters are more like the Evolutionary Learning Progressions than Validation LPs discussed by Duschl et al. (this volume).
5. Studies of classrooms in which teachers skillfully use Facets as an interpretive framework for assessment and instruction have shown evidence of highly responsive teaching (Schoenfeld, 2002; van Zee, Iwasyk, Kurose, Simpson, & Wild, 2001; van Zee & Minstrell, 1997; Vokos, Kraus, Close, Close, & Seeley, 2008) as well as significant student learning gains. A series of studies conducted in secondary science classrooms documented student gains of 15% on average, ranging from 5% to 35%, on end-of-year science assessments compared to previous years and as compared to students in comparison groups (Hunt & Minstrell, 1996; Minstrell, 1984; Minstrell & Matteson, 1994).
6. For a more elaborate examination of the research behind the three components of the system, see Minstrell et al. (2011).
7. We recognize that there are a growing number of innovative resources to support teachers in various aspects of instruction and/or assessment, but these rarely (if ever) provide a complete integrated system.
8. Preparation of this paper was partially supported by grants from the National Science Foundation (Grant numbers 0535818, 0733169, and 0822342) and The US Department of Education, Institute for Education Sciences (Grant number R305A100475). The ideas and opinions presented are those of the authors and do not necessarily represent those of the funding agencies.
9. The authors wish to thank Penn State University, the College of Education, conference staff, and participants in the Waterbury Summit for their questions, comments and suggestions, which helped us to clarify our ideas. Finally, we want to thank our teacher colleagues who continue to help us identify and address the learning needs of their students and fellow teachers.

4

RESPONSE 1

Systems Thinking as a Design Problem

Marcela Borge

The chapters by Litzinger (this volume) and Minstrell, Anderson, and Li (this volume) offer a refreshing and thought-provoking perspective on systems thinking. Together, these chapters reconsider what constitutes necessary STEM educational objectives and focus on improving the methods and sequencing of STEM instruction. They also address the development of students' abilities to think and reason scientifically and present strategies to better integrate assessment with instruction. These are arguably some of the most important issues in need of consideration by STEM researchers, educators, and policy makers.

I respond to the chapters from both a cognitive and human-centered design perspective and begin by speaking to the importance of systems thinking and discussing notable points made in both chapters. I then draw from learning theory to identify additional requirements that could strengthen the proposed educational designs. In doing so, I point out important learning factors that the field as a whole should consider, and I conclude by pushing us to evaluate the educational designs we create from a human-centered-design perspective.

The Importance of Systems Thinking and Scientific Reasoning

Systems thinking is a form of nonlinear reasoning used to understand relationships between multiple variables and outcomes and to guide the design of solutions to complex problems. It requires that individuals be able to synthesize information about complex systems in order to identify all of the components that make up that system and understand how those components interact with each other as part of that system. All of this reasoning is usually done in service of problem solving as a means to identify and argue about claims and trade-offs of potential outcomes stemming from changes in the system. This type of thinking allows people to consider multiple unintended consequences in order to design the best possible solutions for problems at hand.

Though individuals in today's society need to be able to engage in systems thinking, support for the development of this type of reasoning is currently lacking. Our population as a whole needs to be able to engage in systems thinking in order to be able to carefully consider the trade-offs between economic, cultural, and environmental demands. For example, we need to be able to use multivariable information as a means to guide the design of technical solutions and societal regulations to help mitigate or correct the many complex problems that face our society. Unfortunately, higher-order thinking skills associated with systems thinking and other forms of scientific reasoning are not "expected outcomes of education" (Kuhn, Black, Keselman, & Kaplan, 2000). Both young students and adults commonly lack the cognitive abilities to observe and interpret multivariable problems and engage in higher order reasoning (Kuhn, 2009a, 2009b); these are requirements of systems thinking. Given the poor higher-order reasoning abilities that exist across people of different ages, Kuhn (2009a) argues that the development of higher-order reasoning will not occur unless their development becomes a first order goal in classroom environments.

It is imperative that we begin to seriously consider how to make different forms of reasoning central aspects of STEM education and how best to support their development in all children. The chapters presented by Litzinger and Minstrell et al. are good first steps in this direction.

Litzinger's Chapter: Thinking about a System and Systems Thinking in Engineering

Litzinger (this volume) defines systems thinking as a form of holistic thinking that moves beyond the immediate components of a problem to consider the entirety of the potential impacts and outcomes of all of its interrelated and overlapping components. Systems thinking, he argues, includes the ability to understand and reason about how parts of a system can impact each other, the ability to use data to develop an external representation of a system, and the ability to apply the developed understanding of that system as a means to design the best possible solutions to real-world problems.

Others have defined systems thinking using similar terms and discussed its importance as a means of solving multifaceted problems (Dym, Agogino, Eris, Frey, & Leifer, 2005; Richmond, 1993; Senge & Sterman, 1992), but Litzinger's chapter helps the reader fully comprehend the practice of systems thinking. One of the strengths of Litzinger's chapter is the extent to which he unpacks systems thinking, provides examples, and then uses this as a means to propose a set of learning objectives for K-12 education.

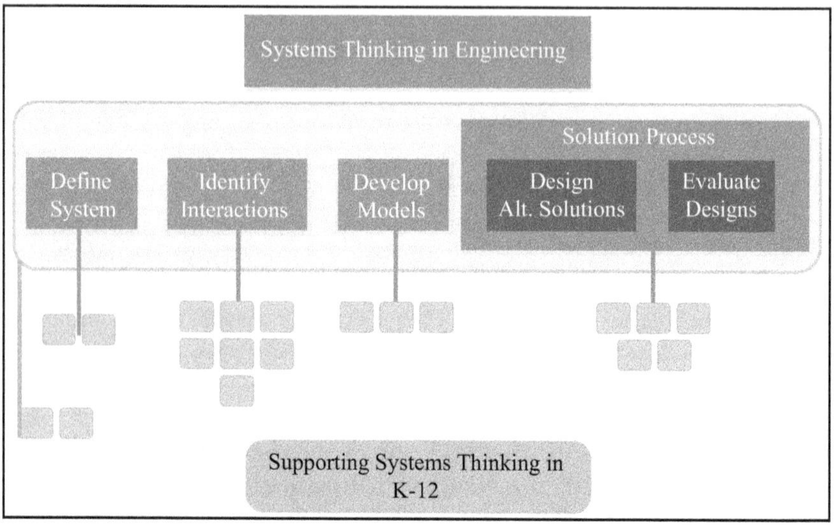

FIGURE 4.1 A visual representation of the argument for specific objectives presented by Litzinger, at the Summit. The author draws on engineering perspectives and professional practice to define systems thinking and identify concrete objectives for it. This is used as the foundation for proposed objectives for K-12 education.

Litzinger begins by defining systems thinking in engineering practice and describing five concrete activities engineers engage in that require systems thinking. These activities coincide with processes for requirements, design, and evaluation. Litzinger then builds on these concrete examples of engineering practice as a means from which to draw learning objectives for younger students (see Figure 4.1). For example, given his definition of systems thinking as requiring the ability to analyze and represent systems, he proposes that K-12 students should be able to identify and create graphical representations of interactions within a system and between a system and its surroundings and then classify the type of interaction as mass, energy, or information. Litzinger concludes his chapter by discussing the challenges and opportunities posed by integrating systems thinking into K-12 education and eloquently speaks about this challenge as a design problem.

Minstrell, Anderson, and Li's Chapter: Diagnostic Instruction: Toward an Integrated System for Classroom Assessment

Minstrell et al.'s chapter (this volume) is a nice complement to Litzinger's in that Litzinger defines what systems thinking is and then Minstrell et al. apply systems thinking to the design of classroom interactions that facilitate deeper learning experiences and conceptual change. Specifically, Minstrell et al. seek to resolve two core instructional problems: (1) the lack of integration between assessment and instruction and (2) the responsiveness of instruction to students' developing understanding of concepts. They do so by proposing a new instructional framework, called BOLT, that changes the focus of assessment from measuring student performance to selecting the path of instruction (see Figure 3.1, Minstrell et al., this volume, p. 52).

Minstrell et al.'s framework prioritizes the educational experiences of students over what they refer to as "teacher moves," or the teacher's predetermined instructional actions. It prioritizes a bottom-up approach to instruction, where instruction begins with students' existing ideas and an assessment of student activity dictates the instructional activities the teacher carries out. The BOLT framework also places large emphasis on classroom sensemaking discourse. The authors contrast their approach to a more traditional top-down model, where instruction begins with the accepted scientific ideas and the teacher chooses the activities students engage in. Such classrooms usually emphasize summative assessments to test students' abilities to correctly restate and apply proper scientific ideas.

The BOLT framework is somewhat similar to the guided assessments and instructional principles proposed by Ann Brown and Joseph Campione as part of their Fostering a Community of Learners (FCL) program (Brown & Campione, 1994; Brown, Metz, & Campione, 1996), a program based on Vygotskian theories of learning and development (Vygotsky, 1978). However, FCL focused more heavily on supporting metacognitive reasoning than is currently emphasized by the BOLT framework. On the other hand, the BOLT framework is far more concrete. Minstrell et al. provide examples and Facets of learning that can aid teacher adoption of the BOLT framework and digital tools to help solve known challenges identified by Brown & Campione (1994), in particular, issues related to teacher competence, limited knowledge capital, and student misconceptions. They do this by introducing a series of tools to support teacher implementation of complex instruction developed as part of a program called the Diagnoser Project.

One of the strengths of Minstrell et al.'s chapter is in envisioning progressive ways to view the purpose of assessment and classroom experiences. Minstrell et al. push us to reconsider the purpose of assessment and instruction. Their position is bolstered by research in cognitive psychology arguing that too much emphasis is placed on short-term learning gains and the use of such gains on assessment of teaching ability; such approaches may have negative impacts on how teachers teach and how students learn (Bjork, 1994). Others maintain that too much time is also spent on the learning of decontextualized facts over the development of cognition (Collins, Brown, & Holum, 1991). Changing the emphasis to evaluating longer-term learning goals and sensemaking rather than short-term performance could have positive impacts on the quality of learning experiences and outcomes.

Reflecting on Both Chapters

Both chapters make important contributions to the field of education, and the arguments presented provide few opportunities for criticism. So rather than focusing on finding flaws with the substance of what was presented, this response will focus on addressing what was missing. It is unrealistic to expect that any one chapter could address all of the factors associated with supporting complex reasoning while also proposing novel educational designs. Nonetheless, I will highlight some key factors that the authors may want to consider to further strengthen the proposed designs.

As Litzinger points out and Minstrell et al. illustrate, improving K-12 education can be perceived as a design problem. Deciding on the concepts, methods, and sequencing that should comprise core objectives for STEM classrooms is a difficult task associated with a great deal of controversy. For example, the research community is divided as to what constitutes higher quality instruction, direct teacher led instruction, or active learning in inquiry or problem based learning environments (Hmelo-Silver, Duncan, & Chinn, 2007; Kirschner, Sweller, & Clark, 2006; Mayer, 2004; Norman & Schmidt, 1992; Schmidt, Loyens, Van Gog, & Paas, 2007). Another source of debate centers on whether instruction should emphasize subject content or thinking processes (Kuhn, 2009a; Schoenfeld, 2002, 2004). To add to this complexity, developing domain-specific capabilities while at the same time meeting the needs of students is truly a challenge. Since we are dealing with a design challenge, it seems reasonable to take a human-centered approach to the design of educational solutions. Toward this aim, I will identify additional requirements not emphasized in the chapters that focus on supporting the *whole learner*.

Requirements for Supporting Systems Thinking

In order to adequately support the complex learning inherent of systems thinking, it is critical to go beyond identifying learning objectives and instructional procedures to considering for whom we are designing: the students. In human-centered design, it is typical to analyze the needs of all the stakeholders in a system, but the educational system is so complex and includes so many different stakeholders that such an approach would likely not be very productive or effective. Thus, it makes sense to begin with our primary stakeholder and identify student needs first. After all, students are the ones for whom the educational system is designed; we strive to support their learning and development.

Meeting the needs of students not only requires attention to cognitive factors, but also metacognitive, social, and emotional factors, as well. Each of these is important to consider from a human perspective. Each one of these factors can affect important learning processes and outcomes (Boekaerts, 1996; Hu et al. 2007; van Merriënboer & Sweller, 2005), and each one can make the case

TABLE 4.1 Requirements for supporting systems thinking

Learning Requirements	*Outcomes*
Integration of general and specific domain skills	Students should be able to develop general skills like metacognitive and collaborative abilities in addition to understanding systems thinking concepts and techniques.
Leveraging local and existing knowledge	Students should be able to learn about important systems concepts and techniques while working on unpacking and understanding a system for which they have existing familiarity. This experience can serve as a bridge to introducing systems that are more complex.
Promoting learner control	Students should be in command of their own learning; able to set learning goals, pay attention to, reflect on, and assess their own learning processes.
Accommodating for individual differences	Students should be able to connect to STEM content and meet rigorous standards regardless of socio-economic background or existing cognitive abilities. In order to do so, it is necessary to create educational environments that meet the needs of a range of students by redefining the role of the teacher, using technology as a means to support and enhance cognition, progressively minimizing overly structured activity, and providing added opportunities for collective, intellectual discourse.

for various learning requirements. As such, there are many requirements one could list, but in this response, I will only focus on four (see Table 4.1).

Requirement 1: Integration of General and Specific Domain Skills

It is important to integrate the learning of domain general and domain specific skills. Domain specific skills refer to skills necessary for competence in a particular knowledge area. For example, being able to form a testable hypothesis is a domain-specific science skill. Domain general skills are skills that are required for competence across knowledge areas. For example, being able to effectively articulate ideas to others is a skill that is required in multiple academic domains.

Systems thinking can be a domain specific skill when used as part of scientific practices defined by Litzinger. However, this domain specific skill comprises metacognition and collaborative problem solving, two constructs composed of a collection of domain-general skills (Osborne, 2010; Schraw, 1998; Wecker & Fischer, 2011). Metacognition has been defined as the ability to monitor and control one's thinking (Brown, 1984), whereas collaborative problem solving has been defined as the ability to maintain a shared problem space in order to negotiate ideas and design joint solutions to the task at hand (Roschelle & Teasley, 1995).

These general domain skills play an essential role in systems thinking and need to be used in context, with appropriate guidance and opportunities for repeated practice (Brown, Collins, & Duguid, 1989; Collins et al. 1991).

Metacognition and collaborative problem solving are considered important aspects of science, design, and engineering (Borge & Carroll, 2010; Dym et al. 2005; Shimoda, White, Borge, & Frederiksen, 2013; White & Frederiksen, 2005; White & Frederiksen, 1998). Scientists and engineers are constantly evaluating thought processes, from the goals they are trying to fulfill to the strategies they use to do so. They are constantly reflecting on methods and procedures and revising approaches to difficult problems. They are also keenly aware of their own cognitive biases, of common pitfalls they must avoid, and they are constantly monitoring for such problems. In order to ensure that they are careful with their approaches and meticulously considering the consequences and implications of their work, they often problem solve with peers, partake in peer reviews or feedback sessions, and work with stakeholders to solve complex problems. As such, they have to be able to communicate their ideas well in order to share, synthesize, dispute, and build on ideas in a strategic and reflective manner with others.

The reason metacognitive and social factors are important to systems thinking is because, at its core, systems thinking is a way of reasoning, decision-making, and problem solving about complex multifaceted problems. For example, complex systems usually include interacting socio-cultural variables that influence how people and groups engage with new designs (Cole & Engestrom, 1993; Engestrom, 2000). Given the intricate nature of such systems, it is important to collect and understand as much information about a complex system as possible in order to avoid negative unintended consequences. Engineers and designers mitigate problems associated with understanding and designing for activity systems by working as part of collaborative teams tasked with identifying, articulating, and negotiating potential problems and design solutions. For this reason, one can argue that to engage in competent systems thinking about complex problems requires the ability to (1) collaborate with system experts and stakeholders, (2) ask questions in different ways to accommodate differing individuals and set design goals, (3) collect and synthesize large amounts of data and reflect on their meaning while being aware of cognitive biases, (4) build models of interactions between system components as part of teams, and (5) communicate assumptions to those involved as part of the system in order to reflect and revise on the models developed. What this implies is that metacognition and collaborative problem solving are not just supplementary activities that can help improve systems thinking but rather are integral to systems thinking.

Given the nature of systems thinking, we need to ensure students are developing abilities to critically reflect and revise collaboratively on the externalized representations they create and on their own learning processes. Otherwise, as Schunk (2012) points out, we run the risk of developing testable knowledge in the absence of practical knowledge. We will develop students' knowledge of systems thinking and how to apply systems thinking when asked to do so, but we fail to develop students' understanding of when and why to apply systems thinking and how to strategically coordinate thinking processes associated with systems thinking.

Let us return to one of Litzinger's learning objectives as a concrete example of Schunk's argument. Being able to identify, create, and classify graphical representations of interactions within, between, and around a system is a notable and necessary learning goal. However, it does not constitute as much competence if students do not know when to draw a graphical representation during authentic practice, why it is useful, how to share it with others, how and why to articulate claims and assumptions, and how to use feedback as a means to iteratively reflect and revise models in order to modify proposed solutions. Developing systems thinking in the absence of such important domain-general skills is not sufficient to constitute a competent systems thinker, and it is difficult to see how we could not include factors associated with metacognition and collaborative problem solving as first order objectives from which to assess the quality of systems thinking.

Requirement 2: Leveraging Local and Existing Knowledge

It is also necessary to develop ways of introducing systems thinking that leverage local and existing knowledge and then build on these introductions as a means to incrementally increase complexity of the subject matter. There are a variety of reasons why we should find ways to leverage students' existing knowledge and provide them with opportunities to make meaningful contributions toward classroom discourse. From a cognitive and developmental perspective, young students are more likely to demonstrate higher levels of reasoning when investigating subjects about which they already have existing knowledge (Schraw, 1998). From a social perspective, when students perceive an individual as having expertise in a domain, students change how they interact with that individual and ascribe leadership specific roles to them (Brown et al., 1993). Feeling valued and needed by the learning community could meet the emotional needs of underrepresented students who might otherwise be disengaged or alienated (Osborne & Walker, 2006). Thus, creating such leadership opportunities may help to encourage participation and a sense of connection for underrepresented students.

As Minstrell et al. argue, it is important to begin with students' existing ideas. However, students' existing ideas can be more than tools to get students to understand what teachers want them to know. Students' existing ideas can be a means to provide students with areas of expertise from which to understand complex reasoning. Students' existing knowledge or intuition is not a bad or deficient thing; it is a rich knowledge source from which students can begin reasoning about complex systems. It is important to be able to draw on existing mental models of systems, the internalized representations, analogies, artifacts, and experiences associated with systems, as a means to facilitate systems learning and reduce cognitive load (Johnson-Laird, 1983; van Merriënboer & Sweller, 2005; Norman, 1988; Sweller, 1988).

Complex learning involves so many different facets that it places extremely high demands on the working memory. In order to minimize cognitive load for such instruction, researchers argue that it should be connected to real-life examples (van Merriënboer & Sweller, 2005). Students' developing knowledge can then be pushed further by following these local, simplified examples with different types of problems or problems of increasing complexity (White & Frederiksen, 1990). In this way, students can be pushed towards the upper ranges of what they can learn and understand with instructional assistance and develop increasingly more sophisticated forms of higher-order reasoning, thereby steadily improving what they will eventually be able to do without guidance (Vygotsky, 1978).

Requirement 3: Promoting Learner Control

Providing students with a sense of agency and concrete strategies for monitoring and controlling their own learning provides huge benefits to students. Learner control is an important affective, motivational, and metacognitive learning factor associated with higher performance (NRC, 1999; White, Frederiksen, & Collins, 2009). Pedagogies guided by these theories highlight the need for learners to have an active role in constructing, evaluating, and regulating their own learning experiences (Bandura, 1993; Boekaerts, 1996; Butler & Winne, 1995; Papert, 1991; White & Frederiksen, 1998; Zimmerman, 1989). In order for students to self-regulate learning, they must have some freedom of choice in classroom environments (Zimmerman, 2005). This means students should set learning goals, closely monitor their own performance, and reflect on outcomes in order to revise learning strategies, behaviors, motivations, and learning experiences (Zimmerman, 2002). Coinciding with Collins & Halverson's (2009) proposal that the role of teachers needs to be reconsidered, promoting learner control would entail that teachers' engagements shift from managing student learning to facilitating the development of core cognitive capabilities that would enable students to manage their own learning.

Instructional methods that emphasize student control over learning processes have been associated with a variety of positive learner outcomes. For example, emphasis on the monitoring and regulation of learning has demonstrated higher rates of learning transfer, desire to learn, and persistence in the face of difficulty (Bandura & Dweck, 1985; Bjork, 1994; Dweck, 1985; Dweck, 1986, 1999; NRC, 1999; Pintrich & de Groot, 1990; Schunk, 2012). Actively involving students in assessing their own thinking processes can also help students to improve their general understanding of science and the epistemologies that serve as foundations for higher order reasoning, and to increase overall metacognitive ability (Shimoda et al. 2013; White & Frederiksen, 1998).

Reflective assessments can improve students' general understanding of science and the practice of inquiry (Shimoda et al., 2013; White & Frederiksen, 1998). Reflective assessments place the student in the evaluator role. Students use explicit criteria (i.e., a set of objectives that compose competence in a domain) to reflect on their work and think about how well they have met these objectives (White et al., 2009). Low achieving students tend to get the most benefit from this form of assessment (White & Frederiksen, 1998). White and Frederiksen (1998) postulate various reasons as to why this might be the case, but chief among them are (1) the likelihood that such experiences provide students with a better understanding of evaluation criteria and (2) that, through repeated cycles of trying, reflecting, and improving work, the students are also developing essential cognitive skills.

Though the BOLT framework proposed by Minstrell et al. emphasizes student reflection, the primary evaluator in their framework seems to be the teacher. Students do not seem to be in control of evaluating, correcting, and modifying their learning processes, and it is not clear to what extent they are made aware of misconceptions, problems, and strategies for learning or to what extent they have explicit control over the course of instruction. It would be possible to enhance learner control in the BOLT framework by encouraging students to choose topics, set learning goals, reflect on learning and performance, revise their learning strategies, and discuss their experiences.

Students can also reflect on their understanding of systems and the extent to which they fulfill systems thinking objectives, such as those proposed by Litzinger. In this way, students can begin to internalize what it means to engage in complex reasoning, problems associated with complex reasoning, and strategies they can use to prevent or mitigate problems with systems they already inherently understand.

Requirement 4: Accommodating Individual Needs

Accommodation of individual needs is another important variable that was not emphasized in either chapter. Accommodating individual needs refers to ensuring that educational designs account for students' cognitive, sociocultural, and economic differences. These differences are important, since the educational system serves a range of students with different backgrounds, experiences, values, and goals.

Sociocultural and economic differences can affect the types of social interactions students have and therefore the types of opportunities they have to learn and develop. Theories of development, cognition, and learning highlight the extent to which children learn by internalizing behaviors they experience as part of sociocultural interactions (Brown et al., 1996; Collins et al., 1991; Vygotsky, 1978).

For example, students with more access to physical and cognitive tools may develop aspects of cognition faster in formal classroom settings than students with a history of less access, as those with access have more experiences augmenting and reflecting on their thinking with the aid of these tools (Pea, 1993).

Social interaction is another powerful sociocultural factor that can vary depending on student background. For example, children from low socioeconomic backgrounds have significantly lower daily verbal utterances spoken in the home, less sophisticated parent-child interactions, and lower expectations from teachers, than children from higher socioeconomic backgrounds (Ginsborg, 2006). The Pygmalion studies, a series of studies spanning over forty years, provide further support to the claim that subtle biases experienced as part of social interactions can have large, collective impacts on student learning and performance. These studies demonstrate that people's a priori expectations affect how they interact with others (for a detailed review, see Rosenthal, 2003). For example, teachers modify verbal and non-verbal communication depending on their expectations of a student's potential ability; they modify the tone and warmth of communication, the frequency of times they talk to students, the amount of material they teach, and the quality of feedback they provide (Rosenthal, 1994). What is more alarming is that the Pygmalion effect has been shown to occur across contexts, in educational, business, military, and health settings. This research suggests that our society establishes extremely unequal cognitive environments for children and that these inequalities are systemic rather than the product of a few isolated experiences. Over time, the subtle differences can accumulate and result in significant developmental differences. As such, it is not surprising that disadvantaged students would withdraw from domains that are highly dependent on the cognitive forms of thought that they are often wittingly or unwittingly denied and in which they are expected to fail (Osborne & Walker, 2006).

What we know about how people learn and the individual differences of children, with regards to ranges of development and opportunities for learning, has implications for design of educational environments and the development of scientific reasoning. Struggling students need to be provided with added social support, technological tools that promote strategic thinking and intellectual discourse, and opportunities to learn in environments that are not overly structured, devoid of agency, or lacking in opportunities to develop metacognitive abilities. Furthermore, given that many children may lack opportunities to engage in sophisticated forms of social discourse, the development of learning communities should also be a first order goal.

All students, regardless of current ability, need opportunities for enrichment and the development of abilities to monitor and regulate their own thinking processes. However, in order to allow students the opportunities to learn subject matter at differing levels of complexity, we need to design curricular activities and standards that will foster the creation of a range of educational opportunities to help students develop differing aspects of higher-order reasoning.

Conclusion: Evaluating Educational Designs from a Human-Centered Design Perspective

Though the chapters presented by Litzinger and Minstrell et al. are well-crafted, thoughtful pieces, I would argue that each could improve on the extent to which they consider how well they are meeting the needs of all students—something our educational system as a whole needs to consider. Our educational system places too much focus on creating learning goals and evaluating student performance in the absence of careful consideration of educational designs and how different designs affect different students.

Currently, the prevailing practice in education is to define the quality of learning by the extent to which students can outperform peers on the explanation and emulation of accepted ideas. We reward students who do better than their peers by giving them good grades and richer learning opportunities and punish students who do worse by giving them poor grades and a reduction in learning opportunities. As educational designers we need to ask ourselves some important questions regarding this practice:

- Is this approach equitable?
- Are we meeting the needs of *all* students?
- Are we providing similar learning opportunities for *all* students?
- Are we creating engaging curriculum for *all* students?

I would encourage us to think more deeply about these educational design questions by examining how we teach design students to create human-centered designs. Human-centered design students must learn to apply cognitive and developmental psychology as a means to design solutions for a wide range of people. They are introduced to the concept that human-computer-interaction is a complex system even at its simplest level—where one individual is interacting with one computer—because people draw on a lifetime of experiences to interpret and use objects as a means to accomplish goals (Cole, 1995; Cole & Engestrom, 1993; Norman, 1988; Vygotsky, 1978). Design students learn that, in order to design effective solutions for people, it is important to investigate the needs and activities of individuals and of the social interactions that constitute their community. They are warned not to fall victim to common fallacies of systems engineering: do not allow personal assumptions to impact how phenomena are interpreted or guide the development of systems without identifying and testing assumptions. These assumptions can lead people to design solutions that meet the needs of those similar to them, while ignoring the needs of those who are too different.

As such, we need to remember that, as educational designers, we bring with us a large set of assumptions about what students are capable of doing, what opportunities are appropriate, how best to assess learning, and what these assessments mean. These assumptions can have huge impacts on the designs we create. These designs amount to the standards we set forth, the assessments we use to measure learning, and the educational system that stems from it. This ultimately sets up the rules to the "educational game" and prescribes the behaviors students must display in order to become valued members of educational and STEM communities.

If the goal of STEM educational reform is to create better, more informed design of standards, curriculum, and pedagogies to make STEM experiences more authentic, accessible, and meaningful for a diverse population of students, then we need to start by thinking about student needs first. Integrating and supporting complex forms of reasoning, such as systems thinking, into classroom experiences would likely benefit many students, and finding better ways of assessing and developing student thinking will greatly benefit many more. However, the educational system as a whole needs to change its perspective on how we support learning and evaluate educational practice if we are to improve educational experiences for *all* students. It is necessary to understand the needs of different populations and use this information as a foundation for the instructional designs we create. We cannot create effective learning environments unless we make better use of the decades of research on cognition, development, social psychology, and design fields as a means to inform our proposed educational designs. A human-centered design approach to educational reform could ensure that we provide the best possible educational design solutions to meet the changing needs of our students and that our solutions work as intended throughout the educational system.[1]

Note

1. I would like to thank Barbara White for her feedback on earlier versions of this draft. I would also like to thank her for being a dedicated mentor and for her profound influence on my work. She spent her life challenging us to test our assumptions about what young children are capable of doing and making science accessible to ALL students.

5

RESPONSE 2

Improving Learning about Systems Requires Designing for Change in Educational Systems

William R. Penuel

In recent years, a wide variety of consensus reports published by the National Academies and other groups have based their recommendations on a growing body of research on how people learn in the disciplines (NRC, 2007, 2009, 2012). In science and engineering education, the new *Framework for K-12 Science Education* (NRC, 2012) drew not only on findings from the learning sciences, it also drew upon empirically grounded conceptions of science and engineering practices developed within science studies. The *Framework's* vision of scientific and engineering practices and the development of scientific understanding is also reflected in the *Next Generation Science Standards: By states, for states* (NGSS Lead States, 2013), which are now being adopted by several states across the country.

Realizing the visions of consensus reports and frameworks and implementing standards is no simple matter, however. It requires the active engagement of researchers in a broader range of policy arenas—including state houses, school boards, district offices, and community coalitions—to effect the kinds of change outlined in these documents. Translating research findings into practitioner-friendly summaries or transportable innovations are insufficient to bring about change; different models for relating research to practice are needed that allow for sustained interaction around problems of practice (Contandriopoulos, Lemire, Denis, & Tremblay, 2010; Donovan, 2013; NRC, 2012). In the case of science and engineering education, this means interaction among scientists, engineers, education researchers, curriculum experts, practitioners, families, educational leaders, and textbook companies.

Making school science more closely resemble the practices of scientists requires more than finding "authentic" projects for students to do inside classrooms. Engaging students directly in the practices of science also means helping communities use science and engineering in service of a better life, especially for youth from nondominant communities (Bang & Medin, 2010; Penuel, 2014; Roth & Lee, 2004). No individual teacher can be expected to create and maintain the kinds of connections to the practices of science or opportunities for community engagement in order to engage students in authentic practice. Currently, the majority of public school systems function better to produce success and failure based on performance on various standardized tests (Varenne & McDermott, 1998) than they do to engage students in meaningful scientific problems and practice. But achieving these goals requires analysis of systems of educational practice, their histories, and an inclusive, participatory approach to design (Bang, Medin, Washinawatok, & Chapman, 2010; Gutiérrez & Vossoughi, 2010).

In this response, I outline what it means to design for educational systems. Drawing on analogies and arguments presented in two other chapters (Litzinger, this volume; Minstrell, Anderson, & Li, this volume), I outline key components of educational systems that need to be redesigned. Then, I develop an argument for designing improvements to educational infrastructures using an approach my colleagues and I refer to as *design-based implementation research* (Penuel, Fishman, Cheng, & Sabelli, 2011). I outline the principles of that approach and offer some questions to guide future research on improvement in engineering education and assessment, the foci of chapters by Litzinger and Minstrell et al.

Designing for Educational Systems

Even though we refer to the US educational system as a system, it is not one in which the components cohere, as called for in the *Framework*. There are many different stakeholders in education, with often conflicting goals and strategies for improvement (Rowan, 2002). Tests are used to sort and select students into different tracks, which in turn enable and constrain access to future educational opportunities in ways that reproduce larger social inequities (Guiton & Oakes, 1995; Oakes, 1985). Organizational norms and routines can just as easily promote stasis as change among practitioners (Horn & Little, 2010; Little, 1990; Warren Little, 2002). At the same time, like the engineering systems that Litzinger (this volume) describes, the educational system is shaped by and shapes a variety of technical, environmental, social, societal, and political processes on multiple timescales (Lemke & Sabelli, 2008). Schools—as one part of this system—are important nodes in a broader network of learning supports for young people, but they are not the only nodes: families and informal learning settings figure prominently in the lives of many young people (NRC, 2009) and shape their imagined and actual futures (Maltese & Tai, 2010; Polman & Miller, 2010).

Litzinger (this volume) cites aerospace engineer Charla Wise as asserting that

> [T]he days of engineers creating designs, structures, and platforms as an independent product without an understanding of the end product, the intended use, and the unintended consequences are over. (Wise, 2010, p. 228, cited in Litzinger, this volume, p. 36)

He goes on to argue, "complex engineering systems do not exist inside a bubble independent of the people who design, build, and use them, nor are engineering systems free from societal and political influences." (p. 4). Litzinger's and Wise's point about engineering applies equally well to those of us who want to improve educational systems. For one, it reminds us that the artifacts that we design—whether visions for improving science and engineering reflected in reports, curricula, or assessment systems—have a life that we do not control. The intended use is often not the actual use, and there will inevitably be many unintended consequences. Sometimes, though, we are enamored by our own visions of design, and we forget the agency of practitioners, both in how we design and in how we characterize what the practitioners do with our design. We include practitioners only peripherally in our design research projects (Ormel, Roblin, McKenney, Voogt, & Pieters, 2012). We claim to be "use-inspired," but we are rarely disciplined by problems of practice as experienced by teachers, school leaders, or informal educators (Donovan, 2013). And even when we succeed in bringing our innovations to scale, they don't necessarily last (Fishman et al., 2011).

Part of the challenge lies in our objectives for design. We typically design innovations for classrooms when we need to be designing for systems. Though we are quite good at designing for "the technical core" of teaching and learning, we attend very little to the infrastructure needed to support our innovations. We treat classrooms as if they are isolated from this larger infrastructure and the politics surrounding education as an annoyance to our ongoing work (Donovan, Snow, & Daro, 2013). This is a mistake.

The key to any system is its infrastructure. Design activities that are likely to have the largest impact on practice are those aimed at redesigning the *infrastructures* of systems and organizations. Most infrastructures are largely invisible; it takes deep investigations to uncover and make visible what the infrastructures do, let alone redesign them (Bowker & Star, 1999; Star & Ruhleder, 1996). But redesigning infrastructures is a core task of changing systems, and many innovations require new infrastructures to be used—for example, electricity requires materials and a means to produce it and a power grid to distribute it across a geographic area. Similarly, changes in educational systems require materials and means to prepare educators, processes for attracting and enrolling learners, and the like. Redesign of educational infrastructures requires great effort—such as when schools redefine attendance patterns or states adopt new standards and assessment systems.

Efforts to redesign infrastructure in educational systems are made in an attempt to meet various goals. Many recent redesign efforts at the state and national level in the US have sought to bring greater coherence and reduce fragmentation of the system of standards, curriculum, assessments, and professional development (Cohen, Moffitt, & Goldin, 2007). At the local level redesign can involve establishing new networks or linkages across different types of activities and organizations, such as making connections between schools and informal settings for learning (de Kanter, Adair, Chung, & Stonehill, 2003). Redesign can also entail the creation of new organizational routines and tools intended to support coordination within organizations (Spillane & Coldren, 2011). Efforts to re-organize district offices are made to better support instructional improvement in schools or to institute new leadership roles such as instructional coaches who are expected to support content area teachers in schools (Honig, 2013; Woulfin, 2012). Similarly, Comprehensive School Reform models offer new infrastructures for supporting instruction inside schools (Peurach, 2011). As Litzinger (this volume) argues, these models sit within larger systems that must be considered in design. To this, I advocate for attention to design at the system level.

Designing for Improvements to Infrastructures

How do we begin to design for improvements to infrastructure? The Minstrell et al. chapter (this volume) on the power of effective formative assessment embedded in instruction contains some important implications when applied to designing for change beyond the classroom. First, we need to engage participants in systems not just as objects to be changed but as co-learners and agents in design. In Minstrell et al. classrooms, teachers build from student ideas rather than seeking to destroy misconceptions the moment they arise. Likewise, we need to build on practitioners' ideas rather than imagining them to be deficient before we begin to engage them. Second, like teachers in BOLT, we need an interpretive framework that guides implementing agents (teachers) in making sense of what's happening and guiding their responses to students. That framework needs to extend beyond our theories of individual learning and encompass theories of how organizations change. Third, we have to imagine as I have argued above, that the entire system is the focus of our design. The BOLT framework is a network with many nodes, many of which are linked together. Its robustness as a framework for guiding improvements in the classroom lies not in its "single bullet" approach, but the many components that have to be coordinated and the appreciation for the emergent character of the system's functioning. Similarly, improving any educational system will require envisioning how to coordinate across different groups, activities, and levels of the system.

The Minstrell et al. framework, however, will not get us to systems change on its own, as the story of the *Contingent Pedagogies* project illustrates. Minstrell and I collaborated on this project in partnership with colleagues in a large urban school district. We accomplished impressive results, too, in a quasi-experimental test of the impact of an intervention designed to improve teachers' formative assessment practice. Teachers in the *Contingent Pedagogies* group employed more varied talk moves and scaffolded richer, evidence-based arguments with students in class, and students in the group learned more than students in the comparison group (Penuel et al., 2013). But two years after the study, there's been little uptake of the tools beyond our initial pilot group, and the district decided to implement an alternative approach to formative assessment in middle school science. We'd failed to engage multiple stakeholders within the system, or to have a consistent enough interaction with leaders at the district level to know how their concerns had evolved beyond our project's goals in such a way that our thinking and project could inform district leaders' decision making.

My colleagues and I (Penuel et al., 2011) have recently articulated an alternative approach to research and development that is specifically aimed at designing for systems in ways that embody and reflect Minstrell et al.'s formative assessment model in BOLT. The approach is called *design-based implementation research* (DBIR). DBIR applies design-based perspectives and methods to address and study problems of implementation. DBIR challenges education researchers to break down barriers that separate those who design and study innovations within classrooms from those who study the diffusion of innovations. It also aims to reconfigure the roles of researchers and practitioners in bringing about systemic change by establishing partnerships that support practitioners in adapting innovations productively to meet the needs of diverse students.

Four principles characterize the approach:

1. A focus on persistent problems of practice from multiple stakeholders' perspectives;
2. A commitment to iterative, collaborative design;
3. A concern with developing theory and knowledge related to both classroom learning and implementation through systematic inquiry; and
4. A concern with developing capacity for sustaining change in systems.

In DBIR, teams form around a focus on persistent problems of practice from multiple stakeholders' perspectives. Design-based implementation research shares a commitment with other forms of design research (e.g., formative experiments, design experimentation) to using research to solve practical problems. In DBIR, multiple stakeholders help identify practical problems that are candidates to become the focus of research and development. Depending upon the endeavor, participants may include researchers, organizational leaders, practitioners, parents and community members, and young people. A diversity of ideas on what problems should be taken up is not only likely across these different groups, it is likely within these groups.

To improve practice, teams commit to iterative, collaborative design. Collaborative design research often focuses on the development and testing of usable tools for improving teaching and learning in specific subject matter domains and settings (Cobb, Confrey, diSessa, Lehrer, & Schauble, 2003). As with other forms of design-based research, the design effort begins with a vision for learning that sets the focus for design (Collins, Joseph, & Bielaczyc, 2004). Other decisions that are critical within design-based research are important within DBIR as well, such as deciding on the composition of the design team, the process of iteration, and what evidence and rationales will form the basis for making changes to designs (Penuel et al., 2011; Penuel,

Tatar, & Roschelle, 2004). Participants may need to include school and district leaders (Cobb & Jackson, 2012), family members (Goldman & Booker, 2009), community members (Bang et al., 2010), and youth (Kirshner, 2007).

The focus of design work is a key distinction between DBIR and most design-based research. A hallmark of design-based research in the learning sciences has been a focus on improving learning environments in classrooms. At the same time, some learning scientists do focus more explicit attention on designing learning opportunities for teachers (e.g., Fishman, Marx, Best, & Tal, 2003), with an eye towards building capacity for teachers to enact innovations. DBIR encompasses such efforts, as well as designs for bringing about system-level improvements (Bryk, Gomez, & Grunow, 2011; Resnick & Spillane, 2006). The focus of design in DBIR may also be on developing supports for implementers of informal education programs, or for interventions that span multiple settings. The key for all of these is that DBIR focuses not only on the design of tools and practices for learners, but also on the design of the necessary means of support for implementing those tools.

As a strategy for promoting quality in the research and development process, teams develop theory and knowledge related to both classroom learning and implementation through systematic inquiry. Though DBIR draws from models of evaluation research, it diverges from them somewhat with respect to this principle. Most evaluation studies are motivated by practical and policy questions, aimed at generating knowledge of specific programs' worth or value. But DBIR aims to develop knowledge that might be useful across a range of settings, either to inform design or to develop deeper understandings of learning of different kinds (e.g., student, teacher, organizational). DBIR also aims to develop theory both for guiding design teams (which include practitioners) and future learning research, both in the ways that design-based research typically does (Edelson, 2002) and in the ways that well-structured case study research does (see Flyvbjerg, 2006).

The kinds of theories that are relevant to DBIR relate specifically to the objects of design, and they encompass theories of learning within and across subject matter domains, informal or everyday learning, teacher learning, organizational change, and leadership. In contrast to design-based research that focuses only on theories related to student learning in domains, DBIR can also contribute to theories of organizations and institutions that guide much contemporary policy research in education, particularly by pointing out how the deployment of new tools (e.g., curricula, technologies) can bring to light new needs for coordination across different system levels and for capacity building (e.g., Stein & Coburn, 2008).

DBIR is concerned with developing capacity for sustaining change in systems. One strategy for promoting sustainability of designs is to develop capacity through intentional efforts to develop organizational routines and processes that help innovations travel through a system. The predominant approaches to building capacity in education have long been focused on improving human capital of implementers—whether through professional development or by selecting and rewarding teachers on the basis of their students' test scores—and on developing and supplying improved material capital (e.g., curriculum) to them. DBIR focuses not only on building the skills of individual implementers—although this may be a positive outcome of DBIR—but on developing the capacity of the entire system to implement, scale, and sustain innovations. Moreover, we include researchers as individual participants and research organizations and universities as institutions that are part of the ecology of supports that must be strengthened to improve excellence and equity.

DBIR for Engineering Education, Classroom Assessment, and Beyond

The chapters in this section take up the topics of engineering education and classroom assessment, arguably disparate topics, made all the more distinct by the fact that the focus of the work is at different levels of the system. But using the DBIR approach, I can pose some of the same questions to the authors, even though I would expect different answers for each. The questions are grounded in the four principles:

1. What are the problems of practice experienced by teachers? If the vision were to enhance their capacity to teach for and assess understanding of systems, how would they name the key problems?
2. Who needs to be at the table when the problems of practice are articulated, and when the search for solutions begins? How will you organize design?
3. What theories and research on student understanding of systems (particularly complex systems) should guide the design effort? What theories of organizational change are relevant to the context?
4. How will the design partnership enhance the capacity of the system to improve?

My sincere wish is that the answer that scholars of engineering education and assessment might give to these questions are not answers to an implicit question many of us try to answer, "How do we get others to do what we know to be good?" But rather, we should be asking a broader question, "How do we change the system together to address the problems we have?" Answering that question will require work that goes beyond designing new courses, curricula, or formative assessment interventions. It will require broader, more long-term partnerships and a willingness to take up the ideas and concerns of people charged with implementing change, at whatever level they are working.

Poster Presentation I

Virtual and Authentic Web-based Ecological Inquiries

X. Ben Wu[1], Stephanie Knight[2], Aubree Webb[2], Melisa Ziegler[2], Janie Schielack[1]

[1] Texas A&M University [2] Pensylvania State University

Introduction

A virtual learning environment based on simulations, Virtual Ecological Inquiry (VEI), and a web-based authentic inquiry based on remote observations of bear behavior (BearCam) were developed, implemented in a large introductory ecology course, and evaluated for their effects on student learning and attitudes.

Process of Inquiry Projects

1. Background exploration and discussion in eLearning

For VEI in Second Life

2. Learning and design in the virtual museum:
   - Explore ecology of Wolong; formulate hypothesis; design field investigation
3. Field sampling
4. Data analysis in the lab at the virtual museum

For BearCam

2. Observing bear distribution and interactions using Still pictures from Larry Griffing (Griffing, 2007) and colleagues' research on grizzly bear behavior, taken at McNeil River Fall in Alaska
3. Developing and refining hypothesis
4. Collection and analyzing data and interpreting results

5. Development of ecological report
6. Calibrated Peer Review (CPR) with a 30-item rubric
7. Revision of reports based on peer reviews

Required group discussions and feedback throughout the process

Implementation of the Inquiry Projects

- Introductory ecology course; two sections in Fall 2012. BearCam: 211 students; VEI: 157 students
- 4-week period: One class period to discuss background and process of inquiry project. Regular in-class discussion at the beginning of each class period
- Online assessments: Pre-test and post-test usig Classroom Test of Scientific Reasoning (Lawson, 1978; 2000) and Test of Science-Related Attitudes (TOSRA) (Fraser, 1981); post-survey of student self-assessment and feedback

Summary of Findings

- Students reported statistically significant learning gains in ability to formulate testable hypotheses, understanding how ecologists conduct research, and ability to evaluate quality of scientific report. But, increase in interest in ecology was not significant for VEI.
- Analyses of reports supported this learing. Avery high percentage of students were able to implement the inquiry skills identified in the rubric. More than half of the students received high overall score on reports.
- Students seem to like the BearCam project more than the VEI project: gave BearCam project significantly higher ratings on their learning and interests with the project, and the overall rating of their experiences.
- No overall difference in TOSRA, $t(155) = 2.738, p > .003$. But, students scored Subscale 2: Adoption of Scientific Attitudes significantly lower from pre to post $t(269) = 3.570, p < .003$. May be due to significant decrease for students in BearCam treatment, $t(155) = 3.313, p < .003$.
- No significant difference between the Lawson's Scientific Reasoning pre-test and post-test for students in both treatments, $t(269) = -1.660, p > .003$. But, significant interaction by gender for post-test: females scored higher for VEI; males scored higher for BearCam.
- Lawson's Scientific Reasoning total score is a significant predictor of students' third exam scores, which has content questions related to the inquiry project ($B = .358$, $t(270) = 6.31$, $p<.001$).
- TOSRA Scale 2: Adoption of Scientific Attitudes is a significant predictor of Lawson's Scientific Reasoning scores ($B = .24$, $t(270) = 4.14$, $p<.0001$).

References

Frascr, B. J. (1981). *Test of science-related attitudes (TOSRA).* Australian Council for Educatonal Research.
Griffing, L.R. 2007. Secret Lives of The Brown Bears of McNeil River. Retrieved from http://griffing.tamu.edu/Site/McNeil/index.htm
Lawson, A.E. 2000. Classroom Test of Scientific Reasoning. Arizona State University.
Lawson, A.E. 1978. Development and validation of the classroom test of formal reasoning. Journal of Research in Science Teaching, 15(1): 11-24.

Acknowlegement

This project was partially supported by grants from National Science Foundation (DUE-0942715 and DRL 0083336).

Theme I References

Abbott, A.D. (1988). *The System of Professions: An Essay on the Division of Expert Labor.* Chicago, IL: University of Chicago Press.

ABET. (2013). *2013–2014 Criteria for Accrediting Engineering Programs.* Retrieved from http://www.abet.org/wp-content/uploads/2015/04/eac-criteria-2013-2014.pdf.

Bandura, A. (1993). Perceived self-efficacy in cognitive development and functioning. *Educational Psychologist, 28*(2), 117–48. http://doi.org/10.1207/s15326985ep2802_3

Bandura, M., & Dweck, C. (1985). *Self-conceptions and motivation: Conceptions of intelligence, choice of achievement goals, and patterns of cognition, affect, and behavior.* Unpublished Manuscript, Harvard University.

Bang, M., & Medin, D. (2010). Cultural processes in science education: Supporting the navigation of multiple epistemologies. *Science Education, 94*(6), 1008–26. http://doi.org/10.1002/sce.20392

Bang, M., Medin, D., Washinawatok, K., & Chapman, S. (2010). Innovations in culturally based science education through partnerships and community. In M.S. Khine & I.M. Saleh (Eds.), *New Science of Learning Cognition, Computers and Collaboration in Education* (pp. 569–92). New York; London: Springer. Retrieved from http://public.eblib.com/choice/publicfullrecord.aspx?p=603372

Bell, B. (2002). *Formative Assessment and Science Education.* New York: Kluwer Academic.

Bennett, R.E. (2013). Preparing for the future: What educational assessment must do. In *To Assess, To Teach, To Learn: A Vision for the Future of Assessment.* Retrieved from http://gordoncommission.org/rsc/pdfs/technical_report_executive_summary.pdf

Bjork, R.A. (1994). Memory and metamemory considerations in the training of human beings. In J. Metcalfe & A.P. Shimamura (Eds.), *Metacognition Knowing About Knowing.* Cambridge, MA: MIT Press. Retrieved from http://search.ebscohost.com/login.aspx?direct=true&scope=site&db=nlebk&db=nlabk&AN=1765

Black, P., & Wiliam, D. (2009). Developing the theory of formative assessment. *Educational Assessment, Evaluation and Accountability (formerly: Journal of Personnel Evaluation in Education), 21*(1), 5–31. http://doi.org/10.1007/s11092-008-9068-5

Boekaerts, M. (1996). Self-regulated learning at the junction of cognition and motivation. *European Psychologist, 1*(2), 100–12. http://doi.org/10.1027/1016-9040.1.2.100

Borge, M., & Carroll, J.M. (2010). Using collaborative activity as a means to explore student performance and understanding. In *Proceedings of the 9th International Conference of the Learning Sciences-Volume 1* (pp. 889–96).

Bowker, G.C., & Star, S.L. (1999). *Sorting Things Out: Classification and its Consequences.* Cambridge, MA: MIT Press.

Broadfoot, P., & Black, P. (2004). Redefining assessment? The first ten years of assessment in education. *Assessment in Education: Principles, Policy & Practice, 11*(1), 7–26. http://doi.org/10.1080/0969594042000208976

Brookhart, S.M. (2009). *Exploring Formative Assessment.* Alexandria, VA: ASCD.

Brown, A.L. (1984). Metacognition, executive control, self-regulation, and other more mysterious mechanisms. In F.E. Weinert & R. Kluwe (Eds.), *Metacognition, Motivation and Understanding* (pp. 60–108). West Germany: Kohlhammer.

Brown, A.L., & Campione, J.C. (1994). Guided discovery in a community of learners. In K. McGilly (Ed.), *Classroom Lessons Integrating Cognitive Theory and Classroom Practice* (pp. 229–70). Cambridge, MA: MIT Press.

Brown, A.L., Ash, D., Rutherford, M., Nakagawa, K., Gordon, A., & Campione, J.C. (1993). Distributed expertise in the classroom. In G. Salomon (Ed.), *Distributed Cognitions: Psychological and Educational Considerations* (pp. 188–228). New York: Cambridge University Press.

Brown, A.L., Metz, K.E., & Campione, J.C. (1996). Social interaction and individual understanding in a community of learners: The influence of Piaget and Vygotsky. In A. Tryphon & J.J. Vonèche (Eds.), *Piaget-Vygotsky: The Social Genesis of Thought*. Hove, East Sussex, UK: Psychology Press.

Brown, J.S., Collins, A., & Duguid, P. (1989). Situated cognition and the culture of learning. *Educational Researcher, 18*(1), 32–42. http://doi.org/10.3102/0013189X018001032

Bryk, A.S., Gomez, L.M., & Grunow, A. (2011). Getting ideas into action: Building networked improvement communities in education. In M.T. Hallinan (Ed.), *Frontiers in Sociology of Education* (pp. 127–62). Dordrecht, Netherlands: Springer. Retrieved from http://public.eblib.com/choice/publicfullrecord.aspx?p=770122

Butler, D.L., & Winne, P.H. (1995). Feedback and self-regulated learning: A theoretical synthesis. *Review of Educational Research, 65*(3), 245–81. http://doi.org/10.3102/00346543065003245

Camp, C.W., Clement, J.J., & Brown, D. (1994). *Preconceptions in Mechanics: Lessons Dealing with Students' Conceptual Difficulties*. Dubuque, IA: Kendall/Hunt.

Cardenas, C., Sosa, R., Moysen, R., & Martinez, V. (2010). Sustaining sustainable design through systems thinking. *International Journal of Engineering Education, 26*(2), 287–92.

Cobb, P., & Jackson, K. (2012). Analyzing educational policies: A learning design perspective. *Journal of the Learning Sciences, 21*(4), 487–521. http://doi.org/10.1080/10508406.2011.630849

Cobb, P., Confrey, J., diSessa, A., Lehrer, R., & Schauble, L. (2003). Design experiments in educational research. *Educational Researcher, 32*(1), 9–13. http://doi.org/10.3102/0013189X032001009

Coffey, J., & Alberts, B. (2013). Improving education standards. *Science, 339*(6119), 489–89. http://doi.org/10.1126/science.1235590

Cohen, D.K., Moffitt, S.L., & Goldin, S. (2007). Policy and practice: The dilemma. *American Journal of Education, 113*(4), 515–548.

Cole, M. (1995). Socio-cultural-historical psychology: Some general remarks and a proposal for a new kind of cultural-genetic methodology. In J.V. Wertsch, P. del Río, & A. Alvarez (Eds.), *Sociocultural Studies of Mind* (pp. 187–214). Cambridge; New York: Cambridge University Press.

Cole, M., & Engestrom, Y. (1993). A cultural-historical approach to distributed cognition. In G. Salomon (Ed.), *Distributed Cognitions: Psychological and Educational Considerations* (pp. 1–46). Cambridge [England]; New York: Cambridge University Press.

Collins, A., & Halverson, R. (2009). *Rethinking Education in the Age of Technology: The Digital Revolution and Schooling in America*. New York: Teachers College Press.

Collins, A., Brown, J.S., & Holum, A. (1991). Cognitive apprenticeship: Making thinking visible. *American Educator, 15*(3), 6–11.

Collins, A., Joseph, D., & Bielaczyc, K. (2004). Design research: Theoretical and Methodological Issues. *Journal of the Learning Sciences, 13*(1), 15–42. http://doi.org/10.1207/s15327809jls1301_2

Contandriopoulos, D., Lemire, M., Denis, J.L., & Tremblay, E. (2010). Knowledge exchange processes in organizations and policy arenas: A narrative systematic review of the literature. *Milbank Quarterly, 88*(4), 444–83.

Córdoba-Pachón, J. R. (2011). Abstracting and engaging: Two modes of systems thinking education. *INFORMS Transactions on Education, 12*(1), 43–54. http://doi.org/10.1287/ited.1110.0072

Coughlan, P., & Ponto, C. (2011). Systems thinking + Design thinking for innovative sustainable solutions. Paper presented at the Systems Thinking In Action Conference, Seattle, WA. Retrieved from www.pegasuscom.com/stia11/slides/d01-coughlin.pdf

Davidson, C. I., Hendrickson, C. T., & Matthews, H. S. (2007). Sustainable engineering: A sequence of courses at Carnegie Mellon. *International Journal of Engineering Education, 23*(2), 287–93.

Davidz, H. L., & Nightingale, D. J. (2008). Enabling systems thinking to accelerate the development of senior systems engineers. *Systems Engineering, 11*(1), 1–14. http://doi.org/10.1002/sys.20081

DeBarger, A., Minstrell, J., Ayala, C., Freed, R., & Vasquez, S. (2013). Using formative assessment to improve learning in high school chemistry. Paper presented at the NARST annual international meeting. San Juan, PR.

de Kanter, A., Adair, J. K., Chung, A.-M., & Stonehill, R. M. (2003). Ensuring quality and sustainability in after-school programs: How partnerships play a key role. *Yearbook of the National Society for the Study of Education, 102*(2), 201–20. http://doi.org/10.1111/j.1744-7984.2003.tb00072.x

de Weck, O. L., Roos, D., & Magee, C. L. (2011). *Engineering Systems Meeting Human Needs in a Complex Technological World.* Cambridge, MA: MIT Press. Retrieved from http://site.ebrary.com/id/10529578

diSessa, A. A. (1988). Knowledge in pieces. In G. E. Forman & P. B. Pufall (Eds.), *Constructivism in the Computer Age* (pp. 49–70). Hillsdale, NJ: L. Erlbaum.

diSessa, A. A. (1993). Toward an epistemology of physics. *Cognition and Instruction, 10*(2–3), 105–225. http://doi.org/10.1080/07370008.1985.9649008

Donovan, M. S. (2013). Generating improvement through research and development in education systems. *Science, 340*(6130), 317–19. http://doi.org/10.1126/science.1236180

Donovan, M. S., Snow, C. E., & Daro, P. (2013). The SERP approach to problem-solving research, development, and implementation. In B. J. Fishman, W. R. Penuel, A.-R. Allen, & B. H. Cheng (Eds.), *Design-based Implementation Research: Theories, Methods, and Exemplars* (Vol. 112, pp. 400–25).

Driver, R., Squires, A., Rushworth, P., & Wood-Robinson, V. (1994). *Making Sense of Secondary Science Research into Children's Ideas.* New York: Routledge. Retrieved from http://public.eblib.com/choice/publicfullrecord.aspx?p=1791219

Dweck, C. (1985). Intrinsic motivation, perceived control, and self-evaluation maintenance: An achievement goal analysis. In R. Ames & C. Ames (Eds.), *Research on Motivation in Education: The Classroom Milieu* (pp. 289–305). Orlando, FL: Academic Press.

Dweck, C. S. (1986). Motivational processes affecting learning. *American Psychologist, 41*(10), 1040–48. http://doi.org/10.1037/0003-066X.41.10.1040

Dweck, C. S. (1999). *Self-theories: Their Role in Motivation, Personality, and Development.* Philadelphia, PA: Psychology Press.

Dym, C. L., Agogino, A.M., Eris, O., Frey, D. D., & Leifer, L. J. (2005). Engineering design thinking, teaching, and learning. *Journal of Engineering Education, 94*(1), 103–120. http://doi.org/10.1002/j.2168-9830.2005.tb00832.x

Edelson, D.C. (2002). Design Research: What we learn when we engage in design. *Journal of the Learning Sciences, 11*(1), 105–121. http://doi.org/10.1207/S15327809JLS1101_4

Engestrom, Y. (2000). Activity theory as a framework for analyzing and redesigning work. *Ergonomics, 43*(7), 960–74. http://doi.org/10.1080/001401300409143

Fishman, B.J., Marx, R.W., Best, S., & Tal, R.T. (2003). Linking teacher and student learning to improve professional development in systemic reform. *Teaching and Teacher Education, 19*(6), 643–58. http://doi.org/10.1016/S0742-051X(03)00059-3

Fishman, B., Penuel, W.R., Hegedus, S., Roschelle, J., Fishman, B., Penuel, W.R., Roschelle, J. (2011). What happens when the research ends? Factors related to the sustainability of a technology-infused mathematics curriculum. *Journal of Computers in Mathematics and Science Teaching, 30*(4), 329–353.

Flyvbjerg, B. (2006). Five misunderstandings about case-study research. *Qualitative Inquiry, 12*(2), 219–45. http://doi.org/10.1177/1077800405284363

Frederiksen, J.R., & White, B.Y. (2004). Designing assessments for instruction and accountability: An application of validity theory to assessing scientific inquiry. *Yearbook of the National Society for the Study of Education, 103*(2), 74–104. http://doi.org/10.1111/j.1744-7984.2004.tb00049.x

Froyd, J., Pchenitchnaia, L., Fowler, D., & Simpson, N. (2007). Systems thinking and integrative learning. Proceedings of the American Society for Engineering Education: *Annual Meeting*. Honolulu, HA.

Ginsborg, J. (2006). The effects of socio-economic status on children's language acquisition and use. In J. Clegg & J. Ginsborg (Eds.), *Language and Social Disadvantage: Theory into Practice* (pp. 9–27). Chichester, England; Hoboken, NJ: John Wiley & Sons.

Goldman, S., & Booker, A. (2009). Making math a definition of the situation: Families as sites for mathematical practices. *Anthropology & Education Quarterly, 40*(4), 369–87. http://doi.org/10.1111/j.1548-1492.2009.01057.x

Guiton, J., & Oakes, J. (1995). Opportunity to learn and conceptions of educational equality. *Educational Evaluation and Policy Analysis, 17*(3), 323–36.

Guthrie, P. (2010). Beyond systems engineering—Educational approaches for the 21st century. In D. Grasso & M.B. Burkins (Eds.), *Holistic Engineering Education* (pp. 93–7). New York: Springer. Retrieved from http://site.ebrary.com/id/10372426

Gutiérrez, K.D., & Vossoughi, S. (2010). Lifting off the ground to return anew: Mediated praxis, transformative learning, and social design experiments. *Journal of Teacher Education, 61*(1–2), 100–117. http://doi.org/10.1177/0022487109347877

Heritage, M. (2010). Formative assessment and next-generation assessment systems: Are we losing an opportunity? In *Paper prepared for the Council of Chief State School Officers*. Washington, D.C.

Herman, J.L., Osmundson, E., Ayala, C., Schneider, S., & Timms, M.J. (2005). The nature and impact of teachers' formative assessment practices. In J.L. Herman (Chair), Building science assessment systems that serve accountability and student learning: The CAESL model. Symposium conducted at the annual meeting of the American Educational Research Association, Montreal, Canada. Retrieved from http://works.bepress.com/michael_timms/28

Hmelo-Silver, C.E., Duncan, R.G., & Chinn, C.A. (2007). Scaffolding and achievement in problem-based and inquiry learning: A response to Kirschner, Sweller, and Clark. *Educational Psychologist, 42*(2), 99–107. http://doi.org/10.1080/00461520701263368

Hokanson, D.R., Phillips, L.D., & Mihelcic, J.R. (2007). Educating engineers in the sustainable futures model with a global perspective: Education, research and diversity initiatives. *International Journal of Engineering Education, 23*(2), 254–65.

Honig, M.I. (2013). Beyond the policy memo: Designing to strengthen the practice of district central office leadership for instructional improvement at scale.

In B.J. Fishman, W.R. Penuel, A.-R. Allen, & B.H. Cheng (Eds.), *Design-based Implementation Research: Theories, Methods, and Exemplars* (Vol. 112, pp. 256–73).

Horn, I.S., & Little, J.W. (2010). Attending to problems of practice: Routines and resources for professional learning in teachers' workplace interactions. *American Educational Research Journal, 47*(1), 181–217. http://doi.org/10.3102/0002831209345158

Hu, H., Real, E., Takamiya, K., Kang, M.-G., Ledoux, J., Huganir, R.L., & Malinow, R. (2007). Emotion enhances learning via norepinephrine regulation of AMPA-receptor trafficking. *Cell, 131*(1), 160–73. http://doi.org/10.1016/j.cell.2007.09.017

Hunt, E., & Minstrell, J. (1996). Effective instruction in science and mathematics: Psychological principles and social constraints. *Issues in Education: Contributions from Educational Psychology, 2*(2), 123–62.

Johnson-Laird, P.N. (1983). *Mental Models: Towards a Cognitive Science of Language, Inference, and Consciousness*. Cambridge, MA: Harvard University Press.

Kirschner, P.A., Sweller, J., & Clark, R.E. (2006). Why minimal guidance during instruction does not work: An analysis of the failure of constructivist, discovery, problem-based, experiential, and inquiry-based teaching. *Educational Psychologist, 41*(2), 75–86. http://doi.org/10.1207/s15326985ep4102_1

Kirshner, B. (2007). Introduction youth activism as a context for learning and development. *American Behavioral Scientist, 51*(3), 367–379. http://doi.org/10.1177/0002764207306065

Kuhn, D. (2009a). Do students need to be taught how to reason? *Educational Research Review, 4*(1), 1–6. http://doi.org/10.1016/j.edurev.2008.11.001

Kuhn, D. (2009b). The importance of learning about knowing: Creating a foundation for development of intellectual values. *Child Development Perspectives, 3*(2), 112–17.

Kuhn, D., Black, J., Keselman, A., & Kaplan, D. (2000). The development of cognitive skills to support inquiry learning. *Cognition and Instruction, 18*(4), 495–523. http://doi.org/10.1207/S1532690XCI1804_3

Lemke, J.L., & Sabelli, N.H. (2008). Complex systems and educational change: Towards a new research agenda. *Educational Philosophy and Theory, 40*(1), 118–29. http://doi.org/10.1111/j.1469-5812.2007.00401.x

Little, J.W. (1990). The persistence of privacy: Autonomy and initiative in teachers' professional relations. *Teachers College Record, 91*(4), 509–36.

Litzinger, T.A., & Lattuca, L.R. (2014). Translating research into widespread practice in engineering education. In A. Johri & B.M. Olds (Eds.), *Cambridge Handbook of Engineering Education Research* (pp. 375–92). New York: Cambridge University Press.

McCloskey, M.E. (1984). Cartoon physics. *Psychology Today, 18*(4), 52–8.

Maltese, A.V., & Tai, R.H. (2010). Eyeballs in the fridge: Sources of early interest in science. *International Journal of Science Education, 32*(5), 669–85.

Mandinach, E.B., Gummer, E.S., & Muller, R.D. (2011). The complexities of integrating data-driven decision making into professional preparation in schools of education: It's harder than you think. Presented at the CNA Education, Education Northwest, and WestEd, Alexandria, VA, Portland, OR, and Washington, DC. Retrieved from http://educationnorthwest.org/resources/complexities-integrating-data-driven-decision-making-professional-preparation-schools

Mayer, R.E. (2004). Should there be a three-strikes rule against pure discovery learning? *American Psychologist, 59*(1), 14–19. http://doi.org/10.1037/0003-066X.59.1.14

Meadows, D.H., & Wright, D. (2008). *Thinking in Systems: A Primer*. White River Junction, VT: Chelsea Green Pub.

Minstrell, J. (1982). Explaining the "at rest" condition of an object. *The Physics Teacher, 20*(1), 10–14. http://doi.org/10.1119/1.2340924

Minstrell, J. (1984). Teaching for the development of understanding of ideas: Forces on moving objects. In C.W. Anderson (Ed.), *Observing Science Classrooms: Observing Science Perspectives from Research and Practice* (pp. 67–85). Columbus, OH: Ohio State University.

Minstrell, J. (1992). Facets of students' knowledge and relevant instruction. In R. Duit, F. Goldberg, & H. Niedderer (Eds.), *Research in Physics Learning: Theoretical Issues and Empirical Studies: Proceedings of an International Workshop* (pp. 110–128). Breman, Germany: IPN.

Minstrell, J. (2001). Facets of students' thinking: Designing to cross the gap from research to standards-based practice. In K. Crowley, C.D. Schunn, & T. Okada (Eds.), *Designing for Science: Implications From Everyday, Classroom, and Professional Settings* (pp. 415–43). Mahwah, NJ: Erlbaum.

Minstrell, J., & Matteson, R. (1994). Adopting a different view of learners. In *In Annual Report video to the James S. McDonnell Foundation, Cognitive Studies for Educational Practice.* A.C.T. Systems for Education, Bellevue, WA.

Minstrell, J., & van Zee, E. (2003). Using questioning to assess and foster student thinking. In J.M. Atkin & J. Coffey (Eds.), *Everyday Assessment in the Science Classroom* (pp. 61–72). Arlington, VA: NSTA press. Retrieved from http://site.ebrary.com/id/10240811

Minstrell, J., & Kraus, P. (2005). Guided inquiry in the science classroom. In National Research Council (Ed.), *How Students Learn: History, Mathematics, and Science in the Classroom* (pp. 475–514). Washington, DC: National Academy Press. Retrieved from http://nap.edu/catalog/10126/how-students-learn-history-mathematics-and-science-in-the-classroom

Minstrell, J., Anderson, R., & Li, M. (2011). *Building on Learner Thinking: A Framework for Assessment in Instruction* (Commissioned paper for the Committee on Highly Successful STEM Schools or Programs for K-12 STEM Education).

Nagel, R.L., Pierrakos, O., Pappas, E.C., & Ogundipe, A. (2011). The integration of sustainability, systems, and engineering design in the engineering curriculum at James Madison University. In *International Design Engineering Technical Conferences & Computers and Information in Engineering Conference.* Washington, DC. http://doi.org/10.1115/DETC2011-47847

National Research Council. (1999). *How People Learn Brain, Mind, Experience, and School.* Washington, DC: National Academies Press. Retrieved from http://site.ebrary.com/id/10041047

National Research Council. (2007). *Taking Science to School: Learning and Teaching Science in Grades K-8.* Washington, DC: National Academies Press.

National Research Council. (2009). *Learning Science in Informal Environments People, Places, and Pursuits.* Washington, DC: National Academies Press. Retrieved from http://site.ebrary.com/id/10315878

National Research Council. (2012). *A Framework for K-12 Science Education: Practices, Crosscutting Concepts, and Core Ideas.* Retrieved from http://www.nap.edu/catalog/13165/a-framework-for-k-12-science-education-practices-crosscutting-concepts

NGSS Lead States. (2013). *Next Generation Science Standards: For States, By States.* Retrieved from http://www.nap.edu/catalog/18290/next-generation-science-standards-for-states-by-states

Norman, D. A. (1988). *The Psychology of Everyday Things*. New York: Basic Books.
Norman, G. R., & Schmidt, H. G. (1992). The psychological basis of problem-based learning: A reviewof the evidence: *Academic Medicine*, (9), 557–65.
Oakes, J. (1985). *Keeping Track: How Schools Structure Inequality*. New Haven, CT: Yale University Press.
Ormel, B. J. B., Roblin, N. N. P., McKenney, S. E., Voogt, J. M., & Pieters, J. M. (2012). Research–practice interactions as reported in recent design studies: Still promising, still hazy. *Educational Technology Research and Development*, *60*(6), 967–86. http://doi.org/10.1007/s11423-012-9261-6
Osborne, J. (2010). Arguing to learn in science: The role of collaborative, critical discourse. *Science*, *328*(5977), 463–6. http://doi.org/10.1126/science.1183944
Osborne, J. W., & Walker, C. (2006). Stereotype threat, identification with academics, and withdrawal from school: Why the most successful students of colour might be most likely to withdraw. *Educational Psychology*, *26*(4), 563–77. http://doi.org/10.1080/01443410500342518
Otero, V. K. (2006). Moving beyond the "get it or don't" conception of formative assessment. *Journal of Teacher Education*, *57*(3), 247–55. http://doi.org/10.1177/0022487105285963
Papert, S. (1991). Situating constructionism. In I. Harel & S. Papert (Eds.), *Constructionism* (pp. 1–11). Norwood, NJ: Ablex.
Pea, R. (1993). Practices of distributed intelligence and designs for education. In G. Salomon (Ed.), *Distributed Cognitions: Psychological and Educational Considerations* (pp. 47–87). Cambridge, England; New York: Cambridge University Press.
Pellegrino, J. W. (2013). Proficiency in science: Assessment challenges and opportunities. *Science*, *340*(6130), 320–323. http://doi.org/10.1126/science.1232065
Penuel, W. R. (2014). Studying science and engineering learning in practice. *Cultural Studies of Science Education*, 1–16. http://doi.org/10.1007/s11422-014-9632-x
Penuel, W. R., Tatar, D. G., & Roschelle, J. (2004). The role of research on contexts of teaching practice in informing the design of handheld learning technologies. *Journal of Educational Computing Research*, *30*(4), 353–70.
Penuel, W. R., Fishman, B. J., Cheng, B. H., & Sabelli, N. H. (2011). Developing the area of design-based implementation research. Paper presented at the Annual Meeting of the AERA, New Orleans, LA.
Penuel, W. R., DeBarger, A., Boscardin, C., Moorthy, S., Beauvineau, Y., Kennedy, C., Leary, H., Allison, K. (2013). Improving learning by improving classroom assessment in earth science: Findings from the Contingent Pedagogies Project. Paper presented at the Annual International Meeting of NARST, San Juan, PR.
Peurach, D. J. (2011). *Seeing Complexity in Public Education: Problems, Possibilities, and Success for All*. Oxford, UK: Oxford University Press.
Pintrich, P. R., & de Groot, E. V. (1990). Motivational and self-regulated learning components of classroom academic performance. *Journal of Educational Psychology*, *82*(1), 33–40. http://doi.org/10.1037/0022-0663.82.1.33
Polman, J. L., & Miller, D. (2010). Changing stories: Trajectories of identification among African American youth in a science outreach apprenticeship. *American Educational Research Journal*, *47*(4), 879–918. http://doi.org/10.3102/0002831210367513
Pourdehnad, J., Wexler, E., & Wilson, D. (2011). Integrating systems thinking and design thinking in systems. *Thinker Newsletter*, *22*(9). Retrieved from http://in2in.org/od/thought/2013-02-ThoughtPiece-Pourdehnad.pdf
Ramage, M., & Shipp, K. (2009). *Systems Thinkers*. Dordrecht, Netherlands; New York: Springer.

Ramaprasad, A. (1983). On the definition of feedback. *Behavioral Science, 28*(1), 4–13. http://doi.org/10.1002/bs.3830280103

Resnick, L.B., & Spillane, J.P. (2006). From individual learning to organizational designs for learning. In L. Verschaffel, F. Dochy, M. Boekaerts, & S. Vosniadou (Eds.), *Instructional psychology: Past, Present, and Future Trends: Sixteen Essays in Honour of Erik de Corte* (1 edition, pp. 257–74). Amsterdam; Boston: Emerald Group Publishing.

Richmond, B. (1993). Systems thinking: Critical thinking skills for the 1990s and beyond. *System Dynamics Review, 9*(2), 113–133. http://doi.org/10.1002/sdr.4260090203

Roschelle, J., & Teasley, S.D. (1995). The construction of shared knowledge in collaborative problem solving. In C. O'Malley (Ed.), *Computer Supported Collaborative Learning* (pp. 69–97). Berlin; New York: Springer-Verlag.

Rosenthal, R. (1994). Interpersonal expectancy effects: A 30-year perspective. *Current Directions in Psychological Science, 3*(6), 176–9. http://doi.org/10.1111/1467-8721.ep10770698

Rosenthal, R. (2003). Covert communication in laboratories, classrooms, and the truly real world. *Current Directions in Psychological Science, 12*(5), 151–4. http://doi.org/10.1111/1467-8721.t01-1-01250

Roth, W.-M., & Lee, S. (2004). Science education as/for participation in the community. *Science Education, 88*(2), 263–291. http://doi.org/10.1002/sce.10113

Rowan, B. (2002). The ecology of school improvement: Notes on the school improvement industry in the United States. *Journal of Educational Change, 3*(3–4), 283–314. http://doi.org/10.1023/A:1021277712833

Sadler, D.R. (1989). Formative assessment and the design of instructional systems. *Instructional Science, 18*(2), 119–144. http://doi.org/10.1007/BF00117714

Scalise, K., & Gifford, B. (2006). Computer-based assessment in e-learning: A framework for constructing "intermediate constraint" questions and tasks for technology platforms. *The Journal of Technology, Learning and Assessment, 4*(6), 4–45.

Schmidt, H.G., Loyens, S.M.M., Van Gog, T., & Paas, F. (2007). Problem-based learning is compatible with human cognitive architecture: Commentary on Kirschner, Sweller, and Clark. *Educational Psychologist, 42*(2), 91–7. http://doi.org/10.1080/00461520701263350

Schneider, C., & Gowan, P. (2011). Deconstructing student work: Investigating teachers' abilities to use evidence of student learning to inform instruction. Paper presented at the Annual Meeting of the AERA, New Orleans, LA.

Schoenfeld, A.H. (2002). Making mathematics work for all children: Issues of standards, testing, and equity. *Educational Researcher, 31*(1), 13–25. http://doi.org/10.3102/0013189X031001013

Schoenfeld, A.H. (2004). The math wars. *Educational Policy, 18*(1), 253–86. http://doi.org/10.1177/0895904803260042

Schraw, G. (1998). Promoting general metacognitive awareness. *Instructional Science, 26*(1–2), 113–125. http://doi.org/10.1023/A:1003044231033

Schunk, D.H. (2012). *Learning Theories: An Educational Perspective* (6th ed.). Upper Saddle River, NJ: Merrill.

Senge, P.M., & Sterman, J.D. (1992). Systems thinking and organizational learning: Acting locally and thinking globally in the organization of the future. *European Journal of Operational Research, 59*(1), 137–150. http://doi.org/10.1016/0377-2217(92)90011-W

Shepard, L.A. (2005). Linking formative assessment to scaffolding. *Educational Leadership, 63*(3), 66–70.

Sheppard, S., Macatangay, K., Colby, A., & Sullivan, W.M. (2008). *Educating Engineers: Designing for the Future of the Field*. San Francisco, CA: Jossey-Bass.

Shimoda, T., White, B.Y., Borge, M., & Frederiksen, J.R. (2013). Designing for science learning and collaborative discourse. In *Proceedings of the 12th International Conference on Interaction Design and Children* (pp. 247–56).

Spillane, J.P., & Coldren, A.F. (2011). *Diagnosis and Design for School Improvement: Using a Distributed Perspective to Lead and Manage Change*. New York: Teachers College Press.

Star, S.L., & Ruhleder, K. (1996). Steps toward an ecology of infrastructure: Design and access for large information spaces. *Information Systems Research*, 7(1), 111–34. http://doi.org/10.1287/isre.7.1.111

Stave, K., & Hopper, M. (2007). What constitutes systems thinking: A proposed taxonomy. In *25th International Conference*. Boston, MA. Retrieved from http://digitalscholarship.unlv.edu/sea_fac_articles/201

Stecher, B., Frohbieter, G., & Greenwald, E. (2011). What teachers learn from formative assessments and how they use this information to guide instruction. Presented at the American Educational Research Association, New Orleans, LA.

Stein, M.K., & Coburn, C.E. (2008). Architectures for learning: A comparative analysis of two urban school districts. *American Journal of Education*, *114*(4), 583–626. http://doi.org/10.1086/589315

Sweller, J. (1988). Cognitive load during problem solving: Effects on learning. *Cognitive Science*, *12*(2), 257–85. http://doi.org/10.1207/s15516709cog1202_4

van Merriënboer, J.J.G., & Sweller, J. (2005). Cognitive load theory and complex learning: Recent developments and future directions. *Educational Psychology Review*, *17*(2), 147–77. http://doi.org/10.1007/s10648-005-3951-0

van Zee, E., & Minstrell, J. (1997). Using questioning to guide student thinking. *Journal of the Learning Sciences*, *6*(2), 227–69. http://doi.org/10.1207/s15327809jls0602_3

van Zee, E.H., Iwasyk, M., Kurose, A., Simpson, D., & Wild, J. (2001). Student and teacher questioning during conversations about science. *Journal of Research in Science Teaching*, *38*(2), 159–90. http://doi.org/10.1002/1098-2736(200102)38:2<159::AID-TEA1002>3.0.CO;2-J

Varenne, H., & McDermott, R. (1998). *Successful failure: The school America builds*. Boulder, CO: Westview Press.

Vokos, S., Kraus, P., Close, E., Close, H., & Seeley, L. (2008). Improving student learning through continuous formative assessment. Presentation at AAPT Summer Meeting. Edmonton, Alberta.

Vygotsky, L.S. (1978). *Mind in Society: The Development of Higher Psychological Processes*. Cambridge, MA: Harvard University Press.

Warren Little, J. (2002). Locating learning in teachers' communities of practice: Opening up problems of analysis in records of everyday work. *Teaching and Teacher Education*, *18*(8), 917–46. http://doi.org/10.1016/S0742-051X(02)00052-5

Wecker, C., & Fischer, F. (2011). From guided to self-regulated performance of domain-general skills: The role of peer monitoring during the fading of instructional scripts. *Learning and Instruction*, *21*(6), 746–56. http://doi.org/10.1016/j.learninstruc.2011.05.001

White, B., & Frederiksen, J. (2005). A theoretical framework and approach for fostering metacognitive development. *Educational Psychologist*, *40*(4), 211–23. http://doi.org/10.1207/s15326985ep4004_3

White, B.Y., & Frederiksen, J.R. (1990). Causal model progressions as a foundation for intelligent learning environments. *Artificial Intelligence*, *42*(1), 99–157. http://doi.org/10.1016/0004-3702(90)90095-H

White, B. Y., & Frederiksen, J. R. (1998). Inquiry, modeling, and metacognition: Making science accessible to all students. *Cognition and Instruction, 16*(1), 3–118. http://doi.org/10.1207/s1532690xci1601_2

White, B. Y., Frederiksen, J. R., & Collins, A. (2009). The interplay of scientific inquiry and metacognition. In D.J. Hacker, J. Dunlosky, & A. C. Graesser (Eds.), *Handbook of Metacognition in Education* (pp. 175–205). New York: Routledge.

Wiliam, D. (2008). Improving learning in science with formative assessment. In J. Coffey, R. Douglas & C. Stearns (Eds.), *Assessing Science Learning: Perspectives from Research and Practice* (pp. 3–20). Arlington, VA: NSTA Press.

Wilson, M. (Ed.). (2004). *Towards Coherence Between Classroom Assessment and Accountability*. Chicago: NSSE; Distributed by the University of Chicago Press.

Wise, C. (2010). Engineers of tomorrow: Holistic-thinking system engineers. In D. Grasso & M.B. Burkins (Eds.), *Holistic Engineering Education Beyond Technology* (pp. 227–41). New York: Springer. Retrieved from http://site.ebrary.com/id/10372426

Woulfin, S. L. (2012). Reading coaches and the relationship between policy and practice. *Reading Research Quarterly, 47*(1), 5–30. http://doi.org/10.1002/RRQ.008

Zhang, J. Q., & Vanasupa, L. (2011). Systems thinking–learning objectives. Retrieved from http://works.bepress.com/lvanasup/30

Zimmerman, B.J. (1989). A social cognitive view of self-regulated academic learning. *Journal of Educational Psychology, 81*(3), 329–39. http://doi.org/10.1037/0022-0663.81.3.329

Zimmerman, B.J. (2002). Becoming a self-regulated learner: An overview. *Theory Into Practice, 41*(2), 64–70. http://doi.org/10.1207/s15430421tip4102_2

Zimmerman, B.J. (2005). Attaining self-regulation: A social cognitive perspective. In M. Boekaerts, P.R. Pintrich, & M. Zeidner (Eds.), *Handbook of Self-regulation* (pp. 13–39). Burlington, MA: Academic Press. Retrieved from http://www.myilibrary.com?id=344782

THEME II

Model-Based Reasoning

6

MODELING AUTHENTIC STEM RESEARCH

A Systems Thinking Perspective[1]

Annmarie R. Ward[2]

Most inservice teachers have never experienced the practices of science and engineering research first hand or have had only limited exposure. The style of teaching with which they are most familiar is the typical transmission style lecture with occasional cookbook verification laboratory exercises, a common feature of large section undergraduate science and engineering courses. In addition, the often held perception that doing science is simply the "Scientific Method" as espoused in major textbooks results in an over simplification of the nature of research that it is a static, linear process (Windschitl, Thompson, & Braaten, 2008) and distracts teachers and students from "productive inquiry" (Tang, Coffey, Elby, & Levin, 2010).

The Next Generation Science Standards [NGSS] (NGSS Lead States, 2013) advocate for a learning progression orientation for developing student understanding of crosscutting concepts and the discourse and practices that embody the research process. Students are expected to acquire understanding of elements of the nature of science through engagement in classroom activities that incorporate these concepts and research practices. Duschl, Bismack, Greeno, and Gitomer (this volume) present strong arguments for the potential of the NGSS approach for enhancing K-12 science learning, and the ability of K-12 students to successfully engage in this type of learning. However, for this to happen, teachers must incorporate the discourse and practices of science and engineering into their science teaching (NGSS Lead States, 2013; NRC, 2007, 2012a). This expectation poses a major challenge for many K-12 inservice teachers. If teachers who have only minimal experience doing research are expected to incorporate the discourse and practices of science and engineering into their teaching, then professional development programs are needed to provide depth of understanding of the research enterprise and how to translate that understanding into classroom teaching, either by implementing existing curriculum or by designing original lessons.

This chapter presents a teacher professional development model that focuses on understanding how scientists and engineers do research and on helping teachers translate that understanding into the classroom. The chapter begins with a description of the Center for Science and the Schools (CSATS) at Penn State and of the teacher-researcher partnership professional development (PD) programs provided by CSATS. Second, is a section about the PD model, including how it was developed, a description of the systems-based framework and the dynamic research processes it represents. This is followed by how CSATS has been using various aspects of the model in its PD programs involving teachers and STEM graduate students and by reactions of teachers and STEM graduate students and faculty to its use. Finally, a discussion of implications and next steps is provided regarding plans for research on using the model with teachers, graduate students, and faculty.

CSATS at Penn State

The challenge of building capacity for teachers to teach in ways set out by the NGSS has been at the forefront of CSATS efforts (csats.psu.edu). CSATS is a STEM education outreach center whose mission is to build mutually beneficial and sustainable relationships between Penn State STEM faculty and K-12 schools to enhance STEM education in Pennsylvania and nationally. Our PD programs have arisen primarily through researcher initiated STEM education grants or broader impacts components of science/engineering research grants. In all of our programs, CSATS personnel work closely with university STEM researchers, including graduate students, to collaboratively develop teacher PD programs that are grounded in faculty members' research areas, connect to K-12 academic standards, model inquiry-based teaching practices and incorporate best practices for teacher PD. Each PD program is unique in terms of its format, STEM discipline area(s), and degree of direct involvement in an authentic research project. "Authentic research" here is defined as research conducted by STEM faculty which may be supported by internal or external grant funding. Alternatively, "classroom research" refers to coordinated classroom activities that mimic authentic research in design, and which investigate a phenomenon for which there is no particular known answer. Characteristics of the CSATS PD programs are summarized in Table 6.1.

A growing focus of CSATS is to optimize teacher learning through teacher-researcher partnerships by devising strategies that allow teachers to (1) examine the dynamic nature of authentic science and engineering research and (2) translate this understanding into their teaching. Since our PD programs have inherent connections with researchers' projects, we use the PD as a context for examining practices of science or engineering research and as a springboard for engaging teachers in the use of these practices when implementing classroom research projects. As can be seen in Table 6.1, graduate students play important

TABLE 6.1 CSATS Teacher Professional Development Programs and Characteristics (TRE = Teacher Research Experience; AY = academic year; GS = graduate student)

Structure	*Characteristics*
AY GS-Teacher partnerships	• Targeted partner school districts; middle and upper elementary school teachers • Varied topics related to teachers' curriculum and grad students' research and content expertise • Weekly PD for graduate students on communication and elements of inquiry-based teaching • Ongoing academic year program involving partnerships between graduate students & teachers • Collaborative design and co-teaching of standards relevant classroom lessons and research projects
Summer TREs and AY workshop series	• Targeted partner school districts; middle and high school teachers • Topics related to nanotechnology, energy sustainability, and health monitoring • 6-week summer TREs co-mentored by grad students; weekly half-day PD workshops • 3–4 academic year one-day PD workshops for all district science teachers • Support for teacher implementation of classroom inquiry-based activities • Faculty and graduate student presentations of individual research projects • PD for graduate students on communication with non-technical audiences and elements of inquiry-based teaching
AY workshop series	• Targeted partner school district; elementary, middle, and high school teachers • Topics linking environmental disruption and disease outbreak • Ongoing academic year workshops with K-12 teachers teaching a variety of disciplines • Faculty and graduate student presentations about aspects of the ongoing research project • Instruction on procedures and techniques used in research project • Support implementing classroom activities and research projects paralleling authentic research
5-day Summer workshops	• Random school districts; middle and high school teachers • Topics related to engineering systems, bioenergy, and biofuels • 5-day summer workshop with sessions focusing on various components of the project

(*Continued*)

TABLE 6.1 Continued

Structure	*Characteristics*
	• Faculty and graduate student presentations and instruction on procedures and techniques used in the research project • Preparation for researchers in communication and aspects of inquiry-based teaching
AY one-day episodic workshops	• Random school districts across PA; K-8 or 7–12 teachers • Academic year one-day workshops; different research project for each workshop • Focus on standards relevant topics related to researchers' areas of research • Faculty and graduate student instruction and engagement of teachers in inquiry-based activities • Preparation for researchers in inquiry-based teaching methods and design of inquiry-based lessons

roles acting as workshop instructors, as co-mentors for teachers during summer Teacher Research Experiences (TRE), and as coaches or co-teachers for classroom activities and research projects. However, the graduate students also vary in their own breadth and depth of research experience. In order to enhance CSATS's PD efforts, we developed the Modeling Authentic Science, Technology, and Engineering Research (MASTER) Model, which is designed to capture the dynamic systems nature of authentic research to help make that accessible to novice researchers.

Designing the MASTER Model

Although the approach for the design of the MASTER Model (henceforth M-model) considered observations and findings from multiple sources, here two are highlighted. The first source involved working with science and engineering faculty to prepare research grant proposals, and for those that were funded, learning about the progress of that research over the life of the grant. Analysis of the structure of 25 federal research proposals (e.g., National Science Foundation, National Aeronautics and Space Administration, US Department of Agriculture) spanning numerous STEM fields and observations during engagement in the projects revealed a recursive pattern that can be represented in terms of a dynamic, complex system of interrelated and interdependent components, subcomponents, and depending on the complexity of the research design, sub-subcomponents. Each component and subcomponent is guided by research questions or goals, underlying sub-questions or sub-goals, ultimately to the point of the individual experiments or tasks. At the heart of the research design is the expectation that ongoing findings from the seemingly separate parts are regularly communicated across the project to inform ongoing progress, allowing next steps as planned or requiring changes. Through this communication process, the system evolves over the course of the research project. Thus the M-model includes both a framework depicting the research design at any one point in time, and the dynamic processes that shape and mold the framework over the time of the research project. The M-model combines and applies the strands of systems thinking and model-based reasoning to promote understanding of how researchers do research.

Also informing the M-model design and its use at Penn State are research in the literature and personal observations about how undergraduate and graduate students become independent researchers. Figure 6.1 presents a flow diagram depicting a typical undergraduate and graduate student progression observed for STEM majors at Penn State. Future science teachers and future researchers share similar early preparation as undergraduate STEM majors (e.g., coursework taught via traditional dissemination style lecture with associated laboratory exercises). By their nature, these early experiences reinforce the simplistic perception of research as the "scientific method," and the separation of research and knowledge building. Such experiences also tend to isolate students from the larger overarching research system and thus prevent them from seeing interconnections among subcomponents. In addition, many research skills are needed by science teachers in order to support student learning via inquiry-based methods. For example, in a study of five preservice high school teachers engaged in a one-year field experience, Crawford (2007) found that these preservice students struggled with "framing questions, grappling with data, creating explanations, and critiquing explanations (including others' explanations in a public forum), . . . important components associated with teaching science as inquiry" (p. 637). However opportunities to develop these skills are limited in undergraduate education.

In a study aimed at understanding how researchers learn to do research, Feldman, Divoll and Rogan-Klyve (2009) examined research proficiencies of honors undergraduate, Master's, and Ph.D. students working on an interdisciplinary research project involving numerous research faculty. Feldman et al. describe methodological and intellectual proficiencies that emerged from the study, and timing associated with appearance of these proficiencies across student levels. They found that advanced methodological proficiencies, such as data analysis and ability to report findings to other researchers and defend them based on data were not evident until early in graduate studies. Intellectual proficiencies associated with research such as designing research questions based on the literature, framing hypotheses, situating individual's research within the context of the broader research space of the lab or the literature, and contributing to the broader knowledge base were not evident until the later stages of Ph.D. programs. This lack of exposure contributes to teachers' limited skills in aspects of research they are expected to help develop in their students. In response to these observations

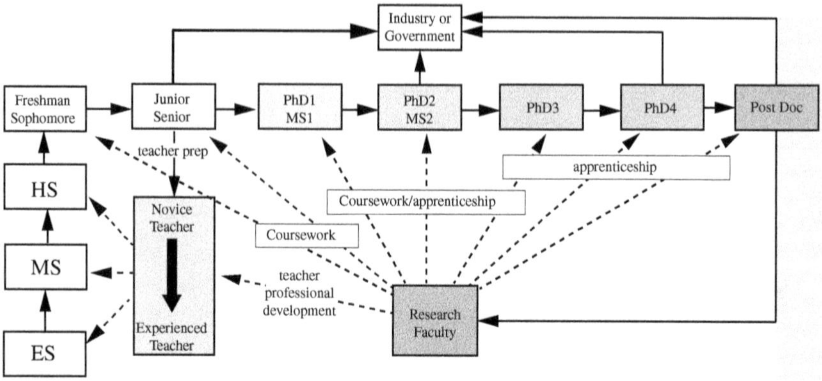

FIGURE 6.1 Traditional educational pathways for STEM students and the role of experienced researchers and teachers (solid blue arrows=pathways; dashed arrows= instruction/mentoring)

and findings, the M-model not only represents the systems nature of research, but also provides opportunities for highlighting methodological and intellectual skills involved in the research process.

Description of the MASTER Model

The M-model consists of a multiple-level systems-based framework and the dynamic processes therein. Figure 6.2 provides a generalized representation of the framework's overall research system subdivided into several interconnected and interdependent recursive units. Black lines represent relationships among subcomponents and components to the overarching research plan; gray arrows represent possible interrelationships and interdependencies among components and subcomponents. The boundaries of the system may vary, depending on the scope of the research, as well as the purpose and context in which the research is done. As such, the framework can represent an entire inter-institutional, multidisciplinary project whose work involves numerous research labs, an individual research faculty's funded research project, and/or the work of an individual graduate student focusing on one piece of the overall project.

Level I of the M-model framework presents the complete planned research project and is intended to capture the development phases of the overall research or engineering design plan. At the initial planning stage, researchers identify a phenomenon or design problem and conduct literature reviews to develop a rationale and identify gaps in knowledge related to the problem. If the idea continues to be viable, the problem is refined to a focused and testable overarching research question or design problem. The next step is development of a research design to address the overarching question or design problem. The plan

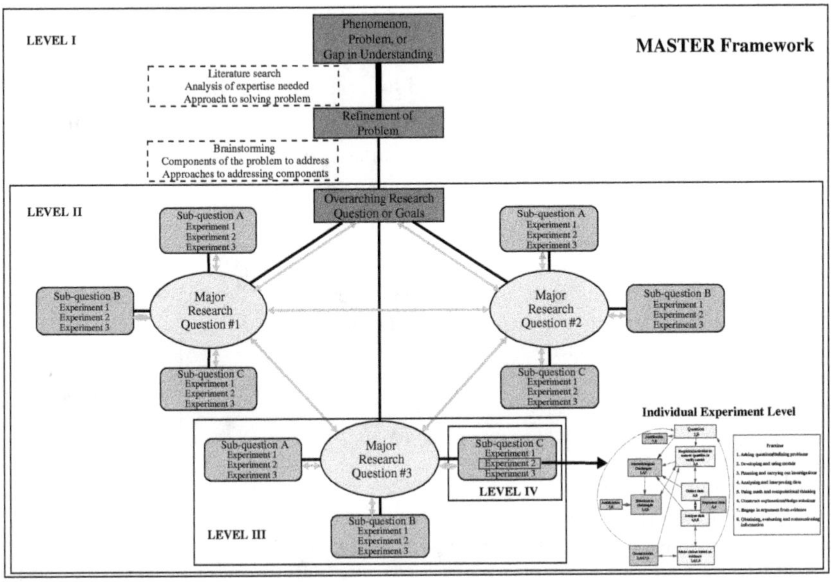

FIGURE 6.2 Template for the MASTER model framework

is comprised of several recursive units, each addressing a different guiding question that represents a different facet of the overarching research question/design problem. Each recursive unit, in turn, is made up of several subcomponents, which may have sub-subcomponents, too. The smallest element in the framework is the individual experiment, several of which may be needed to address the sub-component at hand.

Level II of the system provides a bird's eye view of the entire research process as it unfolds, including interconnections and interdependencies among all elements of the research system (gray arrows). **Levels III and IV** sequentially narrow the boundaries of the system within each recursive unit to consider individual components and subcomponents of the research design. Depending on the scope of the research project, there may be additional levels involved. The finest grained level of focus is the **Individual Experiment**. It is here that measurements are taken, data is obtained, analysis of the data occurs, and patterns of evidence for supporting future claims are generated. As the project progresses, the research design will adapt and evolve due to changes made in procedures and strategies because of new information gained from experiments and analyses of information at the various levels.

Although considered separately in the framework, in practice, these recursive units are interconnected and interdependent across any of the levels. When combined, the findings contribute to solving the overarching problem. In the framework template, all of the arrows are bi-directional and weighted the same. However, when depicting an actual project, they could be used to indicate directions of information flow, and weighted to indicate the extent of the interdependency. The need and opportunity for change arises by virtue of communication within and across components and subcomponents comprising the system. Also, the levels of the framework are described here in numerical order in terms of their hierarchical position within the recursive unit, reflecting how they were derived during the planning stage. However, once the initial research plan is in

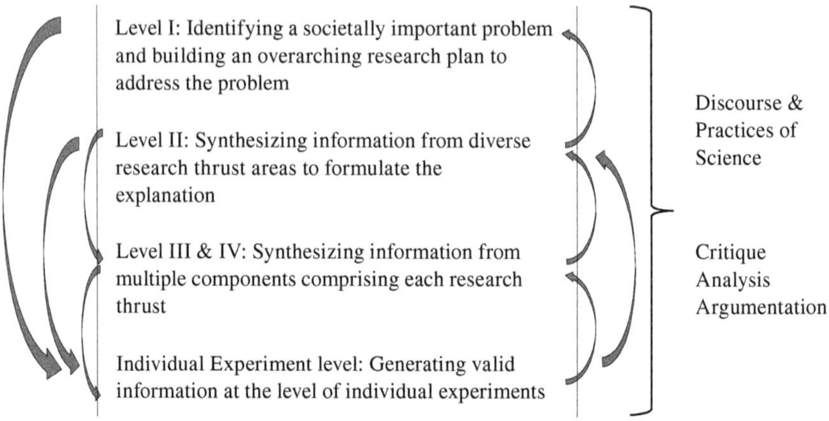

FIGURE 6.3 Summary of flow of activity and information building over the course of the research process

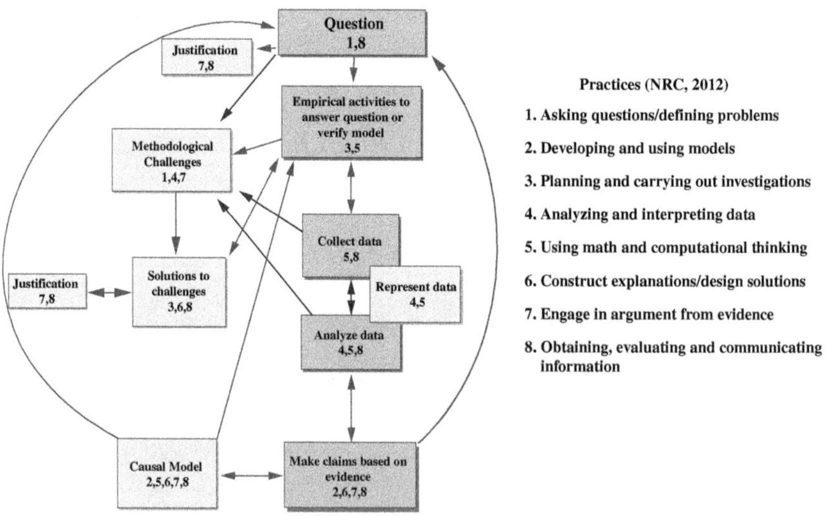

FIGURE 6.4 Level of the individual experiment including connections to NGSS practices of science

place and implementation begins, the sequence becomes reversed and eventually interwoven, as shown in Figure 6.3.

Another important feature of the M-model framework for teacher PD is the detailed depiction of activities within an individual experiment. For experienced researchers, much of what actually happens here becomes routine such that the intricacies of the processes are internalized. Unless the routines are explicitly discussed with teachers, they remain hidden. Figure 6.4 summarizes the kind of hidden thinking inherent when experimenting, and provides links to the science and engineering practices outlined in the *Framework for K-12 Science Education* (National Research Council [NRC], 2012a).

Using the Master Model with Teachers and Graduate Students

Although we have been using the M-model in a variety of PD contexts, each PD utilizes a similar process for introducing the M-model. As a first step, we present and discuss the definition of systems thinking and examples of systems diagrams representing complex natural phenomena such as the water cycle, or food web, or energy cycles, and man-made systems, such as engineered systems, business organizational systems, and governmental systems. We discuss how these diagrams provide mental models of the system, showing interconnections and interdependencies across the system, and enabling (making apparent) predictions of how changes in some component of the system might impact other components and the system as a whole. We then introduce the idea of applying systems thinking to the design and implementation of a research project. At this point, we present the M-model Framework Level I diagram of an active research project showing the recursive pattern of levels and interconnections. If the PD is related to a specific research project, we have one of the research faculty describe the project, referring to the framework diagram. However, if the PD has no relationship to any particular research project, we use an example from a previous project whose components are now public knowledge. The description of the research project includes discussion of the process of refining the guiding question or overarching goal, as well as how findings from various individual tasks combine with others to contribute to answering the overarching question. We also discuss how such new information informs decisions at the subcomponent and component level, such as whether to change or eliminate a particular approach or experiment. We then present and discuss the template for the M-model, as depicted in Figure 6.2.

At this point, our use of the M-model diverges depending on the audience's prior knowledge and experience with research. With teachers we begin by using the M-model framework as a guide for situating the discussion. We involve them in discussions and activities related to various components and subcomponents of the featured authentic research system. Activities include learning about procedures and techniques used in the actual research process, working with actual data from the research project, and developing classroom investigations that parallel aspects of the authentic research project and incorporate the M-model. Discussions include some of the challenges researchers encountered in areas such as data collection and analysis, as well as how findings from the experiments impact other parts of the research system or draw into question underlying assumptions. These activities and discussions are intended to familiarize teachers with an authentic research project and reinforce the system dynamics of the research process. We then have them use the M-model to design and carry out a classroom research project. In some cases, teachers engage their students in classroom projects that parallel the ongoing authentic research project, using the M-model for planning and anticipating interrelationships, and techniques and procedures learned from the PD activities. In the case of the PD program where we work with local school districts, this process is extended into the academic year.

Our use of the M-model with graduate students participating in the PD takes on a different form. The graduate students are all working on a research project intended to be the source of their dissertation work. These projects focus on particular components of a larger research system comprising the advisor's overall research agenda. In order to ground the M-model in their own experience, we ask each graduate student to use the M-model to diagram his or her own research project, indicating how various components of their project interrelate and are interdependent. After they share their diagrams, we ask them to expand the boundaries of their project's research system to incorporate other projects in the research lab, and/or in other labs linked via a common research grant. Once again, using the M-model framework students include areas of interconnections and interdependencies across projects. Finally, we have them expand the boundaries further to consider the broader research spaces to which their research pertains, based on readings of literature and discussions at conferences. We then have them consider how to apply the M-model to the teacher PD activities, such as making presentations about their research and how it relates to the overarching research project, engaging teachers in inquiry-based activities associated with ongoing research of the project, and helping teachers plan classroom research projects.

Teacher, Graduate Student, and Faculty Interactions with the MASTER Model

We are still in the early stages of applying the M-model in our teacher PD programs and have been incorporating different aspects of the M-model with various elements of different programs, so the patterns here represent work-in-progress. Our observations so far are based on informal focus group discussions and interviews with teachers and graduate students who participated in the various programs, as well as anecdotal comments and questions arising during the PD and classroom activities.

Graduate Students/Faculty

- Although their research plans had already been designed, graduate students found that the M-model provided greater clarity of understanding about their projects, and helped them form concrete connections between ideas and topics.
- Use of the M-model helped graduate students organize findings from large amounts of data, enabling them to see outcomes, connections, and holes in the study.
- Graduate students found that the M-model is very useful for conveying ideas to others. Several used it in meetings with their research advisors (who liked it very much), and found that it led to more focused and efficient dialog. Several plan to incorporate the framework into their dissertations to better communicate how they see their projects, connect components, and identify what is missing.

Teachers

- Using the M-model in connection with an authentic research project impacted how teachers think about research. It helped them see how experiments they have their students perform are associated with larger, broader topics.
- Exposure to discussions and activities associated with an authentic research project coupled with the M-model Framework's depiction of the systems nature of the research emphasized the messy, non-linear nature of research and helped teachers move beyond the traditional Scientific Method and recognize the teamwork associated with research, in which many people work toward a common goal.
- The systems thinking aspect of the M-model is helpful in lesson plan development as a framework to guide the design process of the lesson such as what variables should be investigated.

Graduate Student, Teacher, Student Interactions

- When referring to the authentic research project, having a model to represent the overarching research plan and its various components and subcomponents helped teachers, students, and graduate students (1) understand the systems-based organization and complexity of research projects, and (2) situate discussions about the role of the various components and subcomponents, and activities and discussions focusing on individual experiments or tasks.
- The M-model is a helpful tool for graduate students to use when communicating their own research to teachers, and how it fits with the overarching research plan.

- Using the M-model Framework and applying a systems thinking perspective for developing classroom research projects helped teachers and graduate students co-design classroom research and engineering design projects and plan logistics for their implementation.
- The M-model is also helpful while working with students to plan and implement research projects. In the planning stages, information from reading about particular components or subcomponents can be added directly to the framework. During implementation, findings can be added and plans altered depending on those findings.

Implications and Next Steps

The M-model applies the strands systems thinking and model-based reasoning within the context of an authentic research project to enhance teachers' and students' understanding of the nature of research. It is designed to capture the complexities of authentic research in both the planning and implementation phases, in ways that emphasize the systems nature of research and promote development of methodological and intellectual proficiencies associated with research. It provides a representation of the research process that allows various aspects of research to be examined in detail, and combined to represent the whole. There are numerous functions associated with the model, including: (1) a scaffold for applying a systems thinking approach to planning and describing a research project; (2) once completed, a framework or roadmap for situating the various component systems and subsystems comprising the research and their interconnections; (3) a reminder of points of discourse, thinking processes and interactions among researchers that occur at all levels of the research process and which need to be made explicit; and (4) a representation of research as a dynamic, evolving process with internal feedback loops through which findings from individual investigations within components yield new information that builds understanding within and across systems levels.

Duschl et al. (this volume) describe in detail the value of the NGSS approach for providing learning that integrates crosscutting concepts, the discourse and practices of science with content learning, in a way that mimics the work of science and engineering. They also present evidence for the capability of K-12 students to successfully engage in this type of learning when elements are appropriately sequenced and adequate ongoing support is provided. However, preparing teachers to provide that ongoing support remains a key issue associated with the implementation of the NGSS. Borko (this volume) stresses the need for teacher PD to help teachers understand the elements associated with the NGSS approach and be able to use them in their teaching. Borko also provides a plan reflecting best practices for teacher PD that will help teachers understand and implement aspects of model-based reasoning. Such a plan could be applied to a number of individual crosscutting concepts and practices for teachers. However, attention to individual elements in the absence of understanding the big picture of how they fit together may serve to isolate these elements rather than promote their synthesis. The M-model can provide a big picture perspective of how these elements work together in authentic STEM research, and how to translate that perspective into classroom science teaching. Part of its use can include explicit reference and discussion of the individual NGSS elements and how they occur at all levels of the research process. This strategy could be used in combination with more specific attention to individual crosscutting concepts and practices, enabling teachers to understand how the parts fit into the whole.

All of our programs involve connections to authentic research projects and researchers; however, our use of the M-model in PD has varied depending on the context of the program, types of opportunities for graduate student involvement, and the extent of involvement by graduate students and faculty. Our preliminary observations suggest that the M-model is working to enhance teachers' understandings of the systems nature of authentic research and is helping them see how the M-model can be used for designing and implementing classroom research projects. There are a number of research directions we would like to take using more systematic approaches in order to enhance PD in this area. We would like to explore in what ways and to what extent the M-model process influences teachers' understanding of the nature of research and their ability to translate that understanding to the classroom in the form of classroom research projects and model-based inquiry lessons.

A related area of interest is to examine how the M-model can be used most effectively across PD types that vary in how closely associated teacher involvement is with the authentic research project and the degree of graduate student or faculty interaction. Our programs range from six-week summer immersion TREs mentored by STEM faculty and graduate students to one-day episodic workshops led by STEM faculty or graduate students. We would like to understand how teacher understanding from these various PD opportunities differ and how we can maximize learning from each type of experience. For example, a review of the literature on impacts of immersion research experiences by Sadler, Burgin, McKinney and Ponjuan (2010) found mixed results on the value of such experiences to effect change in classroom teaching. Teachers often return to the classroom excited about the experience but unprepared to incorporate what they have learned into the classroom. We hope to investigate in what ways and to what extent implementation of PD based on the M-model helps teachers translate learning from an authentic TRE into their teaching.

We are also interested in exploring use of the M-model as a mechanism for teacher learning about the Nature of Science (NOS). An important goal of the NGSS is for students to develop basic understandings about NOS. These understandings are evidenced by the work that scientists and engineers do, and relate to the methodological and intellectual proficiencies that are acquired over the course of their professional continuum as researchers. Because teachers have only limited exposure to authentic research and apprenticeship with researchers, they have limited opportunities to experience these themes and understandings in action. The M-model, used in conjunction with teacher PD associated with an authentic research project, provides opportunities to explicitly address many of the NOS elements. Having learned how these understandings apply to authentic research via PD that explicitly addresses them, teachers will be better equipped to use them within the context of classroom research or model-based inquiry lessons.

The most surprising outcome from our initial observations has been the impact of the M-model process on graduate students. Our initial purpose for engaging graduate students with the M-model was to familiarize them with the strategy we planned to use for the teacher PD, so they could use it with teachers to help to convey the nature of the research. However, we are finding that the M-model is helping the graduate students on a personal level to conceptualize their own research projects in ways that are different than they were doing previously. Some of these changes seem to be related to advanced methodological, as well as intellectual, proficiencies described by Feldman et al. (2009), such as situating their research within a greater whole, organizing large data sets for analysis, and communicating their research findings in a more organized and efficient way. We are in the process of developing investigations into these and other potential impacts, and how the M-model can be used with faculty/graduate student pairs to enhance graduate students' understanding of the research enterprise. We expect that its value for helping students better understand the nature of research will also apply to undergraduate students as well.

Connecting all of these applications to learning about the nature of research are the strands systems thinking and model-based reasoning that underlie the M-model approach. The model depicts research as a complex system of interrelated and interdependent parts whose boundaries can be expanded or retracted depending on the purpose in which it is being used. Model-based reasoning applied to the complex system allows investigation into the workings of authentic research projects at numerous levels of detail, as well as predictions and implications of changes across the system, causing it to evolve. These are characteristics of systems thinking and model-based reasoning employed in the study of natural phenomena (e.g., Assaraf & Orion, 2005). Based on our preliminary findings, we believe that these strands, when applied to authentic research projects, will lead to deeper understanding of the research enterprise to the benefit of students at all levels.

Notes

1. The author wishes to recognize contributions of CSATS coworker Matthew Johnson for development of Figure 6.4, a compilation of thinking activities involved at the individual experiment level and their connections to NGSS crosscutting concepts and science and engineering practices, and the work of Matthew Johnson and Leah Bug in implementing the MASTER Model in the various CSATS professional development programs.
2. This material is based upon work supported by the National Science Foundation under Grant No. 0909447 and Grant No. 0947962. Any opinions, findings, and conclusions or recommendations expressed in this material are those of the author and do not necessarily reflect the views of the National Science Foundation.

7

MEETING THE STANDARDS FOR STEM EDUCATIONS

Individual and National Needs

Spencer A. Benson

Introduction and Background

In 2002, the National Research Council (NRC), with support from the National Science Foundation (NSF) and the US Department of Education published the results of a two-year study of the Advanced Placement (AP) and International Baccalaureate (IB) programs. The report, *Learning and Understanding: Improving Advanced Study of Mathematics and Science in US High Schools* (NRC, 2002) called for important improvements to advanced study in high school mathematics and science programs. Numerous other reports from government, business, and the academy all stress the need to rethink, reshape and deliver science education in new ways that ensure a science literate citizenry and a workforce that will enable the US to remain a world leader in the STEM (Science, Technology, Engineering and Mathematics) disciplines and to retake leadership in the new, global, knowledge-based economies. Reports such as *Rising Above the Gathering Storm* (NRC, 2010), *An American Imperative* (Johnson Foundation, Inc., 1993), *Bio2010* (Steen, 2005), and *Vision and Change* (American Association for the Advancement of Science, 2011) outline these needs and the consequences of not meeting them. Addressing these challenges requires changes in how we approach STEM education at the national, state, and local levels, from kindergarten through graduate school. In response to this challenge, The *Next Generation Science Standards* (*NGSS*) (NGSS Lead States, 2013) provides descriptions of what K-12 students should know and be able to do and builds upon the recommendations of the *Framework for K-12 Science Education* (NRC, 2012a).

In light of its growing prominence in the American secondary education landscape, the College Board Science AP Programs provide an opportunity to reform STEM curriculums taught at thousands of schools across the United

States and beyond. In 2006—in response to the 2002 National Research Council report—the College Board, with funding from NSF, began redesigning the AP science courses including AP-Chemistry, AP-Physics, AP-Environmental Sciences, and AP-Biology. That work followed a new paradigm for science education in which science is taught *not* as a collection of facts or truths to be memorized, but as a way to understand and make sense of the world around us. It reflects a new paradigm in which course content becomes not the *end* but the *means* for establishing conceptual understandings and engaging in approaches, practices and ways of knowing that are fundamental to practice of science and its roles in society. This chapter will highlight the pathway, progress, and challenges in redesigning one course: the AP-Biology Course and Exam (College Board, 2012c). AP-Biology annually enrolls more than 250,000 students and involves nearly 10,000 science teachers in the US and international locations.

The AP-Biology Course

In the summer of 2007, the AP-Biology redesign commission met and was charged with developing an AP-Biology program that encompasses: (1) enduring understandings, (2) science practices and learner centered instructional practices necessary for the development of student conceptual understandings, (3) integrating new research and advances in biology and student learning, and (4) adopting a learning progression approach for sequencing students' science knowledge learning. In other words, the charge was to develop an exemplary introductory college level biology course that produced enduring understanding and which could serve as a model for universities to emulate.

As a starting point, the course would build upon prior knowledge in biology as described in the College Board Standards for College Success (College Board, 2009). To accomplish this task, a commission composed of exemplary university and high school biology teachers met regularly and developed a draft curriculum framework that was subsequently reviewed, vetted, and refined over a period of five years by university and high school teachers, as well as members of professional societies and the academy. In 2012, the AP-Biology Course and Exam description was released to the public, and the 2012–13 academic year marked the launch of the new AP biology course and the corresponding AP-Biology exam in spring 2013. The new AP-Biology course is organized around four overarching "big ideas" (Table 7.1). Each idea is supported by a set of enduring understandings which in turn build upon a set of essential underlying content (e.g., facts and concepts that students must know in order to understand and be able to communicate their understanding). To assess students, specific learning objectives (LOs) were developed and integrated throughout the revised course.

Each AP-Biology LO describes what students should know and/or be able to do. Each LO has a connection to a specific science practice (SP) (Table 7.2).

TABLE 7.1 2012–13 AP-Biology Curriculum Framework*

Big Idea	*Enduring Understandings*	*Essential Knowledge*	*Learning Objectives*
1. The process of evolution drives the diversity and unity of life.	4	11	32
2. Biological systems utilize free energy and molecular building blocks to grow, to reproduce and to maintain dynamic homeostasis.	5	15	40
3. Living systems store, retrieve, transmit, and respond to information essential to life processes.	5	15	50
4. Biological systems interact, and these systems and their interactions possess complex properties	3	14	27

* see AP-Biology Course and Exam (College Board, 2012c)

The science practices are overarching practices applicable to all of the AP-Science courses (biology, chemistry, physics, environmental science). They were developed and refined by a team of science educator specialists. Each of the seven science practices (Table 7.2) describe in a broad sense what students can do with respect to the practice of science. Each SP description is further refined using action verbs that delineate the nature of the student products that serve as evidence of learning and understanding. The SPs encompass the full range of Bloom's cognitive taxonomy of learning (Anderson & Krathwohl, 2001; Krathwohl, 2002), as well as important learning science practices. In the biological sciences, very often the whole is greater than the sum of the parts and conceptual understanding requires students to recognize and be able to describe content and conceptual connections within and across the big ideas and enduring understandings. The ability to identify, see, and describe such connections requires students to develop both model-based reasoning and a system approach to thinking as described in the Introduction (Duschl et al., this volume).

A major challenge in designing the new course was defining and limiting the essential content students need in order to build conceptual understandings in modern biology. This challenge was especially difficult given the ongoing explosion in new information in the biological science and the mandate to ensure that the student would be equipped to understand the emerging frontiers in biological research. Additional challenges were the tendency for scientists in sub-divisions of biology to argue their content is more essential and the tendency for teachers to resist changes to established instructional models.

An important aspect of the new course was a clear delineation of what content material was beyond the scope of the course. Material designated beyond

TABLE 7.2 Science Practices for AP-Biology*

SP-1.0	The student can use representations and models to communicate scientific phenomena and solve scientific problems.
SP-2.0	The student can use mathematics appropriately.
SP-3.0	The student can engage in scientific questioning to extend thinking or to guide investigations within the context of the AP course.
SP-4.0	Student can plan and implement data collection strategies in relation to a particular scientific question.
SP-5.0	The student can perform data analysis and evaluation of evidence.
SP-6.0	The student can work with scientific explanations and theories.
SP-7.0	The student can connect and relate knowledge across various scales, concepts, and representations in and across domains.

* see AP-Biology Course and Exam (College Board, 2012c)

the scope of the course were not deemed unimportant and should not be taught, but rather as sign posts for instructors to decide the amount of material they needed to teach and include in their course curriculum. The curriculum was designed so that the instructional time needed to teach what would be on the AP exam would *not* exceed 85 percent of the allotted instructional time. This provided teachers a level of ownership and flexibility in designing a course that best addresses the needs of their students and their expertise. Because many of the enduring understandings and biological concepts that make up the course can be taught through a variety of contexts, the course provides a list of illustrative examples to teach enduring understandings and concepts. This adaptive approach allows students to think and understand systems, models and processes, and be able to apply what they have learned to a variety of situations.

The AP-Biology Exam

Each spring more than 170,000 students take the AP-Biology exam. Individual performance on this high stakes exam determines whether students will receive AP credit at the university level. Each college and university sets its own criteria regarding AP credits. Successful completion of AP courses, as measured by exam performance, is a factor considered in admission at many US institutions. The 180-minute exam is administered on a specific day across thousands of sites in the US and abroad. The fact that all AP-students take the same standardized exam is one of the mechanisms that serves to standardize the AP-Biology course across thousands of high schools and school districts in the US and abroad and ensures a fair and level playing field with respect to outcomes assessment. The second mechanism for standardization and fairness is the AP-Audit (College Board, 2012c) in that schools must submit documentation showing the AP-Biology course meets the teaching expectations set forth by the College Board.

The first "new" AP-Biology exam was given on May 13, 2013, and consisted of a 90-minute selected response (SR) section followed by a 90-minute free response (FR) section. The redesigned AP-Biology exam differed from previous exams in a number of ways. The number of SR multiple-choice questions was reduced from 100 to 63 items and the number of answer choices from five to four. A number of the SR questions involved question sets (e.g., several multiple-choice questions that probed understanding of multiple LO using a single question stem). Many of the test's selected response items presented real data, diagrams, and graphs, which required students to interpret, analyze, justify, or construct an answer to matched available choices. The SR section also contained a new part B which consisted of six grid-in questions that required the students to perform mathematical calculation(s) and then bubble in a numerical response on the computer graded test sheet. For the first time ever, students were allowed to use a simple non-graphing calculator with four functions and square root and the exam booklet provided a formula and equation list.

In the FR section of the exam, students wrote responses, which were individually graded by AP teachers and university faculty at the AP-Reading. There is a mandatory (no writing) initial 10 minute reading period where students read and review the questions prior to starting the FR section. The FR section consisted of two long response questions and six short response questions. The long response questions were similar to those present on previous AP-Biology exams and consisted of multiple parts. The short response questions consisted of a single or at most a two-part question, which can be answered in a paragraph. Both types of questions can contain data and/or require quantitative and graphing skills and probe understanding and knowledge across a range of SPs and Bloom's taxonomy levels.

The number of SR and long FR questions was reduced to offset the introduction of the grid-in questions, the short answer questions and the necessity to provide more information and context in the question stems. Because many concepts within the curriculum framework can be taught with various illustrative examples, which are not required knowledge, exam questions that build upon illustrative examples need to provide sufficient background information to enable all students to fairly answer the question. This necessitated an increased reading load for the students. To be sure that the new test did not penalize students due to the increased reading load, prototype test items were extensively tested for reading/test taking times needs during the development phase of the exam.

An important change in the development of the new exam was the use of evidence-centered design (ECD) (Mislevy, Almond, & Lukas, 2003) for the construction of all test items. ECD is based on principles of evidentiary reasoning to make valid inferences regarding what students know and can do. ECD items are written to directly measure content knowledge, conceptual understandings, and application of science practices based on student work (e.g., their answers on the exam). This is in contrast to previous exam development where student knowledge and understanding of content within broad areas of biology (e.g., molecular, genetics, cellular, organ systems, organisms, populations, ecology) served as the basis for test item development and there was no direct linkages to specific LOs or SPs. To ensure scientific accuracy, alignment with the curriculum framework, and equal representation of the four big ideas across all test items, test items were reviewed by a test development committee composed of high school and university teachers who have a deep understanding of the new AP-Biology course. Individual test items were written by independent test item writers, vetted by the Educational Testing Service (ETS), AP-Biology test staff, and then extensively reviewed and refined by the test development committee. Each test item directly aligned with one or more of the 150 learning outcomes and one of the seven science practices as delineated in the AP-Biology Course and Exam descriptions.

In June of 2013, the new exam was scored and the requirements for AP scores of 1 through 5 determined. Panelists who took the exam indicated that the questions effectively balanced content knowledge, quantitative analysis, experimental design, and data interpretation. The initial perception by students and teachers that the multiple-choice part of the new exam was easy proved not to be the case especially for the grid-in questions. The average mean score on the SR sections (multiple-choice plus grid-in) was 61 percent correct, this was only slightly lower than the 63 percent correct observed with the old AP exam (2012). What was striking is that the mean correct score for the quantitative grid-in section was only 36 percent. This, in part, accounts for the slightly lower overall percent correct score for the selected response section. Clearly, the need for increased emphasis on quantitative literacy skills within AP-Biology is underscored by these findings. Further support for this was the observation that the average score on questions linked to SP2 (students can use mathematics appropriately) was lowest (48 percent) for the seven SPs (see Table 7.2). The mean correct score for each of the four big ideas range was: 64 percent, Big Idea 1; 60 percent, Big Idea 2; 55 percent, Big Idea 3; and 65 percent, Big Idea 4. For the two long free response questions, the averages were 36 percent and 37 percent, respectively. For the six short answer questions, the range from 73 percent to 23 percent, with the mean average across all of the short answer questions being 43 percent, slightly above the average for longer free response questions.

Overall, performance on the new AP exam was somewhat below that of previous exams. A plethora of reasons may account for the lower scores: (1) AP teachers are receiving students who are use to pedagogies that focus on coverage rather than on knowledge and enduring understandings, (2) the new exam had a significant increase in quantitative type questions, and (3) many of the questions were designed to require students to demonstrate quantitative analytic and interpretive skills that were in line with the LOs and the SPs. Most college faculty would concur that their own students would be similarly challenged if required to take the redesigned AP exam and program that is focused on student development of enduring understandings and conceptual knowledge rather that memorization of content knowledge.

The AP-Biology Laboratories

The new AP-Biology course is anchored in the belief that student learning is facilitated through inquiry instruction that brings together enduring understandings and essential knowledge with the science practices. While there is a strong expectation that inquiry should be embedded in all aspects of the course (lecture, discussion, projects, etc.), the laboratory component is the area most amenable for easy integration of student inquiry. The new AP-Biology course has a requirement that at least 25 percent of instructional time is devoted to laboratory activities, which is monitored via the AP-Biology course audit (College Board, 2012b). A new AP-Biology Laboratory manual was developed and released in 2013 and contains 13 investigative labs (College Board, 2012a, 2012c). The AP-Biology labs align with the recommendations of *America's Lab Report* (NRC, 2006) and require students to be actively involved in the process of scientific inquiry and developing an understanding of the ways in which science is practiced and knowledge is acquired. The laboratories provide a robust range of inquiry-based activities (from teacher-directed to fully student-directed) that align with the big ideas, enduring understandings, essential knowledge, and science practices. In addition to providing instructional guidance for the laboratory exercises, the laboratory manual contains chapters on: (1) overarching goals for the lab and investigations, (2) learning objectives, (3) guidance in order to help teachers create activities, (4) how to introduce inquiry, (5) how to transition from teacher-directed to student-directed lab investigations, (6) how to modify existing labs to make them more student-directed and inquiry-based, (7) how to manage the investigative experience, (8) safety considerations, and (9) accommodating students with special needs. There are also chapters on written and oral communication from reports to presentations, quantitative skills including statistical tests, standard errors, measuring, modeling, predictions and calculations.

Connecting AP-Biology to the Standards for Science Education: System Thinking, Quantitative Reasoning, and Modeling

Important changes integral to the new AP-Biology course and exam align with the concepts illustrated in Figure 1.1 (see Duschl et al., this volume, p. 4). Among biologists, it is increasingly apparent that 21st century biology will be the century of mathematical biology. Less than a generation ago introductory biology was generally taught almost completely devoid of mathematics or quantitative reasoning. Even though many universities required calculus or other math courses as a prerequisite, the actual use of mathematics within introductory courses was limited or completely absent. Even today, many of the most widely used introductory biology textbooks have few, if any, equations, formulates, or quantitative activities. The need for increased integration of mathematics into biology was recognized and stressed in the 2002 NRC report, and subsequent reports all of which called for an increase in quantitative reasoning in introductory biology. An important component of the new AP-Biology course is an increased emphasis on quantitative literacy in the course, laboratory, and exam. The importance of increasing quantitative reasoning is underscored in the AP-Biology course in the science practices, the inclusion of the equation, and formula list in both the lab manual and the test booklet, the use of calculators for the exam, and the inclusion of the grid-in questions, which required students to solve mathematical problems. In addition to these obvious changes involving mathematics, AP biology teachers are being encouraged through professional development, professional social communities (NRC, 2012c), and sample test questions to integrate quantitative reasoning throughout their AP course. For many high school biology teachers this is a challenge, since their choice of biology as a science may have been in part a reflection of their lack of comfort, confidence, and engagement with mathematics as an undergraduate. Yet, despite the progress made in attempting to bring mathematics into AP-Biology the depth of penetration remains shallow and focuses on the use of a few equations, such as the Hardy Weinberg equilibrium equation. The science practices, which are an important part of the course, implicitly deal with the mathematical aspects of science learning and stress mathematical reasoning applied to biological problems and concepts.

A significant challenge for the AP-Biology community will be to find effective ways to further engrain quantitative reasoning into the course, laboratory, and exam such that it is integral and not viewed as an add-on. This will require an increased emphasis and focus within the professional development activities for AP teachers, a change in perceptions that biology does not require mathematics, eliminating residual negative perceptions or phobias regarding mathematics, a community of practice that embraces and develops new mechanisms and pedagogies that link quantitative reasoning to biological concepts, and the structuring of textbooks that are used for the teaching AP-Biology.

For teachers and student to be successful in the new AP-Biology program, model-based reasoning will need to be introduced, practiced and become interval to how students and teachers think about biological systems. This will require a paradigm shift from the generation-old approach of memorization of facts and system components without understanding the conceptual basis of how the components interact, to students being able to see, recognize and build models that are applicable across a variety of biological systems. Because many of the enduring understandings can be taught using a variety of illustrative examples, students throughout the course will be challenged to learn through the building of mental models, which lead to conceptual understandings applicable to a variety of related overlapping systems. An example of the need for model-based thinking is illustrated by the following: Big Idea 2 *Biological systems utilize free energy and molecular building blocks to grow, to reproduce, and to maintain dynamic homeostasis* is supported by essential knowledge 2.A.1 *All living systems require constant input of free energy*. Teachers can choose to teach Big Idea 2 using a variety of illustrative examples such as the Krebs cycle, glycolysis, Calvin cycle, fermentation, or a system of their choice. Learning success is measured by LO 2.1 *The student is able to explain how biological systems use free energy based on empirical data that all organisms require constant energy input to maintain organization, to grow and to reproduce SP 6.2*. Having teachers focus on model-based reasoning and understanding using one illustrative example rather then trying to teach the many facts that make up metabolism, the student is better equipped to successfully answer exam questions based upon this learning objective. Through the iterative applications of having to learn how and why things operate within biological systems students hopefully will develop the type of system thinking outlined in Figure 1.1 (see Duschl et al., this volume, p. 4).

Connecting the AP-Biology to Educating for Success

The NRC report *Education for Life and Work: Developing transferable knowledge and skills in the 21st century* (NRC, 2012c) looks at cognitive, intrapersonal and interpersonal competencies. While the AP-Biology course, laboratory, and exam focus primarily on cognitive competencies, (e.g., knowledge, information literacy, critical thinking, information literacy, reasoning, and argumentation) a well taught AP-Biology course may help students develop important inter- and intra-personal skills and competencies. The intrapersonal competencies include intellectual curiosity, appreciation for diversity, adaptability, work ethic, perseverance, productivity, and metacognitive skills, among others (NRC, 2012c). The interpersonal competencies include teamwork, the ability to effectively communicate, trust, cooperation, leadership, and self-presentation, among others (NRC, 2012c). The demanding nature of AP science courses require high school students to develop habits and ways of thinking that can directly facilitate their success when they enter higher education. A major transition for students upon entering a university is time management, responsibility for their own learning, and development as a contributing member of the learning community. Data from numerous universities indicate that students who receive high marks on the AP-Biology course (score of 4 or 5) perform as well in second year science courses as students who took the prerequisite resident introductory biology course. One explanation for this is that the high school AP-Biology course is comparable to the resident university introductory biology course. A secondary contributing component is that many of these AP students upon entering the university as first-year students have already acquired the inter- and intra-personal competencies needed to be successful in a second year course as similar to students who have had a year at a university to mature and develop necessary inter-and intra-personal skills needed for university success. It is worth noting that the correlations for success in life, work, and health are stronger for the inter-and intra-personal competencies then for the cognitive competencies. This suggests that a value added benefit for students undertaking challenging AP courses extends beyond content knowledge and may have a role in developing deeper learning and habits of mind that provide life long benefits.

8

RESPONSE 1

Model-Based Reasoning in Professional Development

Hilda Borko

A common theme across the chapters and responses is that to redesign K-12 STEM education in line with the visions set forth in the *Framework* (NRC, 2012a) and the *NGSS* (NGSS Lead States, 2013)—and in particular to integrate the science and engineering practices with the crosscutting concepts and disciplinary core ideas—will require substantial changes in science teaching. This requirement will create, in turn, a need for professional development. Indeed, in a recent *Science* article describing what she sees as "Grand Challenges in Science Education" Suzanne Wilson noted, "The Next Generation Science Standards will require large-scale professional development (PD) for all science teachers" (2013, p. 310) to support changes in teacher knowledge and practice.

In this response, I offer suggestions for designing such large-scale PD efforts, with a particular focus on model-based reasoning. I first describe model-based reasoning and make a case for its centrality in the new vision of science education. I suggest that one important contribution of the chapters could be a set of recommendations for designing the PD needed to achieve this vision. I then offer ideas for a PD program that takes as its starting point the general features of effective PD supported by current research.

Model-Based Reasoning

In the Introduction, Duschl et al. (this volume) proposed a framework for science teaching in which three key elements—systems thinking, model-based reasoning, and quantitative reasoning—are intertwined. Model-based reasoning and quantitative reasoning are the conceptual tools used to characterize components and processes within systems. To engage in systems thinking, a person develops a model that can be used to collect, analyze, and represent data, explain phenomena, and predict outcomes.

Model-based reasoning is an iterative process of analyzing and representing data in ways that support current theories or inform the development of new theories. It involves the use of a variety of representations, some of which may be mathematical. And it incorporates analyses, explanations, and communication about the system and its components. Thus, it draws upon aspects of at least five science and engineering practices: Developing and using models, analyzing and interpreting data, using mathematics and computational thinking, constructing explanations, and communicating information.

Not surprisingly, model-based reasoning is featured prominently in the recommendations for science learning in two recent NRC reports: *Taking Science to School* (NRC, 2007) and *Discipline-Based Education Research: Understanding and Improving Learning in Undergraduate Science and Engineering* (NRC, 2012b): As Duschl et al. (this volume) noted:

> *Taking Science To School* (NRC, 2007) and the *Framework* make the recommendation that science learning needs to be strongly grounded in model-based approaches that focus on the use and consideration of evidence for posing, building and refining models. (p. 14)

Because of its central role in these recommendations, model-based reasoning must also be a central focus of efforts to help teachers acquire the knowledge and skills they need to teach to the new Standards.

The Need for Professional Development

The need for substantial changes in science teaching is similar to mathematics, with most states' recent adoption of the *Common Core State Standards—Mathematics* (*CCSSM*) (National Governors Association Center for Best Practices & Council of Chief State School Officers [NGA], 2010). Both sets of Standards emphasize helping students engage in the disciplinary practices through teaching that integrates content and practices. In the *NGSS*, the focus is on integrating the eight scientific and engineering practices with the crosscutting concepts and the core ideas in the four disciplinary areas. In the *CCSSM*, it is on integrating the eight standards for mathematical practice with all strands of the content standards. To support students in meeting these standards, mathematics and science teachers will need to acquire the knowledge and skills to make significant changes in their practices.

In mathematics and science, PD has been identified as a key to the success of the reform effort. Similar to Wilson (2013), a panel of experts made recommendations for providing mathematics PD to support implementation of the *CCSSM*: "Developing teachers' capacity to enact these new standards in ways that support the intended student learning outcomes will require considerable changes in mathematics instruction in our nation's classrooms. Such changes are likely to occur only through sustained and focused professional development opportunities for those who teach mathematics" (Sztajn, Marrongelle, Smith, & Melton, 2012, p. 3). The panel also identified supporting teachers to foster the practices in their classrooms, and to integrate them with content standards, as central PD goals: "Designers of curricula, assessments, and professional development should all attend to the need to connect the mathematical practices to mathematical content in mathematics instruction" (p. 8).

Because districts have begun to incorporate the Standards into their mathematics curricula, assessments, and expectations for teaching, the demand for PD has already increased. A search for *CCSSM* PD on the Internet produces a multitude of resources and programs, many of which are very recently developed and have no evidence of effectiveness. Undoubtedly, the situation in science will be similar. If this is indeed the case, then one important contribution of the chapters could be a set of recommendations for the role of PD in K-16 STEM education reform, and for assessing the quality and effectiveness of PD resources and programs. What might these recommendations look like, particularly for model-based reasoning?

Features of Effective Professional Development

A growing body of empirical research provides information about the structure and content of PD programs that offer high-quality, high-impact learning opportunities for teachers (Borko, Jacobs, & Koellner, 2010; Desimone, 2009). With respect to content, high quality PD is focused on helping teachers develop the knowledge and skills they need to support students' learning. PD activities are designed to help teachers learn to elicit and interpret students' ideas, examine student work, and use what they learn about students' ideas to inform instructional decisions and actions. With respect to process and structure, high-quality PD programs provide opportunities for teachers to participate actively and collaboratively in a professional learning community with peers. The PD is situated in practice; that is, it engages teachers in inquiry about the concrete tasks of teaching and provides them with the opportunity to make connections between their own learning and classroom instruction. PD activities often use artifacts such as samples of student work and video of lessons to bring the classroom into the PD setting and make instructional practices the object of discussion and reflection. Finally, effective PD is ongoing and sustainable over time. Long-term PD programs enable teachers to engage in cycles of experimentation and reflection. In collaboration with colleagues, they can refine their teaching in ways that increase students' opportunities to engage in the science and engineering practices and develop proficiency with the crosscutting concepts and core content ideas defined in the Standards.

My colleagues and I have argued that although these consensus features are a good starting point for designing a PD program, they provide little information about what PD should look like in practice. We are not the first scholars to make this point. In an analysis of the status of research on PD, referring to the shared agreement on best practices Wayne and colleagues noted, "It is important to recognize that this consensus—although it has endured for more than a decade—lacks sufficient specificity to guide practice" (Wayne, Yoon, Zhu, Cronen, & Garet, 2008, p. 470). They suggested that, "there is much to gain from research that addresses practical questions faced by those who design and adopt PD programs" (p. 476). Taken together these lines of reasoning point to the following situation in science education:

- There is a need for PD that helps teachers learn to teach in ways that integrate the scientific and engineering practices with the disciplinary core ideas and crosscutting concepts and, more specifically, to engage their students in model-based reasoning.
- There is a consensus about general features of effective PD programs.
- However, given the level of generality, these PD features do not provide sufficient information to guide the development of specific programs.

Given this situation, a more refined question to address in developing a set of recommendations for the role of PD in K-16 STEM education reform is: How might we operationalize the general guidelines for effective PD to design a program for model-based reasoning that is composed of "features of PD worth testing"? (Wayne et al., 2008, p. 472).

Professional Development for Model-Based Reasoning: Operationalizing the Features of Effective PD

This section outlines ideas for the design of a school- or district-based PD program for model-based reasoning, using the following guidelines derived from the general features of effective PD:

- *Content-focused*: helps teachers develop the knowledge and skills they need to support students' learning
- *Student-focused*: helps teachers learn to elicit and interpret their students' ideas, examine student work, and use what they learn about students' ideas to inform their instructional decisions and actions
- *Active learning*: provides opportunities for teachers to participate actively in the PD
- *Collaborative learning*: is structured so that teachers have the opportunity to participate collaboratively in a professional learning community
- *Ongoing and sustainable over time*: can be implemented with the district's existing resources, and is aligned with other policies and practices

A PD program might consist of multiple modules, each focused on a particular disciplinary core idea and targeted to a specific grade level or grade bands. Each module includes 4–5 two-hour workshops, led by trained school-based facilitators, to be conducted over the course of one academic semester. Thus, the program could easily be accommodated by a school structure that sets aside two early-release days per month for science PD. The following are ideas for the types of activities included in each module:

1) Content Workshop
   a) Teachers engage in small group science activities that require model-based reasoning.
   b) Activities are designed or selected to be coordinated with the district's curricula and/or learning progressions, and appropriate for participating teachers' students (e.g., the laboratory exercises for an AP Biology course [Benson, this volume]).
   c) Teachers identify the crosscutting concepts and core ideas addressed in the activities and analyze the model-based reasoning in which they are engaged.
   d) Teachers consider how they might design a student-directed inquiry lesson with the activity, and what difficulties or stumbling blocks students might encounter.
2) Analysis of Learning and Teaching
   a) Teachers watch excerpts from videos of students engaged in the same activity and/or examine samples of students' work produced in the activity.
   b) Teachers analyze and discuss the students' understanding and learning, as evidenced in the video clips and work samples.
3) Connecting to Practice
   a) Teachers plan and teach a lesson with the activity.
   b) The lesson is videotaped and/or student work is collected.
   c) Teachers use video clips and student work selected by the facilitator to collectively analyze the lessons they taught, focusing on both student understanding and instructional practices.

Throughout the workshops, the facilitators focus teachers' attention on student understanding and learning, the science and engineering practices, and the crosscutting concepts and disciplinary core ideas. They also attend to developing and maintaining a safe environment that fosters collective participation and collaboration.

Conclusion

Wilson (2013) identified as one of her three Grand Challenges in Science Education: "Identify the kind of PD that will best prepare teachers to meet the challenges of the Next Generation Science Standards" (p. 312). In this response, I have suggested guidelines for developing an approach to PD for model-based reasoning that could be part of the research agenda to address that challenge.

9

RESPONSE 2

"Where Is the Line?"

Brian P. Coppola

My titular question (*Where is the line?*) mainly elicits discomfort among my friends and colleagues who work in the area of teacher education and professional development: "Is there a line past which you will admit that the expectations set for teachers is simply unreasonable?" In my experience, the answers have been highly qualified "maybe" statements: no matter where the line is, it can be relocated, and then teacher education and professional development can provide the means to elevate teachers to any new level. I disagree. I not only think a line exists, but I think we have crossed it.

I am an outsider. By way of full disclosure, my last extensive involvement with precollege teachers was when I graduated from high school in 1974. I have collaborated with my science education colleagues for many years as the "science guy." I have taught organic chemistry in large course settings at the University of Michigan for almost 30 years, mainly to first-year students, and so I interact with many students who have just left the K-12 sector. Since 1989, my department has honored the prior experience of our students who have reasonable mastery of high school chemistry topics by eliminating university-level General Chemistry as a requirement (Coppola, Ege, & Lawton, 1997; Ege, Coppola, & Lawton, 1997). In the organic chemistry course, we promote strongly the value of developing scientific dispositions, particularly through the use of the primary literature as a resource, and built upon the contemporary, mechanistic model of the structure-reactivity relationship (Coppola, 2010).

Let me also say that I am not keen on the way contemporary US education uses the term Standards. I think of Standards as, well, actual standards. I want airplanes built and maintained to strict standards, and I want the expected compliance to meet these standards as 100 percent over the line (and so do you). Yet there are times when 100 percent is unreasonable, and so the line has to be drawn somewhere. For instance, proofing construction in California, for earthquakes, specifies a 6.5 magnitude as an important threshold. As engineering improves, this line can be moved. But there is a point past which the expectation for no damage is simply unreasonable. When I hear my colleagues talk about the science education standards, I hear wishful thinking about theoretical goals that are disconnected from even a minimal set of considerations about the lived reality of *the diverse character and experience of millions of human beings who are teaching in science classrooms across the United States*.

In private conversations, I have heard that wishful thinking is exactly what our Science Standards are: a target, a statement of principle. Perhaps the disconnection from reality disturbs me the most. Pie-in-the-sky "if" statements are not reasonable to call standards: *If we lived in a world where countries satisfied the three pillars of freedom—from poverty, from war, and from prejudice—then. . . .* Is this a realistic statement of an achievable cultural standard? The *NGSS*, and the way my colleagues treat them, strike me this same way: *If we lived in a world where teachers construct science learning environments around the three pillars of science—inquiry through authentic practices, explicit and coherent understanding of interrelated cross-cutting concepts, and discipline-based core ideas—then. . . .* Is this a statement of achievable cultural standard? I think it is beyond reason.

I am equally skeptical that the architecture-du-jour (Learning Progressions) must be the key to the answers we are seeking. It is too premature. Learning Progressions are still too theoretical for me to bet the farm on them, and they still assume a level of skills and literacy about the average teacher that is unjustifiable. I have more to say about this later.

Teaching science as science is certainly a terrific goal, but we are not even asking scientists to do that at the post-secondary level. In the *NGSS*, and through Learning Progressions, we are imposing science on people who, by and large, understand science as an academic exercise, at best. This difference between authentic and theoretical teacher expectations is a simple fact that needs to be taken into account. Whether there is a blogger's summary of rising anti-intellectualism (Williams, 2014) or the comparative criticism of rigid, standards-based systems (Coppola & Zhao, 2012), increasing the distance between the average skill of teachers and the average expectations on teachers makes no sense.

The idea that research-based findings are enough to change behaviors is delightfully naïve, even at the postsecondary level, where, arguably, the scientific credentials of the instructors is more certain (Coppola & Krajcik, 2013, pp. 634–35):

> Singer acknowledges the same dilemma as we have . . . when it comes to the role of evidence in changing behavior. She states, "evidence is necessary but not sufficient to change undergraduate STEM education." This is a critical point. Another apt scientific analogy might be to not confuse thermodynamics and kinetics. To get from A to B, it does not only matter

> how much more favorable the thermodynamic destination is, because you always have to overcome the kinetic barrier (in fact, poor destinations with low barriers are often quite attractive pathways to follow). The futurist, Paul Sappho, says it this way (Sappho, 2013): "Never mistake a clear view for a short distance."

One prescribed solution—and an explicit outcome for this Waterbury Summit—is research, and yet more research, which sustains the professional lifeblood of those who study education from their university chairs, can involve working with a comparative handful of highly supported teachers on their well-funded projects. The research agendas do not come close to even acknowledging the problem of *the diverse character and experience of millions of human beings who are teaching in science classrooms across the United States.*

Finally, because not every child attends college and not every college student studies science, there is a long-standing desire for science education to turn out the near-mythical educated and informed democratic citizen, who makes good decisions, and has developed a shopping-cart's worth of social, psychosocial, and cognitive skills around science.

I get it. I agree with the principles, I really do; as have more than a few generations of scientists and science educators, particularly those in the mid-twentieth century, post-WWII environment of nuclear politics. I also understand that education is a system with lots of moving parts, including its incredibly important context—that is, the cultural perceptions of educators, education and intellectualism in the United States. But these iterations of alarm followed by pronouncements of change have me thinking that I am reading about solutions that are in search of a problem, because the real problems are so damned hard to solve. And if the statement of Standards are not actually standards that can be achieved, and are immediately understood as being so challenging that they cannot be achieved, then there needs to be a prioritization that begins with a set of minimum actual standards—true standards for which we aim for 100 percent achievement by *the diverse character and experience of millions of human beings who are teaching in science classrooms across the United States*. Otherwise, these iterations begin to look like the fresh coat of paint on an old wagon with broken wheels: going nowhere, still.

On the New Standards

Much to the dismay of some of my colleagues, I found myself unable to provide an enthusiastic endorsement of the *NGSS*, despite some general conceptual agreement I have with the larger goals. The short and long of it is that the expectations for teachers have now been set too unreasonably high in an absolute sense. This is because, and I freely admit it, the expectations are higher than what I think I can *generally* achieve in my own first-year university classes, even with the significant advantages I have compared with my pre-college teaching colleagues.

What is the problem being solved? Can any of the issues that the *NGSS* claim to address be attributed to a failing of the previous science standards (NRC, 1996)? The previous standards were created to set a vision, with the aim that the resulting practices would address certain identified inadequacies and/or make improvements. How did that go? What are the explanations that link to why these previous standards did not do their job? Do we understand how to link something like standards to whatever dissatisfaction drives the need for new ones? Is there convincing evidence that those previous standards, and their consequential results, is in any way related to anything that happened? And if so, what? Is there a differential hypothesis on how the *NGSS* can result in a system that does not simply go the way of the previous standards?

If we accept a system of standards, then I argue that we need to think of teacher education and professional development as an intervention, or treatment, in the same way that we do drug testing. One of the standards in medical treatments is the LD_{50}, that is, the median lethal dose: the amount of a drug, or other treatment, that kills 50 percent of the tested population. It is a benchmark against a simple standard: survival. We know that some can tolerate more, and some can tolerate less, but it creates a comparative scale of efficacy (well, terminal efficacy). If we have an educational standard that is coupled with a treatment such as professional development, then one of the things I would really like to know is its ET_{50}, what it takes to get an Effective Treatment (ET), hitting the standard, for 50 percent of a statistically meaningful test population of a given character, with whatever specified criteria are meaningful to understand a teaching context. I would like to know how the professional development treatment to get an ET_{50} varies according to—you can imagine it—a grid of as many critical variables that effect teaching that we know of.

And what about the proposition that Standards-based systems, and their inevitable high stakes testing, is a significant contributor to the real problem itself?

The pro-standards aphorism goes like this: if you do not have a destination, then every road gets there. Interesting. Because right now we have standards and (more or less) none of the roads get there, at least when we talk about *the diverse character and experience of millions of human beings who are teaching in science classrooms across the United States.* I am concerned, here, about the millions of teachers who are left on their own, or perhaps to whatever passes as professional development in their districts, compared with the relatively small handful of teachers who end up associated with highly funded, highly attentive professional development programs—funded by the NSF and directed by many of the participants in this Summit. My question to these professional development (PD) experts during our coffee breaks: do you think that your current PD program, with its teachers under your protective wing, achieves a 50 percent level for Effective Treatment? Can we expect this standard? All of the people I talked with were more comfortable with about a 5–10 percent level. Here, then, is the crux of my concern about why talking about "the line" matters. If the work of these highly funded, academically-based PD programs is good to a (self-assessed) 5 to 10 percent level for a privileged class of teachers, what reasonable hope is there for teachers "in the wild" who find themselves already looking up as the bar is raised farther from where they are?

For two of the pillars, authentic practices and cross-cutting concepts, I think I would swap out for the three scientific dispositions that Camins (2012) suggests ought to characterize the precollege science learning experience: comfort with ambiguity, the search for uncertainty, and learning from failure. For the third pillar, I think that when you peel away all the layers in the new rhetoric, you still end up worried about how teachers and students in America's incredibly diverse educational settings are dealing with learning about phase changes, the Krebs cycle, analytical methods, plate tectonics, glycolysis, and objects rolling down inclined planes, among many others, and how these are constructed and used to explain observable phenomena.

I am worried that by upping the ante on the complexity of what a strained workforce, our teachers, is responsible for, it means that at some point (if not already) the system becomes so strained that it cannot function. What constitutes a responsible foundation for our teachers' understanding of the basic subject matter (phase change, Krebs, etc.) when, particularly, as a technological society we have been so successful in deepening our understanding of these things? And now, one level up, we are going to ask teachers to take their inadequate understanding and design instruction that gets students to understand the distinctiveness of scientific thinking through authentic practices and cross-cutting concepts. And, one more level up, we want teachers to design environments that assume this previous understanding, and tie it to the development of social, psychosocial, and cognitive skills. At some point, the demands on any system can become so severe that it either collapses or it is crippled (famously, at least in *Star Trek III*: the more you overthink the plumbing, the easier it is to stop up the drain).

On a recent trip to New York City, I met with a colleague who was running a summer program for high school students needing additional help in preparing for the Regent's Exams. My colleague was telling me about an experienced teacher who was assisting with the laboratory program, someone who had an undergraduate degree in chemistry from the University of Chicago. In addition to only being comfortable with using fill-in-the-blank worksheets and standing at the electronic balance to do the mass measurements for the students ("write down this number, here, and get out your calculators and enter the other numbers in this order . . ."), the experienced teacher was unaware that she was having them enter a non-sensible value in one of the spaces (mass of crucible = 35.50 g; mass of crucible + hydrate = 3.45 g . . . actually, she was unknowingly accounting for the tare of the crucible in making the measurement and did not realize that this did not match the sheet, or the manipulation of the values that she was directing). When one of the summer students began to ask about the negative mass and the discrepancy, the teacher was unable to do anything except repeat the order to follow her directions. In the daily debriefing, my colleague, who had overheard most of this, asked the teacher about it. The teacher's reply, through tears, was that this was the way she learned how to do this and had no other options. As a faculty member in a department of chemistry, I am a complete stranger to this context, but it makes me believe that there might be some severe disconnects between the theory, implementation, and outcomes from teacher education and professional development.

We do not need to teach everything that we know, and a well-considered approximation is always tied with the need-to-know for any group of students in any given environment. I have been in the audience a few times when Craig Nelson, an evolutionary biologist at Indiana University, poses this question: "What is the shape of the Earth when you teach Architecture 101?" The answer is that, in this situation, we use a flat-earth model. We know, on the curved surface of our planet, that two plumb lines separated by any distance are not ever parallel. However, on the scale of building most buildings, the variation is outside the limits of the needed precision. Courses are like this. While my expertise gives me, say, 10th decimal place precision in my understanding of organic chemistry, one of the terrific challenges in teaching the introductory course is deciding what precision is needed, perhaps 2nd decimal place, and another, greater challenge is constructing a coherent and consistent 2nd decimal place version of the subject. Some faculty instructors have a hard time with these approximations because it seems like they are promoting incomplete heuristics in lieu of "telling the truth." But there is no need-to-know in the 10th decimal place knowledge, in the context of the class, so it does not serve any useful purpose. And as Nelson is quick to point out, my 10th decimal place understanding is not the truth, anyhow, it is just the latest iteration of the flat earth model.

In his essay about the AP Biology upgrade, Benson acknowledges the need to find the expectation boundaries when it comes to the depth of subject matter for which the course is responsible. But even if that expectation for mastery is exactly where a teacher is at, it is not adequate. My level of knowledge and expertise needs to be higher than the instructional boundary in order to be able to bring perspective, alternative entry points, and so on, to instructional design and to the day-to-day challenges of working with students. And subject matter mastery relatively easy compared to things like social, psychosocial, and cognitive goals.

On Learning Progressions

I am worried that tying the *NGSS* to Learning Progressions (LPs) is jumping the gun. I like the LP concept, but if this were a new drug, we would barely be in the Phase One clinical trials. We really do not know enough about how LPs interact with a wide diversity of subjects (students and teachers alike) who are drawn from the diverse array of settings that exist in our country.

Assuming we have a well-constructed and understood LP, LPs still have a pair of coupled weaknesses. First, the details about a LP come from some aggregate, average behavior of the student population(s) that was used to construct it, which means that information about an average is being used to predict the character of individuals. If it is the 4th grade, then this is what you should be able to do. Second, the concept of a progression begins to more formally create expectations that can never be based on the actual experiences of any given set of students. That is, not only is this what we are supposed to be doing in 4th grade, but we also need to be able, in a highly formalized sense, to expect the mastery of the lessons from grades 1 to 3. What are the error bars on what students at any given level should and/or should not be able to do, and how does the absolute placement on this LP depend on the school, the teachers, the setting, and so on. Without advancing to the Phase Two and Three clinical trials, we cannot know how this LP treatment interacts with all the different clinical features of the wide array of "subjects" on which it is supposed to act.

In an essay on the distinctiveness of both disciplinary expertise and embedded understanding of disciplinary dispositions, when comparing the expertise of pre-college and post-secondary educators, this difference in what can be expected emerges (Coppola & Krajcik, 2013, p. 628).

> Discipline is a continuum of increasing focus as one moves through education. Although the distinction is fuzzy-edged, one characteristic of post-secondary science education is the shift from introductory survey classes with multiple topics (e.g., High School and General Chemistry) to classes that specialize according to a more specific disciplinary expertise (e.g., Organic Chemistry, Physical Chemistry, Cellular and Molecular Biology, or an Organismal and Population Biology) and the increased chance that the instructor is also a person who is versed in the art of that specific area (an Organic Chemist, a Physical Chemist, a Molecular Biologist, or an Organismal Biologist). The common impression from survey classes, and one that is never a goal, is that scientific knowledge reduces to a bundle of facts. Ideally, disciplinary context provides an integrated understanding, where the "thinking, feeling, and behaving" part of understanding is an indispensible feature of the story. Progressing through the K-20 spectrum, the larger cross-disciplinary ideas about science, as a way of knowing, become highly contextualized through the more detailed study through the lenses of disciplinary focus.
>
> K-12 and post-secondary instructors generally differ greatly in both the time with, and depth of, their prior experiences carrying out work in the discipline. An educational environment designed and implemented by a scientist whose expertise involves years of first-hand generation of data in the discipline carries the possibility, at least, that contextual, practical, epistemological, theoretical, and experimental issues are part of the knowledge

and cannot be easily separated from them. As we have said, this is a fuzzy-edged distinction, in that there are certainly K-12 teachers who understand and can convey disciplinary insights from having only limited experiences; and there are post-secondary educators whose Ph.D. does not guarantee that they are aware of, or capable of conveying, the deeper understanding from their fields.

Designing Learning Environments around Model-Based Reasoning

Two of the six articles in the August 2013 Special Issue of the *Journal of Research in Science Teaching* include excellent examples of how model-based reasoning can be used to re-conceptualize the traditional introductory course at the university level. In one case (Dauer, Momsen, Speth, Makohon-Moore, & Long, 2013), a group of biologists took a specific system, the "Gene to Evolution" model, and used it as a framework to link concepts from molecular-level processes to population-level outcomes. The subject matter in this class spiraled around the development of this model. In the second case (Ding, Chabay, & Sherwood, 2013), an introductory physics course in Mechanics is described and studied. In this program, all of introductory physics is handled through learning how to identify and apply one of three principles (the Momentum Principle, the Energy Principle, and the Angular Momentum Principle). The vast landscape of endless equations that introductory physics classes are known for collapses to the judicious use of these principles.

In 1989, my colleagues and I undertook a large-scale revision of the entire undergraduate chemistry curriculum at the University of Michigan. The centerpiece of this change, and one that still thrives, is the elimination of General Chemistry for the vast majority of the non-engineering students. Based on an adequate high school background, first-term first-year students begin their study of college chemistry with organic chemistry, and, in particular, a re-thought version of the organic chemistry subject matter, based on a modern model of relating structure and reactivity. In fact, just to make the distinction clear at every turn, we do not even call the course "Organic Chemistry;" we call it "Structure and Reactivity" (Coppola et al., 1997; Ege et al., 1997).

The types of assessments we use in this class have driven the instructional design. Because the structure/reactivity model in this subject area is so well behaved, and because organic chemistry does not rely on numerical/mathematical representations, we are able to use the most up-to-date journal articles as the basis for writing examination questions. The power of the contemporary structure/reactivity model in organic chemistry means that hundreds of thousands of new and unfamiliar examples and observations can be understood by the application of a handful of principles (structure: connectivity, stereochemistry, conformational analysis; reactivity: four transformations involving polar sigma and pi bonds—complexation, substitution, addition, elimination). Exam questions, which usually carry the citation (once again, to signal as clearly as possible that recognition and recall of specific examples is meaningless), present students with data or other pertinent structural or experimental information. As opposed to recognizing and recalling a specific factoid, we ask students to make sense of new and unfamiliar data and to provide an explanation that is based on making meaning from this evidence. Perhaps not surprisingly, "Don't forget to read the question" becomes a familiar admonishment for those who struggle with what we are doing. Up to 30 to 40 percent of the time, exam problems have more than one answer that is completely consistent with the data that are given. Do not for one moment believe that our students are automatically comfortable with this level of uncertainty and ambiguity; they are not. But it is something they learn over the year. Enrollments in the fall term are approximately 1500, and the Drop/Fail/Withdraw (grades of "D" or "F," or those who withdraw) rates are exceptionally small (about 6 to 8 percent).

At every turn in the design and implementation of these classes, my colleagues and I are drawing from our own deep and connected understanding of the subject and its principles. As stated explicitly previously (Coppola & Krajcik, 2013, p. 628): "disciplinary context provides an integrated understanding, where the 'thinking, feeling, and behaving' part of understanding is an indispensible feature of the story. . . . An educational environment designed and implemented by a scientist whose expertise involves years of first-hand generation of data in the discipline carries the possibility, at least, that contextual, practical, epistemological, theoretical, and experimental issues are part of the knowledge and cannot be easily separated from them."

Conclusion

My first-hand experience, drawn from over 30 years of teaching tens of thousands of students, tells me that model-based reasoning is as powerful in designing learning environments as it is in the normal, everyday practice of working with it as a scientist. Indeed, I cannot work without it.

In the classroom, my young colleagues have often reflected on their first experience teaching in the *Structure and Reactivity* courses this way (paraphrasing): *I think what we are doing is simply teaching the science, and not teaching how to do a set of chemistry problems.* Comparatively speaking, we provide a great deal of support (some might call it professional development) for the instructors who are teaching these classes for the first time, including materials and being part of an experienced team that talks regularly with one another about teaching these classes.

There is another side to this coin: What we do is not easy. And it is worthwhile noting that we are in no way under the illusion that we are asking students to actually construct a new model on which to do model-based reasoning! *This is really hard.* We are aiming at developing students' abilities to understand, use, and apply an actual model with new and unfamiliar, yet authentic, data in both lecture and laboratory settings.

I have a difficult time, then, knowing what I know about my one hard-earned and deeply understood authentic model, understanding how a group of middle school children can both learn the science *and* construct *and* use a scientific model that has deep, lingering, and meaningful effect. Hypothetically, at least, achieving any level of sophistication with model-based reasoning—in classroom practice, scaled up, and in diverse settings—relies on a level of discipline-centered literacy and experience with experimental science that, to the best of my understanding, is not held by the majority of school science teachers. I worry that it becomes the straw that could break the back of that teacher in the New York City program I mentioned earlier, and I am absolutely sure she is not alone.

I understand that this is a provocative statement and let me be clear: I am NOT beating up on teachers. I also understand that there are teachers out there with excellent backgrounds who could approach this, or who could be trained to do this. But if we are going to make this expectation a national standard—a real standard—then a certain degree of realism has to accompany this declaration. This clear view not only lacks a short distance to its resolution, but the mountain of challenges sitting in the way is high and non-trivial. I am looking at the expectations in the *NGSS*, and I am wondering if I achieve them in my own teaching, or rather, what it would take to do so.

I wish that all of my clever friends and colleagues who worked so long and hard on the *NGSS* had taken all of that time and thought through about a hundred paradigm-breaking ways for the United States to improve the situation for education in our country. Just to stir the pot a bit more, Yong Zhao and I suggested (2012), for instance:

> *Incentivize the teaching profession.* Even in this era of budget austerity, we need creative, strong, visible, compelling, and cost-effective ways to make the teaching profession more appealing. One drastic measure would be to make primary and secondary teaching an income-tax-free profession.
>
> *Reintegrate the disciplines and teacher education.* While the United States will never return to the Normal school system, some way of putting teeth into the requirement for our disciplinary and education faculties to work together on this problem is needed. To this end, we should simply require, as a condition of accreditation, a meaningful collaboration between college disciplinary units (chemistry, physics, and so on) and schools of education in the early identification, recruitment, and preparation of future teachers, including programs for engaging precollege students and putting them on this path.

> *Resist any temptation to standardize and overly regulate higher education in the name of accountability.* There is an increasing effort to impose government regulations and external standards upon colleges. These seemingly responsible actions will inevitably bring more regimentation, standardization, and testing, ruining what has made American higher education the envy of the world.

The United States is not Finland, where things are so culturally uniform, and the level of diversity is so low, that it is Finnish citizens who speak a different dialect who are considered the chief minority group. Implementing change is easier, there, than here. The United States is not China, where so many high school teachers are the product of their Normal Universities, in which their teacher preparation is integrated with graduate level research in the discipline. Although China has been the poster-child for Standards-driven education, its experience ought to serve as a warning buoy for the United States: one of the reasons they are starting to scoot ahead of us on measures of technology and innovation is because they have begun to abandon some Standards-based practices at the same time, ironically enough, that we have been moving in that direction. Historically, I think that the absolute level of disciplinary expertise held by their teachers has ameliorated some of the intrinsic problems with standardization, precisely because the teachers could default to their deep understanding of science and experimental practice.

And so I am disheartened when the challenges-cum-strengths of education in the United States are not automatically taken together when I read in this area, because it usually seems to be solutions that are simply disconnected from the real challenge, again and again, of *the diverse character and experience of millions of human beings who are teaching in science classrooms across the United States.*

If I were to set a research agenda for those interested in science education, I think I would aim at the following targets:

(1) Clump, cluster, and/or prioritize the *NGSS* into a rational, accessible, nested set of ideas, a subset of which constitutes an actual minimum standard that is rationally accessible by 100 percent of teachers. I think this is a critical exercise in Reality. Call them the *Reality Standards.*

(2) Think about the ET_{50} measurement, and start thinking in terms of what it takes to get 50 percent of some given test population to an absolute level of success on the *Reality Standards.* We can then learn, I think, how the treatment needs to vary wildly, according to the given population, in order to achieve the same outcome (this is the anti-magic bullet principle). In a sense, this is a statement of the pedagogical content knowledge for effective professional development.

Poster Presentation II

Cultivating Systems Thinking and Engagement through an Intercollege Minor in Sustainability Leadership, August 2013

Susannah Heyer Barsom, Ph.D., David R. Riley, Ph.D.
Sustainability Institute, The Pennsylvania State University

Sustainability

Definitions of sustainability are varied and personal. Also, sustainability is broadly interdisciplinary.

Sustainability Leadership Competencies

Sustainability Competencies

- Ability to formulate, articulate, and revise one's own working definition of sustainability, and apply its concepts in multiple settings
- Capacity for systems thinking in sustainability – ability to identify system connections and boundaries
- Proficiency in using sustainability measures and metrics in analyses and decision-making
- Ability to articulate and apply in context the ethical dimensions of sustainability

Leadership competencies

- Ability to articulate a defined self-concept, including one's connections to community and to place
- Ability to implement and facilitate effective interdisciplinary and collaborative approaches to problem solving in sustainability
- Proficiency in describing and responding to different styles of leadership and group dynamics
- Effectiveness in facilitating change toward sustainability in multiple settings

Essential Elements

- Systems thinking and practical leadership
- Opportunity to practice addressing sustainability problems in a non-academic setting
- Both key elements are emphasized in the Sustainability Leadership minor

STEM in Sustainability Leadership Minor

- Includes STEM concepts, social sciences, humanities, and arts.
- Includes the three dimensions in the National Research Council (NRC)'s science education *Framework* (NRC, 2012): *crosscutting concepts* is prominently featured; *practices and core ideas* are also addressed.

Program Design and Requirements

- Program basis can be found in Frank Rhodes's words: the "concept of sustainability could provide a new foundation for the liberal arts and sciences." (Rhodes, 2006, p.1)
- Flexibility built in to ensure the program is accessible to students in any major, on any University campus.
  - Student may focus to varying degrees on the science or mathematics of sustainability, depending on the
  - student's interests, or on the strengths of the campus.
  - All students have exposure to these topics and opportunities to apply knowledge in a practical situation.

Requirements: the 18-credit minor has three parts.

Part 1: Prescribed Coursework [3 credits] - Foundations of Leadership in Sustainability.

Part 2: Elective Coursework [12 credits]

(at least 3 credits at the 400 level; at least 3credits from leadership courses.)

- 9 credits in sustainability coursework; 3 credits that provide immersive* sustainability experience

Part 3: Capstone Coursework or Project [3 credits]

- Complete a capstone course in sustainability or an independent study approved by minor committee,
- which includes an applied or research experience in sustainability leadership.
- At least 9 credits of the minor degree program must come from *outside* his/her major department.

**Immersive sustainability experiences provide opportunities to engage and learn in depth about sustainability challenges and solutions. Courses meet the requirement for an immersive sustainability experience when they involve:*

a) the investigation of a sustainability challenge facing a community,
b) an inquiry process that helps define a sustainability challenge and/or assess implementation of a proposed solution,
c) structured opportunities for student reflection and self-assessment, and
d) regular contact with a faculty member to facilitate integration of scholarly content with the immersive experience.

References

NRC (2012). *A framework for K-12 science education: practices, crosscutting concepts, and core ideas.* Washington, DC: National Academy Press.

Rhodes, F. H. T. (2006). "Sustainability: The Ultimate Liberal Art." *The Chronicle Review* 53 (9): B24.

Theme II References

American Association for the Advancement of Science. (2011). *Vision and Change in Undergraduate Biology Education*. Washington, DC: AAAS.

Anderson, L.W., & Krathwohl, D.R. (Eds.). (2001). *A Taxonomy for Learning, Teaching, and Assessing: A revision of Bloom's Taxonomy of Educational Objectives*. New York: Longman.

Assaraf, O.B.-Z., & Orion, N. (2005). Development of system thinking skills in the context of earth system education. *Journal of Research in Science Teaching, 42*(5), 518–60. http://doi.org/10.1002/tea.20061

Borko, H., Jacobs, J., & Koellner, K. (2010). Contemporary approaches to teacher professional development. In P.L. Peterson, E.L. Baker, & B. McGaw (Eds.), *International Encyclopedia of Education* (Vol. 7, pp. 548–56). Oxford, UK: Elsevier. Retrieved from http://public.eblib.com/choice/publicfullrecord.aspx?p=634860

Camins, A.H. (2012, August 8). Three core values of science, engineering and how ed reform contradicts them. Retrieved from http://washingtonpost.com/blogs/answer-sheet/post/three-core-values-of-science-engineering-and-how-ed-reform-contradicts-them/2012/08/07/130a9e7a-de4e-11e1-af1d-753c613ff6d8_blog.html

College Board. (2009). *Science Standards for College Success*. New York: College Board.

College Board. (2012a). AP Central—AP Biology Lab Manual Resource Center. Retrieved March 28, 2015, from http://apcentral.collegeboard.com/apc/members/courses/teachers_corner/218954.html

College Board. (2012b). AP Course Audit. Retrieved March 28, 2015, from http://.collegeboard.com/html/apcourseaudit/

College Board. (2012c). Changes to AP Biology—Advances in AP—The College Board | Advances in AP. Retrieved March 28, 2015, from https://advancesinap.collegeboard.org/stem/biology

Coppola, B.P. (2010). Structure and reactivity at the University of Michigan. In J. Ryan, T. Clark, & A. Collier (Eds.), *Assessment in the Disciplines: Assessment in Chemistry* (Vol. 5, pp. 175–99). Tallahassee, FL: Association for Institutional Research.

Coppola, B.P., & Zhao, Y. (2012, February 5). US education in Chinese lock step? Bad move. *The Chronicle of Higher Education, 58*. Retrieved from http://chronicle.com/article/US-Education-in-Chinese/130669/

Coppola, B.P., & Krajcik, J.S. (2013). Discipline-centered post-secondary science education research: Understanding university level science learning. *Journal of Research in Science Teaching, 50*(6), 627–38. http://doi.org/10.1002/tea.21099

Coppola, B.P., Ege, S.N., & Lawton, R.G. (1997). The University of Michigan undergraduate chemistry curriculum 2. Instructional strategies and assessment. *Journal of Chemical Education, 74*(1), 84. http://doi.org/10.1021/ed074p84

Crawford, B.A. (2007). Learning to teach science as inquiry in the rough and tumble of practice. *Journal of Research in Science Teaching, 44*(4), 613–42. http://doi.org/10.1002/tea.20157

Dauer, J.T., Momsen, J.L., Speth, E.B., Makohon-Moore, S.C., & Long, T.M. (2013). Analyzing change in students' gene-to-evolution models in college-level introductory biology. *Journal of Research in Science Teaching, 50*(6), 639–59. http://doi.org/10.1002/tea.21094

Desimone, L.M. (2009). Improving impact studies of teachers' professional development: Toward better conceptualizations and measures. *Educational Researcher, 38*(3), 181–99. http://doi.org/10.3102/0013189X08331140

Ding, L., Chabay, R., & Sherwood, B. (2013). How do students in an innovative principle-based mechanics course understand energy concepts? *Journal of Research in Science Teaching, 50*(6), 722–47. http://doi.org/10.1002/tea.21097

Ege, S.N., Coppola, B.P., & Lawton, R.G. (1997). The University of Michigan undergraduate chemistry curriculum 1. Philosophy, curriculum, and the nature of change. *Journal of Chemical Education, 74*(1), 74. http://doi.org/10.1021/ed074p74

Feldman, A., Divoll, K., & Rogan-Klyve, A. (2009). Research education of new scientists: implications for science teacher education. *Journal of Research in Science Teaching, 46*(4), 442–59. http://doi.org/10.1002/tea.20285

Johnson Foundation, Inc. (1993). *An American imperative: Higher expectations for higher education. An open letter to those concerned about the American future. Report of the Wingspread Group on higher education.* Racine, WI: Wingspread Group on Higher Education, Johnson Foundation, Inc. Retrieved from http://eric.ed.gov/?id=ED364144

Krathwohl, D.R. (2002). A revision of Bloom's taxonomy: An overview. *Theory Into Practice, 41*(4), 212–18. http://doi.org/10.1207/s15430421tip4104_2

Mislevy, R.J., Almond, R.G., & Lukas, J.F. (2003). *A Brief Introduction to Evidence-centered Design.* Princeton, NJ: Educational Testing Service.

National Governors Association Center for Best Practices & Council of Chief State School Officers. (2010). *Common Core State Standards for Mathematics.* Washington, DC: National Governors Association Center for Best Practices, Council of Chief State School Officers.

National Research Council. (1996). *National Science Education Standards observe, interact, change, learn.* Washington, DC: National Academies Press.

National Research Council. (2002). *Learning and Understanding Improving Advanced Study of Mathematics and Science in US High Schools.* Washington, DC: National Academy Press.

National Research Council. (2006). *America's Lab Report Investigations in High School Science.* Washington, DC: National Academies Press.

National Research Council. (2007). *Taking Science to School: Learning and Teaching Science in Grades K-8.* Washington, DC: National Academies Press.

National Research Council. (2010). *Rising above the Gathering Storm, Revisited: Rapidly Approaching Category 5.* Washington, DC: The National Academies Press. Retrieved from http://ncbi.nlm.nih.gov/books/NBK259118/

National Research Council. (2012a). *A Framework for K-12 Science Education: Practices, Crosscutting Concepts, and Core Ideas.* Washington, DC: National Academies Press. Retrieved from http://www.nap.edu/catalog/13165/a-framework-for-k-12-science-education-practices-crosscutting-concepts

National Research Council. (2012b). *Discipline-based Education Research Understanding and Improving Learning in Undergraduate Science and Engineering.* Washington, DC: National Academies Press. Retrieved from http://site.ebrary.com/id/10606364

National Research Council. (2012c). *Education for Life and Work: Developing Transferable Knowledge and Skills in the 21st Century.* Washington, DC: The National Academies Press.

NGSS Lead States. (2013). *Next Generation Science Standards: For States, By States.* Retrieved from http://www.nap.edu/catalog/18290/next-generation-science-stan dards-for-states-by-states

Sadler, T.D., Burgin, S., McKinney, L., & Ponjuan, L. (2010). Learning science through research apprenticeships: A critical review of the literature. *Journal of Research in Science Teaching, 47*(3), 235–56. http://doi.org/10.1002/tea.20326

Sappho, P. (2013). *Never mistake a clear view for a short distance.* Retrieved from http://www.saffo.com/idea1/

Steen, L. A. (Ed.). (2005). *Math & Bio 2010: Linking Undergraduate Disciplines.* Washington, DC: Mathematical Association of America.

Sztajn, P., Marrongelle, K., Smith, P., & Melton, B. (2012). *Supporting Implementation of the Common Core State Standards for Mathematics: Recommendations for Professional Development.* North Carolina State University College of Education.

Tang, X., Coffey, J. E., Elby, A., & Levin, D. M. (2010). The scientific method and scientific inquiry: Tensions in teaching and learning. *Science Education, 94*(1), 29–47. http://doi.org/10.1002/sce.20366

Wayne, A. J., Yoon, K. S., Zhu, P., Cronen, S., & Garet, M. S. (2008). Experimenting with teacher professional development: Motives and methods. *Educational Researcher, 37*(8), 469–79. http://doi.org/10.3102/0013189X08327154

Williams, R. (2014, July 7). Anti-intellectualism and the "Dumbing Down" of America. Retrieved from http://www.psychologytoday.com/blog/wired-success/201407/anti-intellectualism-and-the-dumbing-down-america

Wilson, S. M. (2013). Professional development for science teachers. *Science, 340*(6130), 310–13. http://doi.org/10.1126/science.1230725

Windschitl, M., Thompson, J., & Braaten, M. (2008). Beyond the scientific method: Model-based inquiry as a new paradigm of preference for school science investigations. *Science Education, 92*(5), 941–67. http://doi.org/10.1002/sce.20259

THEME III

Quantitative Reasoning

10

QUANTITATIVE REASONING IN MATHEMATICS EDUCATION

Directions in Research and Practice

Heather Lynn Johnson

Imagine that a bottle is being filled with water that is being dispensed at a constant rate[1]. How might one measure the amount of water in the bottle at any given time? How might the amount of water in the bottle relate to the height of the water in the bottle for differently shaped bottles? How might one relate the changing amount of water to the changing height of water? To respond to these questions, an individual would need to make sense of both the height of the water and the amount of the water as attributes that could be measured. Importantly, an individual would not necessarily need to measure (or know) particular numerical amounts to make sense of the situation. For example, dispensing water into a very wide bottle or a very narrow bottle would result in different relationships between the height and amount of water in the bottle. An individual's responses to the questions would depend on how that individual interprets the context of the problem situation (Johnson, 2014). For example, the sound of the water filling into the bottle, bottles with which an individual might be familiar, or dispensers an individual has used might play a role. These introductory questions are intended to provide examples of prompts that could foster individuals' quantitative reasoning.

This chapter stemmed from conversations that took place in the Quantitative Reasoning Working Group of the International STEM Research Symposium: Quantitative Reasoning in Mathematics and Science Education held in Savannah, GA in Spring 2012. Goals of the Working Group, facilitated by Johnson, were threefold: To identify high priority research problems and methods related to quantitative reasoning; To determine critical themes based on those problems and methods; and To identify future actions for the group. Broadly, conversations revolved around three themes: distinguishing between quantitative reasoning (QR) and quantitative literacy (QL), fleshing out interrelated constructs of quantity and quantification, and addressing roles of context in quantitative reasoning. In the discussion that follows, each of these themes is situated within a larger discussion focusing on quantitative reasoning in mathematics education.

To organize this chapter, first relationships between the constructs of quantity and quantification are examined, a working definition of QR is provided, and distinctions between QR and QL are made. Second, the role of QR in multiplicative, algebraic, and covariational reasoning is examined, arguing for a dynamic perspective of context in tasks designed to foster QR, and highlighting learning trajectories in mathematics education K-12 that foster QR. Next, the inclusion of QR in educational policy documents and practitioner resources is discussed, providing subsequent examples of research based tasks fostering students' QR in K-16 mathematics education. The chapter concludes by arguing how quantification and measurement could forge links between mathematics and science education, supporting the use of different, yet compatible, lenses to investigate shared problems.

Background

Quantity and Quantification

Nearly three decades ago researchers defined a quantity as a pairing of a number and a unit (Nesher, 1988; Schwartz, 1988), meaning that quantities are numbers with additional characteristics (a unit). However, as suggested by the example of the bottle problem, a numerical amount could indicate a measure of a quantity, but a particular measure of a quantity does not need to be known to reason about the quantity (cf. Thompson, 1993). For example, an individual would not need to measure (or know) particular amounts of height and water to make relationships between them. Unlike previous definitions, Thompson (1994) defined quantities as "conceptual entities," asserting that a quantity entails an individual's conception of some attribute of an object (e.g., height of water) as being possible to measure. Importantly, from this perspective, the entity of quantity is inextricably linked to an individual's conception of the measurability of some attribute. When referring to quantity in this chapter, it is an entity that an individual conceives as being possible to measure.

While a quantity is an object, quantification is a process. Thompson (2011) defined quantification as a three part process, involving "settling what it means to measure a quantity, what one measures to do so, and what a measure means after getting one" (p. 38). For example, quantification related to the bottle problem would involve deciding what it would mean to measure amounts of height and water, what one would measure to determine amounts of height and water, and what measures of amounts of height and water would mean in this situation. Building from Thompson's research, Mayes, Peterson, and Bonilla (2013) defined quantification as a mathematical process extending beyond assigning a numerical value to a quantity, involving "conceptualizing an object and an attribute of it so that the attribute has a unit measure" (p. 6). The unit of measure indicated by Mayes et al. may be a standard or non-standard unit of measure. Importantly, as illustrated by both definitions, the process of quantification is distinct from the process of measurement, with quantification setting the stage for subsequent measurement (Thompson, 2011). When referring to quantification in this chapter, it is a process of conceiving of an attribute that could be measured, deciding how one might measure that attribute (which could include selecting a unit of measure), and making meaning from measures that would result.

Quantitative Reasoning (QR)

Quantitative reasoning (QR) entails both the objects of the reasoning (quantities) and the operations (Piaget, 1970) that an individual could enact as thought processes or physical acts (Thompson, 1994a). For example, an individual could envision what attribute the height of water in a bottle would represent or physically investigate heights of water in different bottles. In contrast, Mayes et al.'s (2013) definition of QR focuses on the how an individual might apply her knowledge: "mathematics and statistics applied in real-life, authentic situations that impact an individual's life as a constructive, concerned, and reflective citizen" (p. 6). For example, an individual could apply her knowledge to determine an ideal bottle shape given particular constraints. A key distinction between these definitions involves what is claimed to be central to QR: application (Mayes et al.) vs. quantification (Thompson). When referring to QR in this chapter, it is a process composed of operations on objects (quantities), for which quantification is central. This is not to say that application of knowledge could not be involved in QR, but that quantification should be central to QR.

Quantitative Literacy (QL)

Steen (2001) defined quantitative literacy (QL) as a "habit of mind," involving a variety of elements, including solving problems involving different contexts, making meaning from numbers, and using numbers to measure different items. More recently, the American Association of Colleges and Universities (AAC&U) (2010) defined QL as follows: "Quantitative Literacy (QL)—also known as Numeracy or Quantitative Reasoning (QR)—is a 'habit of mind,' competency, and comfort in working with numerical data." Notably, working with numbers plays a key role in both Steen's and the AAC&U's definitions of QL.

Some use QR and QL synonymously (e.g., American Association of Colleges and Universities, 2010; Dingman & Madison, 2010), while others identify QL as a component of QR (e.g., Mayes et al. 2013). Mayes et al. (2013) defined QL as the "use of fundamental mathematical concepts in sophisticated ways for the purpose of describing, comparing, manipulating, and drawing conclusions from variables developed in the quantification act" (p. 6). Importantly, Mayes et al.'s definition indicates that QL results from quantification, which may or may not include working with numbers. Here, it is argued that work with numbers does not capture the scope of QL, which results from quantification and does not necessitate work with numbers.

Quantitative Literacy (QL) and Quantitative Reasoning (QR)

Quantification and measurement are distinct processes at the heart of QR and QL, respectively (See Figure 10.1). Simply put, measurement involves actual measuring (or knowing actual amounts of measure), whereas quantification involves conceiving of how an attribute of an object might be measured and what those measures might mean. For example, an individual might be able to use numerical amounts of city and highway miles per gallon calculations to make comparisons between the fuel efficiency of different cars (measurement). Notably, fuel efficiency comparisons would not necessarily require one to envision how one might measure miles per gallon (QR), rather they would involve being able to use determined amounts of miles per gallon to make informed decisions (QL). In contrast, an individual determining what it might mean to measure miles per gallon (quantification) could form and interpret relationships

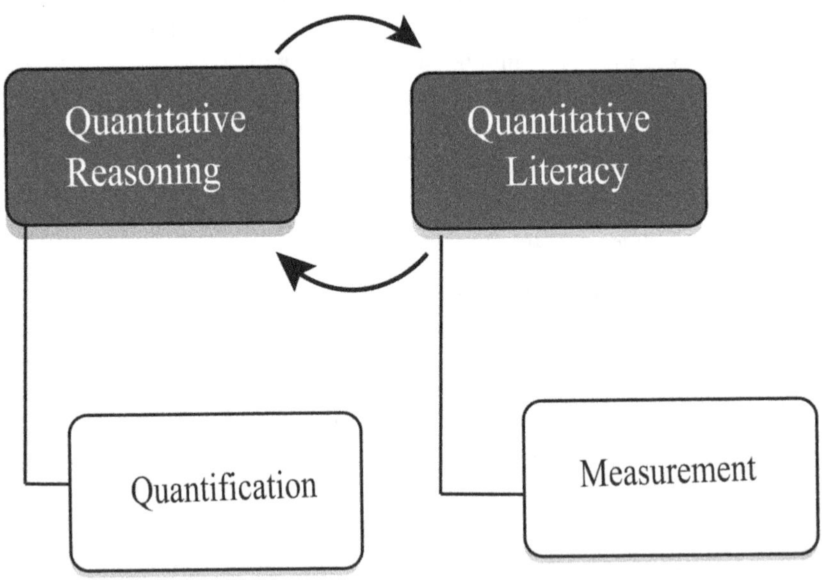

FIGURE 10.1 A reflexive relationship between QR and QL

between the quantities of miles and gallons. Similarly, an individual determining what it might mean to measure miles per gallon (QR) may or may not use those relationships to make informed decisions about automobile travel (QL). Figure 10.1 illustrates a reflexive relationship between QR and QL, such that QL results from QR, and subsequent QL could afford an individual's further engagement in QR.

QR and Multiplicative, Algebraic, and Covariational Reasoning

QR can form a foundation for multiplicative reasoning (Thompson & Saldanha, 2003), algebraic reasoning (Carraher & Schliemann, 2007; Mason, 2008; Smith & Thompson, 2008), and covariational reasoning (Saldanha & Thompson, 1998). Here the focus is on relationships between QR and these three forms of reasoning because of the key role multiplicative, algebraic, and covariational reasoning play in K-16 mathematics. Multiplicative reasoning is essential for students' development of algebraic reasoning (Greer, 1994). Algebraic reasoning encompasses more than reasoning about generalized arithmetic (Carraher & Schliemann, 2007), and QR can support students development of algebraic reasoning that extends beyond generalized arithmetic (Smith & Thompson, 2008). Covariational reasoning is essential for students' developing understanding of the mathematics of change (Carlson, Jacobs, Coe, Larsen, & Hsu, 2002).

QR and Multiplicative Reasoning

Researchers investigating students' reasoning in the early grades have made distinctions between the types of units of which a quantity could be composed (Behr, Harel, Post, & Lesh, 1994; Steffe, 1992, 1994). Two such types of units are single units and composite units. Composite units are entities composed of single units (Steffe, 1994). For example, a tower composed of four unit cubes represents a composite unit because it is an entity composed of single units (unit cubes). To reason multiplicatively, an individual needs to be able to operate on composite units (Behr et al., 1994). For example, consider a situation involving seven towers containing four cubes in each tower. An individual reasoning multiplicatively could operate on composite units (towers) to determine the total number of cubes needed to create seven towers with four cubes in each (cf. Tzur et al., 2013). Individuals operating on composite units in this way would be using QR because they would be envisioning towers as "things" that could be measured by the number of unit cubes contained in each tower. As illustrated by this example, a strong foundation of QR could lay groundwork for operations on composite units, which are at the heart of multiplicative reasoning.

QR and Algebraic Reasoning

QR can support algebraic reasoning that may not depend on symbolic expressions and help students from the early grades through post-secondary to make sense of symbolic expressions in terms of related quantities (Madison & Dingman, 2010; Smith & Thompson, 2008). For example, an individual making sense of the bottle problem might realize that for a wide bottle, the height of water would change more slowly in relation to the increasing amount of water than would the height for a narrow bottle, thereby allowing that individual to make sense of relationships represented by symbolic expressions. To be clear, symbol manipulation is not mutually exclusive from QR. Rather, QR can provide a meaningful basis from which generalizations could emerge and then need to be represented with abstract symbols (Smith & Thompson, 2008).

QR and Covariational Reasoning

Covariational reasoning involves reasoning about quantities that change together (Carlson et al., 2002), thereby extending QR to include forming and interpreting relationships between quantities (e.g., the height and amount of water in the bottle filling situation). Covariational reasoning can take different forms, depending on how an individual envisions the ways in which the quantities change together. Two different forms of covariational reasoning are chunky and smooth (Castillo-Garsow, Johnson, & Moore, 2013). Chunky covariational reasoning involves envisioning change as having happened (e.g., envisioning a bottle being filled in two-ounce increments, then comparing the change in the height of the liquid in each of those increments). In contrast, smooth covariational reasoning involves envisioning change as continuing (e.g., envisioning a bottle being filled from a soda dispenser and coordinating change in the height of the liquid with continuing change in the volume of the liquid). Envisioning change as continuing is more powerful than envisioning change as having occurred (Carlos Castillo-Garsow et al. 2013), in part because it affords the consideration of variation in the intensity of a change (Johnson, 2012), and is involved in robust quantification of rate of change (Johnson, 2015a). Thus, QR could provide a meaningful foundation on which the mathematics of change could be developed.

A Role of Context in Tasks Fostering Students' QR

Tasks designed to foster students' QR have incorporated situations that have realistic contexts. As such, asking whether or not a student is familiar with a context seems a natural question. However, students' familiarity with the situation described by a task is only one aspect of the multifaceted interactions that can occur as students work on a task. When referring to context in this chapter, a dynamic stance is taken, meaning that context is not something external to an individual, but rather that context "constitutes an individual's conception of a given situation" (Johnson, 2014, p. 339). For example, two students encountering the filling bottle problem could have different conceptions of the given situation, depending on how they conceived of quantities involved.

A dynamic stance on context proved useful for the analysis of secondary students' work on an adapted version of the bottle problem (Johnson, 2012, 2015a, 2015b). The adapted bottle problem incorporated a graph that represented the amount of liquid in the bottle as a function of the height of the liquid in the bottle. Given the graph, students were asked to discuss how the volume of liquid in the bottle was changing as the height of the liquid in the bottle was increasing. Students had different conceptions about what the filling bottle problem entailed. Some students made and interpreted relationships between the varying quantities of volume and height. Two distinct types of relationships include (1) making comparisons between associated amounts of volume and height (Johnson, 2015b) and (2) coordinating changes in volume with continuing change in height (Johnson, 2012).

A key distinction between the relationships lay in how students formed and interpreted relationships between the changing quantities—comparison or coordination (Johnson, 2015a). Students making Type 1 relationships compared amounts of change in volume and height that *occurred* on particular intervals (chunky covariational reasoning). In contrast, the student making Type 2 relationships coordinated change in volume *continuing* change in height (smooth covariational reasoning). Despite the bottle problem's intentional focus on the quantities of height and volume of liquid in the bottle, students might consider other quantities or relationships. Consider this student's response: "I'm thinking as the water bottle, you are filling it up and as the rim comes, you can hear it filling up faster" (Johnson, 2010). Although this student was working on the same printed task as students forming and interpreting relationships between height and amount of water, she drew on her auditory experiences related to filling a bottle as she wrestled with the problem of actually filling a bottle. Taken together, these student responses suggest that a student's conception of a situation represented by a task might afford or constrain his or her QR.

Students' perspectives of quantities involved in a task may explain in part why students may not seem to transfer knowledge to a novel task situation (Lobato & Siebert, 2002). Using a task requiring students to investigate the steepness of a ramp, Lobato and Siebert (2002) identified multiple quantities that a student could use to quantify the steepness of a ramp. For a student to interpret that constructing a ratio between height and length of a ramp would be an appropriate activity in this situation is neither transparent nor trivial. Because students working on the same printed task can interpret the task differently and potentially reason about different quantities, conceiving of the task context as a "textual cover story" does not account for how students might draw on QR in different task situations (Lobato & Siebert, 2002).

Working with undergraduate precalculus students, Moore and Carlson (2012) found that students' images of a problem's context were not static. When solving a problem involving constructing an open box from a sheet of paper, students' interpretations of the work to be carried out did not necessarily align with the researchers' interpretations of the work to be carried out. During interactions with a researcher, students adjusted their image of the problem's context, and their emerging images influenced their QR (Moore & Carlson, 2012). Taken together, the findings of Johnson, Lobato and Siebert, and Moore and Carlson suggest that students' perspective of the task context is nontrivial and may influence QR.

Learning Trajectories Involving QR

Beginning with Simon's (1995) articulation of a hypothetical learning trajectory, learning trajectories (LTs) have become more pervasive in mathematics education and can be interpreted in different ways (Daro, Mosher, & Corcoran, 2011). Here, when referring to LTs, Clements and Sarama's (2004) definition is used:

> descriptions of children's thinking and learning in a specific mathematical domain and a related, conjectured route through a set of instructional tasks designed to engender those mental processes or actions hypothesized to move children through a developmental progression of levels of thinking, created with the intent of supporting children's achievement of specific goals in that mathematical domain (p. 83).

In this section, aspects of LTs (or groundwork supporting the development of a LT) related to geometric measurement in the early grades, multiplicative reasoning in the intermediate grades, and exponential and quadratic function in the middle-early secondary grades are addressed. In doing so, the goal is to illustrate how QR can play a role in students' understanding of K-12 mathematics.

Geometric Measurement: LTs

Sarama and Clements (2009) developed LTs on how early elementary students might develop understanding of length, area, and volume measurement. QR is rooted in the LTs; each LT begins with students' identification of length, area, or volume as an attribute of an object that could be measured and then indicates how students could develop the ability to make comparisons between amounts of length, area, or volume. Importantly, it is only after students engage in quantification that the LTs address how students might begin to determine particular amounts of length, area, or volume (measurement). Notably, to determine particular amounts of length, area, or volume, first students would need

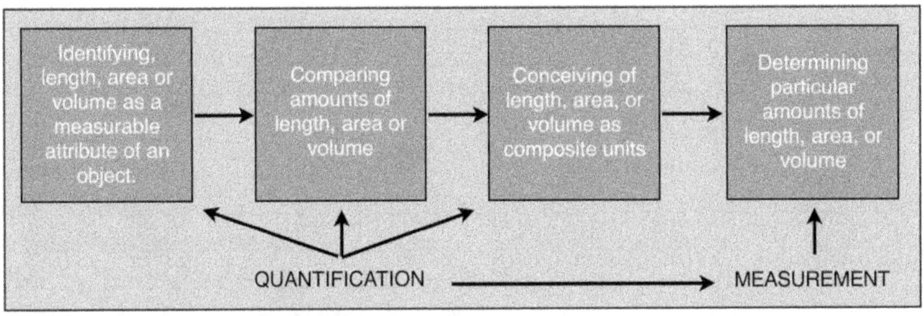

FIGURE 10.2 Sequence of operations for learning trajectories related to length, area, and volume measurement (Sarama & Clements, 2009)

to conceive of length, area, or volume as composite units. Figure 10.2 illustrates a general sequence of the operations for LTs related to length, area, and volume measurement.

Figure 10.2 illustrates how the LTs are rooted in students' quantification of length, area, and volume. The identifying portion of each LT involves determining what it means to measure length, area, or volume. By comparing amounts of length, area, or volume, a student could begin to determine what one might measure to determine an amount of length, area, or volume. Conceiving of length, area, or volume as a composite unit is necessary for determining what a measured amount of length, area, or volume might mean. Notably, measurement—involving the determination of particular amounts of length, area, or volume—occurs after quantification.

Multiplicative Reasoning: Laying Groundwork for LTs

Tzur et al. (2013) proposed a set of six schemes (von Glasersfeld, 1995) and operations (Piaget, 1970) related to students' development of multiplicative reasoning with whole numbers. Schemes are goal directed mental activities that an individual could enact in anticipation of a particular result (von Glasersfeld, 1995). The schemes and operations build on students' envisioning of whole numbers as composite units. As students progress through the schemes, they develop an ability to operate on composite units. The set of schemes indicates students' operations on composite units would develop as follows: (1) coordinating equally sized composite units, (2) additively operating on composite units, (3) comparing collections of differently sized composite units, (4) coordinating additive operations on composite units with a segmenting operation on a group of single units (5) segmenting a set of objects into groups of equally sized composite units (determining the number of groups), then (6) partitioning a set of objects into groups of equally sized composite units (determining the amount in each group) (Tzur et al. 2013).

Tzur et al.'s (2013) set of schemes is rooted in students' quantification of composite units. By developing the operations involved in this set of schemes, students could interpret whole number multiplication problems in terms of quantity. For example, the problem 7×6 could be envisioned as a unit composed of 7 single units replicated 6 times. This is not to be confused with an image of multiplication as repeated addition. Conceiving of 7 single units replicated 6 times would require an individual to conceive of 3 different units: the number of single units in a composite unit (7), the number of composite units (6), and the total number of single units (42). In contrast, repeatedly adding 7 would not necessarily require a student to consider anything other than single units. A purpose of this example is to communicate the central role of quantity in envisioning operations on composite units. In addition, the example highlights a critical distinction between multiplicative and additive reasoning: multiplicative reasoning requires a transformation of units, while additive reasoning involves operation with the same kind of units.

Quadratic and Exponential Function: LTs

QR is an important aspect of students' reasoning about functions as relationships (Ellis, 2011a; Lobato, Hohensee, Rhodehamel, & Diamond, 2012). Recently, groundwork for a LT rooted in QR has been accomplished for quadratic functions (Lobato et al., 2012), and a LT rooted in QR has been proposed for exponential functions (Ellis et al., 2013).

Lobato et al. (2012) investigated 24 eighth grade students' quantitative reasoning related to quadratic function, with the goal of supporting students' understanding that the rate of change in a quadratic function is constant. Students participated in 15 hours of classroom instruction in a small group setting of 7 to 9 students and in an individual follow up task-based interview. Results of the study include 5 pivotal intermediate conceptions (PICs) related to quadratic function. PICs build on students' conception of quantities. The first PIC is: "Students comprehend a quadratic motion situation as containing a set of elapsed distances as quantities—entities that can be mentally operated upon (e.g., compared), interpreted in terms of their meaning in context, and assigned correct units of measure" (p. 93). Lobato et al. argued that conceiving of a quadratic situation in terms of quantities was critical for students' progress in the conceptual domain of quadratic function.

Findings reported by Ellis (2011b) provide evidence to support Lobato et al.'s argument. Using the teaching experiment methodology (Steffe & Thompson, 2000), Ellis conducted a study with 6 middle school students. Given a task involving a table of values of side length and area for a growing square with dimensions 1×1, 2×2, 3×3, 4×4, 5×5, and 6×6, students determined first and second differences. During a class discussion, a student used another student's diagram of a growing square to make sense of the second difference as the increase in the number of new squares being added to the growing square as the side length increased by one unit. Particularly powerful was the student's ability to relate the amount of growth in new unit squares being added to the growing square to the amount of growth in the side length. Interestingly, the student was not able to determine this relationship prior to using the diagram. By interpreting the diagram in terms of the quantities involved in the growing square (side length and unit squares), the student could determine a relationship between amounts of change in quantities.

Recently, Ellis et al. (2013) proposed a LT related to exponential function, resulting from a teaching experiment study with three middle school students. The LT indicates three types of reasoning involved in developing understanding of exponential function: pre-function, covariation, and correspondence, with the development of the different types of reasoning being not entirely sequential. Although Ellis et al. indicate that pre-functional reasoning develops prior to covariation and correspondence reasoning, they found that the covariation and correspondence reasoning developed in a non-linear fashion, and students could shift between covariation and correspondence forms of reasoning to make sense of situations involving exponential function. Central to the study were task situations involving a growing plant. Like the students interpreting the growing square reported in Ellis (2011b), these students interpreted numerical amounts in terms of measurable attributes of the growing plant. Together, this collection of LTs provides some indication of how QR might foster K-12 students' development of important mathematical concepts.

Educational Policy Documents and QR

Principles and Standards for School Mathematics

Over a decade ago, the National Council of Teachers of Mathematics (NCTM) published *Principles and Standards for School Mathematics* (PSSM) (National Council of Teachers of Mathematics, 2000). The *PSSM* organizes the standards into standards for mathematical content (Number & Operations, Algebra, Geometry, Measurement, and Data Analysis & Probability) and standards for mathematical process (Problem Solving, Reasoning & Proof, Communication, Connections, and Representations). Quantitative relationships are included as one of the Algebra Standards to be addressed in K-12: "Use mathematical models to represent and understand quantitative relationships" (p. 90). Expectations related to this standard included modeling problems using multiple representations (K–8) and identifying quantitative relationships in function situations (9–12). Quantity appears in expectations at the 6–8 level in the Number and Operation Standard: "understand and use ratios and proportions to represent quantitative relationships" (p. 214) and the Algebra Standard: "use graphs to analyze the nature of changes in quantities in linear relationships" (p. 222). Although quantity and quantitative relationships were mentioned, they were not central to the standards. Furthermore, QR is not included in a later publication that addressed high school students' mathematical reasoning: *Focus in High School Mathematics: Reasoning and Sense Making* (NCTM, 2009).

Common Core State Standards for Mathematics

The *Common Core State Standards for Mathematics* (*CCSSM*) include standards related to quantity and QR (National Governors Association Center for Best Practices & Council of Chief State School Officers, 2010). The *CCSSM* is organized into standards for mathematical practice and standards for mathematical content, with "Reason abstractly and quantitatively" being one of the eight standards for mathematical practice. The standards for mathematical content are organized by grade level for K–8, then by conceptual categories for grades 9–12. Number and quantity is one of the conceptual categories for grades 9–12. In addition, quantities are related to number and operation in grades K–5, ratio and proportion in grades 6–8, and algebra and function in grades 8–12. Despite a greater inclusion of quantity and QR than *PSSM*, the *CCSSM* define quantities as "numbers with units," and do not distinguish between quantification and measurement. Below are the *CCSSM's* definitions for quantity, QR, and quantification:

- *Quantity*. "In real world problems, the answers are usually not numbers but quantities: numbers with units, which involves measurement" (p. 58).
- *QR*. "Quantitative reasoning entails habits of creating a coherent representation of the problem at hand; considering the units involved; attending to the meaning of quantities, not just how to compute them; and knowing and flexibly using different properties of operations and objects" (p. 6).
- *Quantification*. "In high school, students encounter a wider variety of units in modeling. . . . They also encounter novel situations in which they themselves must conceive the attributes of interest. For example, to find a good measure of overall highway safety, they might propose measures such as fatalities per year, fatalities per year per driver, or fatalities per vehicle-mile traveled. Such a conceptual process is sometimes called quantification" (p. 58).

As argued previously, defining quantity as a pairing of a number and unit conflates number and quantity. With such a definition, quantities do not arise without numbers. However, quantities such as length or area can be envisioned without determining particular numerical amounts. Although the *CCSSM's* definition for QR distinguishes the meaning of quantities from the computation of quantities, the standards do not adequately explicate how such distinctions may be developed. Despite the *CCSSM's* example of finding an appropriate measure as part of the conceptual process of quantification, more elaboration is needed to address the scope of quantification, including attributes one would measure to arrive at those given measures, and what the measures would mean in terms of the situation. Although the *CCSSM* includes QR, they do not emphasize the forming and interpreting relationships between quantities that change (covariational reasoning), which limits the scope of QR as a coherent thread supporting students' mathematical meaning making (Thompson & Carlson, in press).

Educational Resources Connecting Research to Practice

Connecting to the CCSSM's Standards for Mathematical Practice

In 2013, NCTM published *Connecting the NCTM Process Standards & the CCSSM Practices*, which is intended to provide a "roadmap to help teachers navigate these practices" (Koestler, Felton, Bieda, & Otten, 2013, p. ix). In unpacking the mathematical practice of reasoning quantitatively and abstractly, Koestler et al. explicitly indicate that students can reason quantitatively without assigning specific measurements to those quantities. This is a useful step for supporting teachers' making sense of quantities as more than pairing numbers with units. Drawing on research reported in Johnson (2012) and tasks described by van Dyke and Tomback (2005), Koestler et al. argue that a common theme is "the clear emphasis on the relationships among the various quantities *before* turning attention to the specific numerical values" (p. 26). Importantly, this publication begins to elaborate teaching practices that could provide students opportunities to reason about quantities as more than just pairing a number and a unit.

NCTM'S Essential Understanding Series

NCTM's *Essential Understanding* (*EU*) series, edited by Zbiek, has a twofold purpose: to help teachers (1) develop their mathematical knowledge and (2) support their students' development of big ideas of mathematics. The *EU* series addresses mathematical topics PreK–12 that are difficult to learn and challenging to teach, and identifies big ideas and essential understandings for the different topics. The big ideas relate to the areas of mathematics, and the essential understandings provide specifics to support the big ideas. The discussion that follows addresses a role of quantity and QR in the *EU* series across PreK–12.

PreK–2: number and numeration

One big idea defines number as an extension of quantity: "*Number is an extension of more basic ideas about relationships between quantities*" (Dougherty, Flores, Louis, & Sophian, 2010, p. 10). One essential understanding related to this big idea involves comparing quantities without assigning particular numerical amounts. Another essential understanding related to this big idea distinguishes quantity from physical objects: *"Physical objects are not in themselves quantities. All quantitative comparisons involve selecting particular attributes of objects or materials to compare"* (p. 11). By focusing on quantities as attributes of objects (not objects themselves) and distinguishing quantities from particular numerical amounts, Dougherty et al. (2010) highlight a central role for quantity in PreK–2.

3–5: algebraic thinking

One big idea explicitly involves QR: "*Quantitative reasoning extends relationships between and among quantities to describe and generalize relationships among these quantities*" (Blanton, Levi, Crites, & Dougherty, 2011, p. 39). One essential understanding related to QR addresses the use of comparison to make relationships between quantities. Examples indicate how comparisons can be done without assigning numerical values to quantities or determining particular amounts of quantities. Another big idea relates quantity and functional thinking: "Functional thinking includes generalizing relationships between covarying quantities . . ." (p. 47). An essential understanding related to functional thinking explicitly describes functions as tools for expressing covariation: *"Functions can be viewed as tools for expressing covariation between two quantities"* (p. 48). Researchers have argued that a covariation perspective on function is a more intuitive approach to function than a correspondence approach (Confrey & Smith, 1994, 1995; Thompson, 1994b). Blanton et al. indicate how such a view of function can be developed in the early grades.

6–8: ratio, proportion, and proportional reasoning

Two essential understandings underlying the big idea of proportionality articulate how understanding ratio involves coordinating covarying quantities and making multiplicative comparisons between covarying quantities (Lobato & Ellis, 2010). Lobato and Ellis provide examples of tasks and questions that focus on these essential understandings. For instance, students could additively or multiplicatively compare amounts of orange juice and water in different mixtures. To illustrate, consider the sample question provided by Lobato and Ellis:

> Does a batch of orange juice made with 2 cans of orange concentrate and 3 cans of water taste equally orangey, more orangey, or less orangey than a batch made with 4 cans of orange concentrate and 6 cans of water? (p. 22)

A student reasoning additively might determine that the first batch has 1 more can of water than orange concentrate, while the second batch has 2 more cans of water than orange concentrate. The student could use this to conclude (incorrectly) that the first batch is more orangey, because there is less difference between the amounts of water and orange concentrate. In contrast, a student reasoning multiplicatively might determine that doubling the first batch would result in the second batch or that both batches have 1½ times more water than orange concentrate, meaning that both batches would taste the same. These examples illustrate the importance of how students make relationships between amounts of quantities—to reason meaningfully about ratios, students need to make multiplicative rather than additive comparisons.

9–12: functions

One big idea characterizes functions in terms of relationships between quantities: *"Functions provide a means to describe how related quantities vary together. We can classify, predict, and characterize various kinds of relationships by attending to the rate at which one quantity varies with respect to another"* (Cooney, Beckmann, & Lloyd, 2010, p. 23). Here the big idea indicates a covariation perspective on function. They distinguish this perspective from a correspondence perspective, arguing for the usefulness of a covariation perspective for making sense of the mathematics of change. Cooney et al. (2010) identify a rate of change of a function as providing a "mechanism to describe and quantify the covariation between two variables" (p. 25). Consequently, the covariation of variables is an attribute quantified by a rate. Envisioning rate as quantification of covariation of variables is far more powerful than envisioning rate as a procedure to be executed based on a formula. By forming and interpreting relationships between quantities, students could develop a need for different types of functions, such that different types would indicate different kinds of relationships between quantities.

Research-based Tasks Focusing on QR

Researchers working at the elementary (Tzur et al. 2013), secondary (Johnson, 2013), and post-secondary (Dingman & Madison, 2010) levels have designed and implemented mathematical tasks having the potential to engender students' QR. The intent of this section is not to provide an exhaustive list of tasks focusing on QR, but rather to highlight examples of tasks fostering students' QR in K–16 mathematics education.

Elementary: Please Go Bring for Me (PGBM)

Tzur designed a platform game, Please Go Bring for Me (PGBM), to be used for the purpose of developing elementary students' multiplicative reasoning (Tzur et al., 2013). PGBM requires students to engage in coordinated counting activities, beginning with physical manipulatives, then drawn representations of those manipulatives, followed by symbols representing those manipulatives.

The game is played in pairs with one student playing the role of "sender" and the other playing the role of "bringer." The game begins with the sender asking the bringer to produce an amount of equally sized towers composed of cubes (e.g., 4 towers containing 3 cubes each). The sender then asks the bringer four key questions: "(1) How many towers did you bring? (2) How many cubes are in each tower? (3) How many cubes are there in all? (4) How did you figure it out?" (Tzur et al., 2013, p. 88). The questions are central to the task because they focus on the different types of units involved in the PGBM game: the amount of towers, the amount of cubes per tower, and the total amount of cubes. When students can play the PGBM game proficiently with manipulatives, a teacher would introduce sequential constraints including: covering physical manipulatives, then drawing pictorial representations of physical manipulatives, then using symbols rather than pictorial representations. Results from research conducted by Tzur, Johnson, McClintock, and Risley (2012) suggest that the strategic use of representations involved in PGBM supported students' ability to operate on composite units, and hence provided students opportunities to develop their multiplicative reasoning. Such use of representations is grounded in QR, because it fosters students' envisioning of towers as "things" that could be measured by the number of unit cubes contained in each tower.

Secondary: A Filling Triangle

Using Geometer's Sketchpad Software (Jackiw, 2009), Johnson (2013) developed dynamic, interactive computer environments suitable for middle school students. One environment linked a pictorial representation of a right triangle—filling with area—with a dynamic graph relating the changing quantities of area and height (see Figure 10.3). To vary the area and height of the triangle, students could click and drag on point D (see Figure 10.3) or press the action button Animate Point D. If a student pressed animate point when AD was greater than zero, the area would "fill" until point D reached the maximum height of the triangle, then reset AD to zero and began filling area again until the initial length of AD was reached.

By linking dynamic representations of geometric objects with dynamic graphical representations, the goal was to provide students working on the task with the opportunity to interpret graphical representations in terms of related quantities. Animations were designed to continually "fill" the objects with area. It is known that continuous covariational reasoning is important for students to develop (Thompson, 2008, 2011), and even successful mathematics students working at advanced levels may not necessarily engage in continuous covariational reasoning (Carlson et al. 2002). However, research has shown that secondary students can engage in continuous covariational reasoning (Castillo-Garsow, 2013; Johnson, 2012). Tasks requiring students to make predictions about relationships between covarying quantities prior to viewing an animation may provide opportunities for

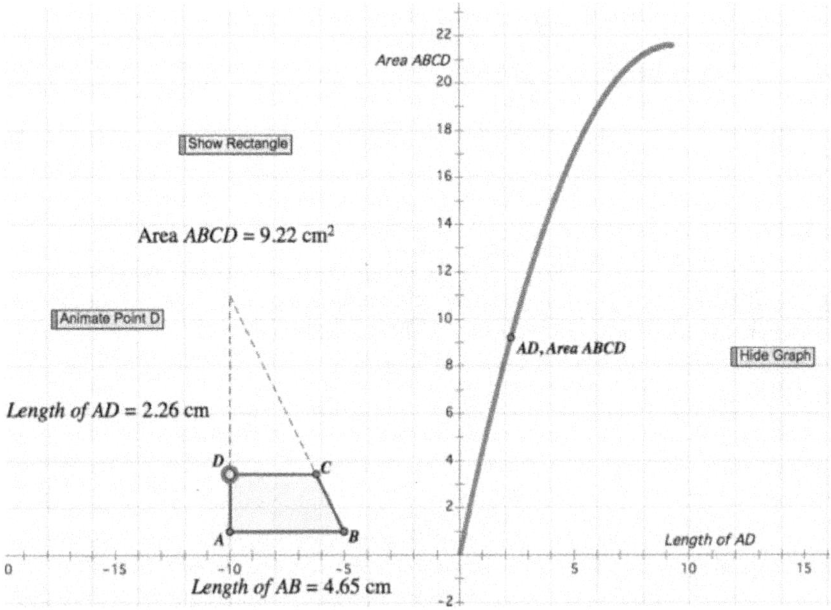

FIGURE 10.3 Filling triangle environment

middle grades students to engage in continuous covariational reasoning (Johnson, 2013). Yet, more needs to be known regarding how students might develop such reasoning.

Post-Secondary

Researchers at the University of Arkansas (Dingman & Madison, 2010) designed and implemented a QR-focused mathematics course titled Quantitative Reasoning in the Contemporary World (QRCW). QRCW was implemented on a pilot basis in 2004, and by 2009 QRCW was being implemented to nearly 500 students. QRCW incorporated contextually rich tasks requiring students to reason about mathematical ideas to make sense of the situations. One such task required students to compare the fuel efficiency of a gasoline and hybrid car. After engaging in comparison, students were then required to respond to a newspaper article arguing that the cost of creating more fuel-efficient cars could outweigh savings in fuel efficiency. Based on their interactions with students in QRCW, Dingman and Madison argued that students' experiences in secondary mathematics did not seem to engender the kind of reasoning required for tasks used in QRCW. Were elementary and secondary students to have more opportunities to engage in QR through the kinds of tasks described in this chapter, they could be better prepared to engage in the QR necessary for tasks used in QRCW.

Discussion: Connections to Introduction

In the Introduction, Duschl, Bismack, Greeno, & Gitomer (this volume) argued that QR could serve as a link between mathematics and science education. Although Duschl et al.'s argument addressed only QR, it could be extended to include both QR and QL. Mayes et al.'s (2013) definition of QL, which stated that QL results from quantification, provides an example of how QR and QL could be integrated to build bridges between mathematics and science.

During the QR Working Group of the International STEM Research Symposium, Duschl proposed that a focus on measurement could forge relationships between mathematics and science. While Duschl's proposition included only measurement, it could be extended to include the reflexively related constructs of measurement and quantification. This is to say that quantification precedes measurement, and subsequent measurement could support further quantification. Importantly, working from results of measurement may or may not incorporate the process of quantification. Thompson (2011) argued that expanding a definition of quantification could bridge mathematics and science education. Mayes et al.'s (2013) inclusion of quantification as part of a framework for QR in science education provides a beginning example that could support such bridging.

To conclude, distinctions between QR and QL could have implications for the design and implementation of tasks focused on QR. Moving too quickly into computation of amounts of quantities (or measuring particular amounts of quantity) can result in too little attention to the process of quantification and too much attention to the interpretation of the results of measurement. During the QR Working Group of the International STEM Research Symposium, the group did not come to consensus regarding definitions of QR or QL. A focus on quantification and measurement as salient aspects of QR and QL could provide a way to forge links between mathematics and science education in this area.[2]

Notes

1. This problem situation is derived from the well-known bottle problem developed by the University of Nottingham's Shell Centre (Shell Centre for Mathematical Education (University of Nottingham), 1985).
2. I would like to thank Hilda Borko, Richard Lehrer, and Robert Mayes for their thoughtful comments related to a prior version of this chapter. I am grateful for the insights that resulted from the conversations.

11

TEACHERS' USE OF DATA, MEASUREMENT, AND DATA MODELING IN QUANTITATIVE REASONING[1]

Anthony J. Petrosino

Introduction

This chapter is a contribution to the Summit session on Quantitative Reasoning and is situated in the broader Summit goals regarding the redesign of K-12 education for alignment of curriculum-instruction-assessment; STEM teacher education; post-secondary STEM education; and the role of informal STEM education on the publics' engagement.

The chapter begins with an overview of the role of data modeling in quantitative reasoning with an emphasis on data variability as a context for designing meaningful investigative activities in K-12 STEM classrooms. Next, key components of an instructional sequence are presented with a focus on scaffolding measurement, variability, and modeling in the pursuit of designing and implementing experiment. Then the chapter shifts from students' engagements to preservice teacher education and professional development. Here the focus is a discussion of attempts on how to take reform pedagogies to a larger scale. The chapter finishes with a discussion that examines how quantification and measurement could forge links between integrating science and mathematics education.

Background

Data Modeling

The *Framework for K-12 Science Education* (NRC, 2012) draws on research from the two paradigms researching science and engineering practices science studies and learning sciences. Moreover, a number of other recent reports have recommendations that are based on an established and ever growing body of research on how people problem-solve, design, reason, and learn in disciplinary contexts

(NRC, 2007, 2009, 2012). Such consensus notions of scientific practices and of how understanding develops are reflected in the 3 dimensional design of *Next Generation Science Standards: By States, for States* (NGSS Lead States, 2013).

We all seem to have an inborn disposition to classify and generalize. Such abilities are foundational in our ability to comprehend and understand the world around us (Medin, 1989). Nonetheless, no matter how much we describe objects and events, their variability is at least as important as their similarity. This point is emphasized by Gould (1996) who noted that modeling natural systems or "full houses" (such as the disappearance of 400 hitters in baseball or Cope's Rule of increasing size) often requires taking into account the variability in the system. (Cope's Rule or Cope's Law is named after American paleontologist Edward Cope and postulates that population lineages tend to increase in body size over evolutionary time.) One example now widely familiar from both the popular and professional press (e.g., Weiner, 1994) is the account of how small variations in beak morphology led to dramatic shifts in the population proportions of various kinds of Galapagos finches. Moreover, these shifts occurred as the environment fluctuated over relatively brief decadal periods of time. From this perspective, understanding variability between and within organisms and species is fundamental to understanding a big idea like "diversity" (NRC, 1996).

Yet, given its importance in understanding the world around us, variability receives little attention in school instruction or in teacher professional development. Learners are given few conceptual tools to reason about variability, and even if they are provided, the tools are often rudimentary. Typically, these tools consist of brief exposure to a few formulas or algorithms (e.g., for calculating the mean or standard deviation), with little to no emphasis on the bigger picture of data modeling. More specifically, students do not get opportunities to engage in or be exposed to data modeling contexts that allow them to develop their own questions, consider qualities of measures and attributes relevant to a specific question or context, and then go on to represent data and make inferences about their original questions (Lehrer & Romberg, 1996).

Superficial Enactments of Inquiry

When students do not have the skills, language, or tools for reasoning about distribution and variability, they find it difficult to cognitively move beyond superficial enactments of inquiry or investigative activities. Unfortunately, such superficial enactments dominate much of the current practices in general education and in STEM education specifically.

Consider a hypothetical investigation about model rocket design. Students spend some time looking at various components or key design features of rockets. They narrow in on a few design attributes such as the number of fins, the material used, or the shape of the nose cone. After some discussion, they decide to focus on nose cone design. In a short period of time, they launch several model rockets that have rounded nose cones and measure the height of each rocket as it reaches apogee. Next the students measure the launch heights of a second group of rockets with pointed nose cones. The children find, to little surprise, that the heights are not identical. Yet, what do these results mean? How different must the results of the rounded and pointed cone model rockets be in order to permit a confident conclusion that the nose cone shape matters? In short, *when is a difference really a difference*?

Unfortunately, much of current science instruction falls short in providing students with the tools to push beyond simple comparison of outcomes (Chinn & Malhotra, 2002). If the numbers are simply different (in this case, the heights at apogee), the treatments (rounded and pointed cones) are assumed to differ in an important way, a conclusion that no practicing scientist or engineer would comfortably support without some analysis of other possible sources of variability for the heights at apogee.

Because most students and teachers do not have the skills, language, or tools for effectively leveraging ideas about sampling, distribution, or variability, teachers have little choice but to employ the use of overly simple "investigations" or "hands-on" activities of very robust phenomena that are already well understood. A typical outcome of this approach is that students spend their time "investigating" a predetermined question posed by the teacher (or the textbook), often following cookbook like procedures that are defined well in advance. This kind of science education undertaking shares some superficial components of the kind of "inquiry" valued in the science standards, but falls well short on its underlying purposes or epistemology (Duschl, 2008). Consequently, many teachers pursue "canned" investigations not for lack of enthusiasm for this kind of genuine inquiry, but because more is needed than simply encouraging students to develop and investigate their own questions. What is required is that students and teachers alike have the requisite tools for interpreting the results in a disciplinary honest manner.

For example, students rarely participate and teachers do not necessarily receive training in how to effectively scaffold the creation, evaluation, and reflective revision of questions about phenomena of personal interest. Moreover, students often do not puzzle through the challenges or see the problematic issues related to instrumentation, or what Pickering (1995) described as achieving a "mechanic grip" on the world—by deciding how to parse the world into measurable attributes (heights of model rockets) and then coming to agreement about best ways of measuring them (standard conditions, measurement devices, units of measure).

Instruction in Variation and Measure

Some initial science education work on data modeling drew from the historical ontogeny of distribution, which suggests that reasoning about error variation may provide a reasonable grounding for instruction (Konold & Pollatsek, 2002; Porter, 1986; Stigler, 1986), and from previous research with students who were modeling density of various substances (Lehrer & Schauble, 2000; Lehrer, Schauble, & Petrosino, 2001). In this work, students obtained multiple measurements of weights and volumes of a collection of objects, and discovered that measurements varied from person-to-person and also by method of estimation. For example, students found the volumes of spheres with water displacement, the volume of cylinders as a product of height and an estimate of the area of the base, and the volume of rectangular prisms as a product of lengths. Differences in the relative precision of these estimates were readily apparent to students (e.g., measures obtained by water displacement were much more variable than those obtained by product of lengths). Also evident were differences in measurements across measurers (e.g., one student's estimate of one-third of a unit of area might be taken as ó of a unit by a different student). Students estimated true values of each object's weight and volume in light of variation by recourse to an invented procedure analogous to a trimmed mean (extreme values were eliminated and the mean of the remaining measures was found as the "best guess" about its value). This work showed clearly that the contexts of measurement afforded ready interpretation of the center of a distribution as an indicator of the value of an attribute and variation as an indicator of measurement process. However, further investigation was needed in order to engage students in examining the structure of variation (e.g., its potential symmetry), and to investigate sources of variation other than person and instrument.

Data Modeling and Experiment

In Petrosino, Lehrer, and Schauble (2003), a teacher and his class of fourth graders worked with researchers to engage in coordinated cycles of planning and implementing instruction to put these tools in place, and studying the forms of student thinking—both resources and barriers—that emerged in the classroom. During the 8-week design study (Petrosino et al. 2003), researchers engaged in making a broader scope of attribution about potential sources and mechanisms of variability in light of the processes involved in measuring. Students measured first the lengths of the school's flagpole and a pencil, and later the height of model rockets at the apex of their flight. These measures were selected for their potential for highlighting ideas about variability and structure. Multiple sources of random error were identified and investigated. To distinguish between random (wind, individual differences of measuring, etc.) and systematic variation (cone design), students conducted experiments on the design of the rockets (rounded vs. pointed nose cones).

The goal of the investigations was to determine whether the differences between the resulting distributions in heights of the rockets were consistent with random variation, or instead, if the difference could be attributed to the shape of the nose cone. During the investigations, students identified and discussed several potential contributions to measurement variability (including different procedures for measuring, precision of different measurement tools, and trial by trial variability). The expectation for a "true" measure served to organize students' interpretations of the variability that they observed.

Although the researchers could have employed a variety of indicators of variation, the choice of difference scores (the difference of a case from the median) facilitated a measurer-eye's view of deviation from center as an indicator of precision. The context and practice of measurement helped students appreciate the need to coordinate the distribution of differences and the original measures so that difference distributions could be more readily appreciated as a rescaling of the original distributions. For example, students expected that most measurements would be "a little over or a little under," and so had similar expectations about differences from the center.

Difference scores maintained emphasis on the case-to-aggregate relation, as opposed to other indicators of spread, which tend to obscure this relationship. For example, students were readily attracted to the range, but this statistic had the disadvantage of relying on the two most extreme measurement values—"bad" measurers. This ready coupling of spread with agency helped present the distributions as sensible, so that qualities like symmetry and a related aspect of shape of the distribution (i.e., skewness) came to be understood as anticipated outcomes of measurement processes (Lehrer & Schauble, 2007). This ready correspondence between process and characteristics of distribution is apt to be harder to maintain in other contexts. For example, what is the meaning of the difference between an individual's height and the average height in the population, or transitions in variation over time in such populations? Porter (1986) suggested that the leap from error at the case level to such "natural" variation among a population distribution represented a major advancement in statistical understanding. Gould (1996) and Mayr (2001) would agree.

Instruction and activities focusing on measurement distribution help set the stage for considering systematic effects (i.e., round vs. pointed nose cones) in the presence of random effects due to error. Like scientists, these fourth grade novices had to consider whether an outcome could be considered a treatment effect in light of multiple sources of error variation, such as variation due to trial-to-trial and individual differences. In short, these young novices had to engage in the kind of reasoning at the core of experimental testing, what Mayo (1996) calls "arguing from error" (p. 7) and decide when a difference was really a difference. It was only after careful consideration and characterization of these sources of variation that the learners moved to consider whether their conjecture could be sustained. Here students noticed that the upper tails of a distribution constructed by pooling both effects (launches with round and pointed nose cones) was populated exclusively by the less preferred and counter intuitive (rounded) alternative (see Figure 11.1).

Despite ideas about pointed cones "cutting" through the air, consideration of the joint distribution suggested, if anything, the opposite effect: Rounded

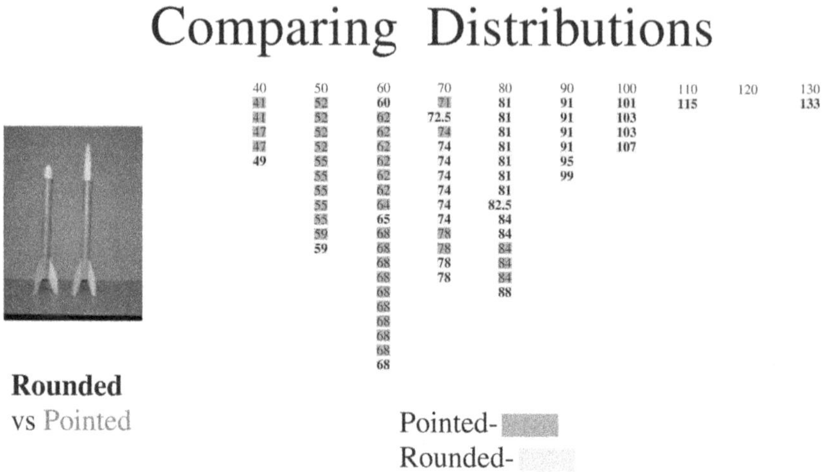

FIGURE 11.1 Comparison of rounded and pointed nose model rockets

nose cones might very well be superior. (The ideal nose shape for a rocket that minimizes its drag and maximizes its altitude depends on how fast the rocket is designed to travel. Most model rockets that you can buy in a store fly well below the speed of sound. As a result, the ideal nose for a model rocket is not the pointed cone but the more rounded parabola-shaped cone.) Although studies of scientific reasoning often offer compelling reason to believe that students of this age cannot readily coordinate theory and evidence (Kuhn, Black, Keselman, & Kaplan, 2000), I conjecture that such studies often do not provide students with the tools that professional practice suggests are important. In this case, armed with an emerging repertoire for reasoning about distribution, students proved capable of coordinating data with conjecture, even though the conclusion that they reached was one that they initially found "insensible." Their puzzling about this resulting "insensibility" set the stage for continued inquiry, again a hallmark of sound scientific practice.

Through a series of assessments and analysis, students in this fourth grade class outperformed fourth-graders in a national NAEP sample on items designed to assess interpretation of data from displays (tables and graphs) and performed comparably to seventh and eighth graders in an instructional study of students' statistical thinking in the areas of reasoning about experiment, using statistics to compare distributions, relations between precision and variability and comparing distributions (Petrosino et al. 2003).

Data Modeling and Natural Variation

In subsequent design research, Lehrer and Schauble (2004) worked with the same teacher in the fifth grade to model changing distributions of plant populations in studies of natural variation. In this and in subsequent related efforts, students constructed models of random and directed variation to make statistical inferences about the effects of different factors on plant growth. Statistical inference here was based on comparisons between simulated model behavior of statistics (an empirical approximation to a sampling distribution) and observed measures of plant growth in different conditions.

Studies like Petrosino et al. (2003) and Lehrer and Schauble (2004) take a fair amount of investment in resources to insure the teacher has the knowledge and skills needed to help learners in their developing data modeling abilities. In fact, informal professional development of the teacher was ongoing during the 8 weeks of the design experiment as was more longer term professional development with a larger group of teachers focused on their reflections of their own students mathematics and science investigations with data (Lehrer & Schauble, 2002). But if the vision of the NGSS or other reform efforts are to be realized, professional development (both in service and pre service) needs to be conducted more efficiently and perhaps as a more central commitment to formal teacher education. To that end, the following study was conducted in order to examine design principles and instructional sequences for professional development in the service of teachers data modeling. In this specific study, experienced teachers worked with researchers to engage in coordinated cycles of (a) planning and implementing instruction and (b) studying the forms of student thinking—both resources and barriers—that can and do emerge in the classroom. Although presented here as separate phases, these two forms of activity were, in fact, concurrent and coordinated. The development of teacher thinking was conducted in the context of particular forms of instruction; and in turn, instructional planning was continually guided by and locally contingent on what we (teachers and researchers) were learning about student thinking concerning aspects of data modeling.

Methodology

The methodology for this study was inspired by an approach called *design-based implementation research* (DBIR) (Penuel & Fishman, 2012). Primarily, DBIR utilizes design-based ideas and methods to study issues of implementation and theory building (see also Barab & Squire, 2004; Brown, 1992). DBIR presents educational researchers with a challenge to reduce barriers between disciplines of learning science research that isolate those who design and study innovations within real world classrooms from those who study the scale and diffusion of innovations. As a secondary feature, DBIR also aims to rethink the roles of researchers and practitioners in bringing about systemic change in practical ways for practitioners[2]. According to Penuel and Fishman (2012) four principles embody the approach: (1) A focus on persistent problems of practice from multiple stakeholders' perspectives, (2) a commitment to iterative, collaborative design, (3) a concern with developing theory and knowledge related to both classroom learning and implementation through systematic inquiry, and (4) a concern with developing capacity for sustaining change in systems.

Design-based implementation research shares a commitment with other forms of design research like *design experiments* (Cobb, Confrey, diSessa, Lehrer, & Schauble, 2003), which also attempt to build theoretical understanding by working in actual classrooms. DBIR best describes the current study due to the persistent problem of lack of discipline informed data modeling education for K-12 teachers, the collaboration with teachers to help inform the instructional sequence, an attempt to build knowledge about teacher understanding about data modeling, and an overall concern for implementation in potentially dozens (if not hundreds) of classrooms

Participants

Subjects for this study were inservice teachers from a highly selected nationally recognized teacher education program. The 21 teachers included 12 women and 9 men between the ages of 22 and 45. All were teachers with 2 to 20 years of experience teaching in a K-12 setting. Teachers came from 4 different states, were mostly secondary STEM teachers, and taught in 15 different school districts ranging in size from major metropolitan districts to small single school rural districts. During the intervention, teachers engaged in a number of different inquiry activities, including model rockets, toy helicopters, making ice cream, using meter sticks to conduct reaction time experiments, and comparing pulse rates. All activities were informed by the instructional sequence for data modeling which consisted of structuring variation in measure as distribution, accentuating precision of measure, instrumentation and spread, and putting distribution thinking in service of experiment (Petrosino et al. 2003).

Procedure

Interviews were conducted to probe teachers' conceptions of individual and collective forms of inquiry. Although we regularly observed teachers engaged in data-based argument, the interview process provided opportunities to learn how teachers were evaluating practices of data modeling. Finally, we wanted to know more about how teachers could reason about experiment as a form of data modeling.

The interview and activity addressed the following themes:

- Design of inquiry: How did students design to learn? Did they design experiments and understand the logic of experimentation? What role did they think replication plays in inference?
- Measurement: What role did measurement play in student inquiry? Were measures integral to questions or considered mere adjuncts?

- Deficit model of experimentation—can young students conduct meaningful investigations?
- Epistemology and the nature of science: How did students characterize their activity? What did they think of the roles and functions of the activity structure of the research meeting? How did they evaluate the challenges of extended, self-directed inquiry when contrasted to more packaged forms of inquiry, like the science kits that comprised their district's science instruction?

Analysis

Interviews were read by the team members and characterized teachers' positions with respect to the four central themes above. We focused the initial characterization of the four themes with a select sub-sample of 6 teachers. We then applied the emerging analytic framework to other interviews selected at random. The resulting adjusted scheme guided our interpretation of all the protocols. We characterized each protocol independently and resolved disagreements consensually.

Results and General Themes

In order to demonstrate the range of teacher performance on the interviews, we first summarized the responses from teachers. The range and forms of reasoning by these teachers were recorded and analyzed. We present findings for the entire sample.

Findings

Design of Inquiry

The primary mode of inquiry was experiment. A valid experiment provided some control of variable when appropriate, which was decided upon by investigating how the experiment was actualized in practice and by additional probes centering on the logic of the valid experiment. For example, Mr. A captured a typical explanation from an "anti cookbook" approach to inquiry:

> Teachers are sticking to experiments that are well understood because they believe students don't have the ability to understand sampling, distribution, or variability. When teachers pose a question and students follow procedures, there is no room for inquiry or further exploration of the topic.

This notion of understanding what inquiry is by stating what it is not was a consistent theme throughout the interviews. For instance, Mr. B compared the sentiment of many other teachers:

> When students investigate something using questions and steps that are already preplanned, they are not really learning through inquiry.

A number of other teachers were fairly self critical in terms of their own comfort levels with inquiry, investigations, and conducting authentic inquiry in their own classrooms. The notion of authentic inquiry seems to include a requirement of novelty or newness for the student as well as the teacher. Mr. C explains:

> Most teachers do not go past their comfort zone when it comes to teaching science, or conducting science activities. Students as well as teachers should conduct science experiments where both the teacher and students are learning something new as they go. Not where the teacher already has an idea of what is going to happen because she's performed the experiment numerous times.

This idea of newness as a preferred component for inquiry resonated with over 15 of the 21 teachers.

While some of these activities reminded the teachers of activities performed in class when they were in school or even activities they conducted as teachers, there was a cogent and noticeable contrasting case with the inquiry activities conducted as part of our activity. For instance, Mr. D explains:

> The experiments we conducted with our students led to inquiry and high levels of student engagement because they were able to experiment, collect data, and make assumptions based on their experiments. In high school and middle school, my science experiences were not like those mentioned in the article (in class). There was little application to the knowledge I gained and thus little information was retained.

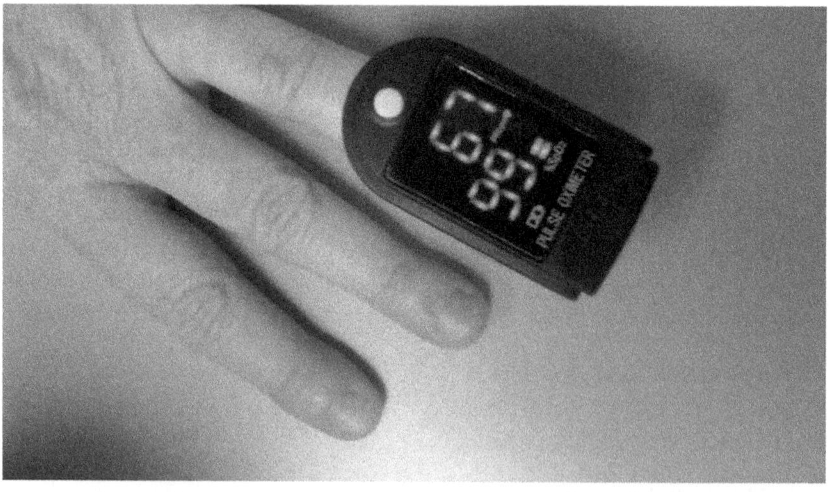

FIGURE 11.2 Teachers collected pulse rate data and dissolved oxygen levels

Measurement

All the teachers coordinated measure with the question. For example, if the length of the meter stick that was displaced during the reaction time activity was measured, teachers played a role in developing an answer to the inquiry at hand. Typically, as in Lehrer and Schauble (2004), measurement played a role in refinement of the inquiry question (e.g., does sight influence reaction time?).

Deficit Model of Experimentation

Conducting effective inquiry is more than heuristics and strategies that students (or teachers) need to learn; much needs to be mastered to understand the very enterprise itself (Lehrer et al. 2001). Teachers and students do not necessarily understand an experiment as a way of modeling, investigating, or manipulating a range of situations that are similar in theoretical and important ways. Some of these difficulties have led researchers to conclude that inquiry and experimentation is something that should be delayed or postponed for later in a student's career (Klahr, Fay, and Dunbar, 1993). But, with the proper scaffolding and support, teachers found it quite amazing what students were able to do in terms of inquiry activities. Mr. E explains:

> I was blown away from what the 4th grade students were conducting . . . the level of math they were utilizing was amazing. To see that students that were in 4th grade were able to do math that was meant for a middle school level. If we engage and have more in depth discussion with teachers being the guide not the answers, then students will amaze us beyond our expectations. Over time, as the teachers worked with actual students, there was a clear transformation that took place and a deficit model of student ability was slowly replaced by a more empowered model of student inquiry with the proper supports to conduct inquiry activities in a class setting.

Mr. F explains in some detail:

> It gives the teacher an opportunity to learn something new and along with the students become a scientist in search of answers, evidence, and data that can explain their claims or ways of thinking. By doing this, teachers are deviating from the idea that experiments have to be fun and replaced with experiments being valuable to the students and their education . . . During all our activities, the techniques we used to teach our students was mainly through the experimentation itself. The students were expected to form their claims and support them with evidence (argumentation).

Epistemology and Nature of Science

I examined the teachers' descriptions of science in the classroom, their impressions of what goes on in science classrooms, what should go on in science classrooms, and their responses to a contrasting case between extended inquiry and more constrained "cookbook" science often conducted in classrooms. I found near unanimity across the group of teachers. Teachers' descriptions of the day-to-day conduct of science in the classroom strongly conveyed themes of a text-based approach to science instruction and general lack of discipline informed enactments of activities. Mr. G explains this quite well:

> We need to use argumentation, representation, and modeling in our experiments to make them more relevant and engaging to the students so they can actually learn what it is they are doing and why they are doing it. We might believe the students see these experiments as representations of the real world when they actually don't.

Notions of hands-on activities persist with the teachers although with increasing articulation and sophistication. In this excerpt, we hear from Mr. H who explains the importance of hands-on, manipulation of the materials and activities of science class but with key components of data modeling and argumentation to portray a more discipline specific conception of inquiry and experiment:

> Hands-on-learning is student based inquiry. Students are interacting with materials, making observations, discussing, collecting and analyzing their own data, and creating their own hypotheses. When they test these hypotheses, they are able to confirm or reassess their hypothesis. True hands-on activities allow for kids to gather their own data to support their own ideas. In cases where evidence and data does not support their hypothesis, as happened with the model rockets, students change their ideas or misconceptions. The ability to interact with the materials makes the learning more memorable and effective, and either validates or alters their understanding.

Discussion: Connections to the Introduction

A common theme across the sessions at the Waterbury Summit was that in order to rethink K-12 STEM education in a manner that is consistent with the NGSS and *STEM Integration in K-12 Education: Status, Prospects, and an Agenda for Research* (NRC, 2014) there will need to be significant innovations in science teaching as well as professional development. In the Introduction, Duschl et al. (this volume) propose an adoption of a systems thinking approach to STEM education. As part of this rethinking, they propose model-based reasoning, systems thinking, and quantitative reasoning as interwoven core cognitive skills.

Systems thinking incorporates the capability to understand and articulate a system or its components with an ultimate goal of being able to make predictions about that system and other surrounding systems. In order to fully engage in systems thinking, a person must develop a model. This model will help facilitate collection, analysis, and representation of data as well as explain phenomena and make informed and reasoned predictions. When this model-based reasoning is joined with quantitative reasoning, a learner posses the tools needed in order to characterize parts and processes of systems. Duschl et al. suggest that we adopt a systems thinking approach to science. Within the framework they proposed, systems thinking, model-based reasoning, and quantitative reasoning are interwoven essentials. They characterize systems thinking as the ability to explain a system or its components, and then to predict the impacts the system will have on surrounding systems. To engage in systems thinking, a person develops a model, which can be used to collect, analyze, and represent data; explain phenomena; and predict outcomes. Model-based reasoning (Ward, this volume) and quantitative reasoning (Johnson, this volume) are the tools used to characterize components and processes within systems.

Perhaps most reverent to the emphasis of this chapter is that Duschl et al. emphasize that model-based reasoning is featured prominently in the recommendations for science learning in *Taking Science to School* (NRC, 2007). This report makes the recommendation that science learning needs to be strongly grounded in model-based approaches that focus on the use and consideration of evidence for posing, building, and refining models. Thus, although the NGSS emphasizes the interweaving of practices, it seems that learning to participate in the practice of modeling is the linchpin for the related practices of developing quantities, constructing data, and explaining the functioning of natural systems (Lehrer & Schauble, 2015). Furthermore, Duschl (2008) and others (Windschitl, Thompson, & Braaten, 2008) explain that helping teachers integrate epistemic practices such as systems thinking, model based reasoning, and quantitative reasoning into their professional practice is a significant challenge now and in the near future.

These new practices exemplified in current reform documents such as the *Framework* (NRC, 2012) and the NGSS are complex and yet hold promise though will require a transformation in how professional development for pre- and in-service teachers of science is conceived, designed, and implemented[3]. In addition, more empirical work, detailed examples, and implementations to scale are needed to address the question of how to support teachers in shifting their pedagogical practices to be more in line with quality science education in light of the NGSS. I would further Duschl et al.'s (this volume) recommendation and propose that the field needs to think much more in how to assure that department chairs, supervisors, directors, principals, and superintendents recognize and understand the types of pedagogies advocated for in the reforms so well articulated in the Introduction.

Like Duschl et al. (this volume) and others (Lehrer & Romberg, 1996; Lehrer & Schauble, 2004), I refer to data modeling as a conceptual bridge from hands-on investigative activities (inquiry) to inference-making and experimentation (Petrosino et al., 2003). Data modeling is rooted in understanding measure, error, and structuring of data. In fact, measure is critical to the creation of quantified variables thereby serving as the core components of models. The goal for students and teachers alike is to develop ideas about units and scale, and how unit and scale are integrated to produce measurable quantities is fundamental to modeling in science.

In terms of professional development, analysis of sources of error, measures of center, variability, and spread in data modeling tasks are all part of a quantitative literacy currently lacking in teacher preparation and seems to fall into a void between preparation for science educators and mathematics educators. While these critical ideas are rooted in fairly basic arithmetic skills that most middle school students are expected to have mastered, teachers and students alike struggle with the application of these ideas within a discipline specific context. While this area may offer great leverage for integration in STEM teacher preparation and professional development going forward, it is currently an unrealized potential. To that end, a data modeling focus on analysis of variance and in multiple measures shows promise for setting the stage for model-based inference in teacher education.[4]

Notes

1. This chapter is a revised version of a paper presented by A. Petrosino and R. Lehrer at the Waterbury Summit, The Pennsylvania State University, August 7–9, 2013.
2. Teachers involved with this work were part of a program specifically designed to implement reform minded pedagogy not only in their own classrooms, but as part of a nationwide dissemination effort involving over 50 colleges and universities.
3. A growing body of empirical research provides information about the structure and content of professional development programs that offer high-quality, high-impact learning opportunities for teachers (Borko, Jacobs, & Koellner, 2010; Desimone, 2009).
4. The work described in this chapter was supported in part by funding from the US National Science Foundation under grant number MSP 0831811 made to the author.

12

RESPONSE 1

Quantitative Reasoning in STEM Disciplines

Robert Mayes

This response provides brief comments on Quantitative Reasoning (QR) lead chapters (Johnson, this volume; Petrosino, this volume), then offers comments on the Introduction (Duschl et al, this volume). My responses are guided by the research and development conducted through the Pathways Project[1], an NSF supported initiative to develop environmental literacy learning progressions. The *Pathways* QR team led efforts to incorporate QR into the science learning progressions, culminating in the formation of a QR learning progression. Hence, the response is from the perspective of QR as defined in this project.

Johnson (this volume) begins by addressing the fundamental issue of what we mean by quantitative reasoning. The literature is full of definitions of quantitative reasoning and overlapping or related terms such as quantitative literacy, numeracy, and mathematical literacy. Johnson builds on Thompson's statement about "the promise of research on quantitative reasoning as a foundational component of students' learning of important mathematical ideas and contributing to their capacity to participate in important mathematical practices" (Thompson, 2012). This is a psychology of mathematics perspective of QR, which is rich in building theory on how students develop conceptual understanding in mathematics. Johnson ties this view of QR to underpinning development of multiplicative, algebraic, and covariational reasoning. The *Pathways Project* team built on definitions that emphasize applications of mathematical knowledge in real world contexts. Also included were definitions of quantitative literacy and quantitative reasoning provided by Steen (2004), the National Numeracy Network (2011), Program for International Student Assessment (OECD, 2000), and the Association of American Colleges and Universities (2010).

The concepts of quantity and quantification discussed by Johnson and based on Thompson's (2011) work impacted the inclusion in our QR progression of what we called the Quantitative Act (QA). This was one of three progress variables in our proposed learning progression for QR. Thompson's conception of quantification as a process that requires only conceiving a quantity as being possible to measure (not actually having to measure it) we found compatible with determining attributes of variables (objects) within a real-world science context. We also took from Thompson's definition of quantification the idea of the attribute having a unit measure. This is compatible with the importance of units and dimensional analysis in the sciences. However, as Johnson points out, we may have moved beyond her notion of QR when we included identifying standard or non-standard unit of measure, since this implicates potential measurement. Measurement is so central to science that we felt compelled to include the act of measuring in our QA progress variable, but it is a point of contention needing further discussion. I agree with her statement that the act of measuring the quantity is preceded by the quantification process. We consider the act of measurement part of quantitative literacy (QL), which focuses on the ability to reason with the quantities that are created through quantification. The act of quantification includes the identification of the objects within the context and attributes by which they can be compared, contrasted, and combined. The comparison of variables requires QL.

Johnson does an excellent job of summarizing the difference between our definition of QR and that of Thompson. I agree with her distinction, but feel that both definitions have an important place in the study of how mathematics serves as a support or barrier in the study of other disciplines. The work being done on the psychology of QR in mathematics is fundamental to improving students' ability to construct viable variables that can serve as the building blocks for models in real-world science situations. This is analogous to the microscopic scale in science where hidden mechanisms that explain an event are revealed. Our work on QR represents the landscape scale where we are attempting to determine overarching patterns of QR in becoming a scientifically literate citizen. Johnson's comment that inclusion of the act of quantification (QA) in a QL definition expands the QL construct is, I hope, a correct one. Whether QL is part of QR or a compatible activity serving different purposes, I believe is based on the QR definition you hold. From our interdisciplinary view of QR within real-world contexts, QA enables the student to move from context to quantitative account; QL allows them to move from quantitative account to quantitative interpretation (QI) of a given model; or to quantitative modeling (QM) where they build their own models.

Johnson's discussion of the role of context in QR is well stated and very important for our work as well. The research-based tasks focusing on QR identified in her chapter are not complex authentic real-world tasks. Yet even these more controlled tasks have shown that how the student interprets the problem context and quantities is nontrivial. The influence of context on QR and science understanding when engaging students in real-world science tasks is vastly more complex. Much more research needs to be done on the interaction of real-world contexts and QR. Such work is fundamental in understanding how students learn in interdisciplinary STEM situations.

A comment on learning trajectories in mathematics and learning progressions in science is important to make here. Though trajectories and progressions have a lot in common, they also represent somewhat different perspectives on student learning of complex concepts. For example compare the Sarama and Clements (2009) trajectory cited in Johnson's chapter to science learning progressions. The Consortium for Policy Research in Education report *Learning Progressions in Science: An Evidence-based Approach to Reform* (Corcoran, Mosher, & Rogat, 2009) identified learning progressions as a promising model that can advance effective adaptive instruction teaching techniques and thereby change the norms of practice in schools. We hypothesize that QR is essential for data-based and modeling approaches to learning the sciences, and that developing a QR learning progression will support a change in STEM teaching practices. While learning progressions in science have incorporated some components of QR (Louca, Zacharia, & Constantinou, 2011; Pluta, Chinn, & Duncan, 2011; Schwarz et al., 2009; Stefani & Tsaparlis, 2009; Taylor & Jones, 2009), there presently is no research in science education on a progression examining either the quantification act or the trajectory of quantitative literacy, quantitative interpretation, and quantitative modeling supporting science as model-building. Science progressions that integrate significant quantitative components include modeling in science (Schwarz et al., 2009), data modeling and evolution (Lehrer & Schauble, 2002), and atomic molecular theory (Smith, Wiser, Anderson, & Krajcik, 2006).

In conclusion, Johnson discusses the potential links between mathematics and science that could result from expanding the definitions of QR and QL. Such interdisciplinary STEM connections are essential for students who will have to make decisions about global grand challenges impacting their future. QR is an essential tool for students if they are to model real-world interdisciplinary STEM problems.

Petrosino (this volume; Petrosino et al., 2003) highlights the ways teachers think about inquiry and measurement when engaging in various activities that are informed by an instructional sequence for data modeling. This relates to our *Pathway* efforts to create a learning progression for QR. Our QR progression includes three progress variables: the quantification act (QA), quantitative interpretation (QI), and quantitative modeling (QM). Data modeling is defined as a set of practices including posing questions, generating measures and data representations, structuring and visualizing measures, and making inference in light of uncertainty. Petrosino also refers to data modeling as a conceptual pathway from inquiry to inference. Data modeling is grounded in understanding measure, error in measure, and in structuring of the resulting quantities as data. I agree that measure is essential to creation of quantified variables, which serve as the building blocks of models. Thompson's conception of quantification, which we incorporate in the QA progress variable, is a mental construct for an object within a context that includes both measure and attributes. Lehrer (2013) identifies the need for students to develop conceptions of the nature of units and scales, and how unit and scale are coordinated to produce measured quantities. The conceptions of analysis of error, measure of center, variability and spread in data modeling tasks we have included as elements of quantitative literacy within the QA progress variable. These conceptions are based in arithmetic skills that middle school students are expected to have mastered, yet as Lehrer (2013) points out and we have found, students struggle with applying these ideas within a context. A procedural approach to learning mathematics is likely the cause. Lehrer's (2013) data modeling focus on analysis of variance in repeated measures shows promise for setting the stage for model-based inference. The data based goal of model-based inference relates to our QI and QM progress variables. The student generated repeated measures in their example provides a table model ripe for interpretation (QI). The participant structure of students inventing displays of the measures constitutes QM, and the Display Review an excellent structure for QI.

Lehrer (2013) discusses the transformation of the learning trajectory for data modeling into a viable learning progression. The literature uses the terms learning trajectory and learning progression at times interchangeably and at times for different concepts. In addition, mathematics education appears to be more focused on learning trajectories, while science education pursues learning progressions. I think of learning trajectories as being of shorter duration (perhaps within a semester rather than across multiple years) and having more specific focus then a learning progression. The learning trajectory for data modeling created by Lehrer crosses over the three components for QR in our progression. The difference among measures, multiple shapes of same data, and statistics-as-measure components fall within our QA component. The measure of uncertainty falls within our QI component and the modeling measurement process with chance within QM. His inference in light of uncertainty returns to QI, indicating the cyclic nature of the interplay between QI and QM. I view the data modeling learning trajectory as an excellent case of how a trajectory plays out within and across a learning progression. I had not thought of learning trajectories in this way. Lehrer and Petrosino summarize this in their Summit Presentation: "Learning progressions include prospective pathways of conceptual development and specific realizations of these we call learning trajectories."[2]

Lehrer (2013) proposes that learning progressions are more than trajectories, they end up in the classroom as trading zones in which different realms of educational practice intertwine. Creation and implementation of the data modeling learning progression required significant collaboration among software designers, psychometricians, learning researchers, and teachers. Our *Pathways* work on a quantitative reasoning learning progression follows a similar pattern of engagement. We began by interacting with scientists, science educators, and mathematics educators around issues of developing environmentally literate citizens. Lehrer speaks of the wrenching work of aligning disparate communities. We found this to be the case even within the learning researcher community. There was a great deal of conversation and effort exerted over agreeing on the interdisciplinary interactions between science and mathematics. We debated and sometimes even agreed on very basic questions. What is quantitative reasoning within the context of environmental science? What mathematics is important for an environmentally literate citizen to possess: quantification, numeracy, measurement, calculation, representation, interpretation, or modeling? Our science and science education colleagues identified three overarching strands for environmental science: biodiversity, the carbon cycle, and the water cycle. We developed the QR learning progression around the science strands, developing QR interviews and QR assessments that supported development of the learning progression. In a parallel process, we also interacted with the psychometric community on building the science assessments. Once the learning progressions in science were developed, the project moved to creating teaching experiments and professional development workshops to assist teachers in using the progressions to improve teaching and learning, as Petrosino discusses (this volume). The QR research team reviewed the teaching experiments and integrated quantitative reasoning into the experiments. The outcome of this intense interaction with our science colleagues was a hypothesized QR learning progression for science. The QR learning progression was refined by conducting student interviews and administering diagnostic assessments with students across the middle and high school grade bands.

Interaction with the teaching community can be challenging and classroom implementation is often sporadic. Essential components of moving from progression to classroom implementation include the revision of design documents into educative curricula that include teacher notes, thought-revealing questions, students' ways of thinking, mathematical appendices supporting science contexts, and extensions (Lehrer, 2013). The method of video annotated construct maps and formative assessment provide a means of moving teachers' discourse practice away from what they term the initiate-respond-evaluate cycles. Given the interplay of these innovative strategies, the trading zone metaphor for learning progressions is well taken. The collaboration of disparate communities is essential for learning progressions to have an impact in the classroom. Such interdisciplinary undertakings are challenging.

Reaction to the Introduction

The Introduction (Duschl et al.; this volume) based on the drivers of systems thinking, model-based reasoning, and quantitative reasoning provides a strong frame for future research and change in STEM education. But the challenges to incorporating these drivers into school STEM are significant. Ken Robinson (2013) outlines some of the systemic barriers in his work on changing education paradigms. Countries around the world are reforming schools with two goals in mind: economic development and cultural preservation. The problem, according to Robinson, is that they are doing so by alienating students. He claims the current school system was designed and conceived in the intellectual culture of the enlightenment to serve the economic circumstance of the industrial revolution. Public education was built on two twin pillars: economic (develop a work force) and intellectual (intelligence judged as the ability to reason deductively and knowledge of the classics). These pillars are still influential in education today, causing a view of learners as academic or non-academic and education systems modeled on the interest of industry and in the image of industry: schools are organized on factory lines with ringing bells, separation of subjects into silos, and educating children in batches based on their age (their date of manufacture). If we are to move towards systems thinking, model-based reasoning, and quantitative reasoning, we will need to break down this production line mentality which includes increased national conformity and standardized testing.

Robinson espouses a move in the opposite direction, with a focus on developing divergent thinkers. He characterizes divergent thinking as an essential capacity for creativity, the ability to see lots of possible answers to and ways of interpreting a question, to think laterally and not just in linear or convergent ways, and to envision multiple answers not just one. Robinson's perspective on the education system supports the Waterbury Summit focus on systemic change, but it also promotes the idea of complex adaptive systems (CAS) as an important goal for curricular change. We need to get past what he refers to as the gene pool of education to think differently about human capacity, move beyond old conceptions of academic, non-academic, vocational, abstract, and theoretical; and realize the power of learning in groups. I believe we need to concern ourselves with both systemic change in our schools and the role of complex adaptive systems in changing instructional practice in the classroom. CAS is inherently interdisciplinary and requires a significant shift in how both teachers and students view the world. It requires a shift from a linear, silo subject, mechanistic view to a recursive, interdisciplinary, systems view.

Duschl et al. (this volume) propose that students need to move from experiment through modeling to theory building. QR is essential to this progression because it enables students to move from qualitative accounts, which may be based on authority (of teacher or text) to quantitative accounts, which enable a data-informed argument. If QR is to play an essential role we need to come to agreement on a definition of QR that serves the interdisciplinary STEM approaches we are proposing. Research on the role QR plays as both a support and barrier for CAS and model-based reasoning is needed. This is echoed by Duschl et al. in that "developing quantitative reasoning and model-based reasoning in the science classroom is an important step toward developing students' abilities to analyze and model quantitative data to support an understanding of systems thinking" (this volume, p. 5). I would expand this statement from science education to include STEM education, considering the impact of these approaches on not only science, but technology, engineering, and mathematics education, too. In addition, the movement from experiment to theory through modeling opens the door to inclusion of new 21st century skills and new science paradigms. Research on modes of reasoning is needed, including model-based reasoning, QR, and engineering design. We need to consider how to incorporate two new paradigms of science, computational science and data-intensive science, into interdisciplinary STEM. These new paradigms allow for the integration of the TE in STEM into science and mathematics education, with the computational science focus on simulations, visualizations, and representations of large data based science, and the data intensive science focus on analysis of large data sets.

Duschl et al. (this volume) cite the call by the NGSS to have STEM education further the understanding of the world around us (this volume). I believe that to engage students in STEM we need to have them focus this understanding of the world on the contexts and places familiar to them. Most students are capable of learning STEM, but as Robinson points out they have no faith in the education system, which leads them to disengage. Student motivation and engagement is an issue that we need to consider in our research. In our projects with teachers, we are using tenets of place-based education (PBE) and problem-based learning (PBL) to engage teachers and students. We establish collaborations between research institutes and school districts, providing STEM experts that work with interdisciplinary professional learning communities of STEM teachers to identify critical issues impacting the region that students can study in their place.

Grand challenges in STEM identified in National Research Council reports provide research topics that can be studied in place, then expanded to regional, national, or global scales. The issues of authenticity, collaboration, and inquiry raised by Duschl et al. (this volume) are supported by the approaches of PBE and PBL. One goal of such collaborations is to have teachers engage in interdisciplinary teaching. Interdisciplinary STEM teaching is an issue at the K-12 and college level. In our work, 6–12 grade science and mathematics teachers have difficulty in communicating about QR because they use different terminology, as well as concerns about subject content coverage. How do we restructure schools so that STEM teachers have the opportunity to collaborate on interdisciplinary teaching, with modeling and QR as bridges between subjects?

The issue of discipline-based education highlights some of the competing forces in the reform of STEM education: "Research on learning is moving away from a focus on general principles of learning to a focus on developing domain specific knowledge" (Duschl et al.; this volume, p. 10). There is a potential conflict here with interdisciplinary approaches. We need to be clear that domain specific does not mean a specific science domain, such as chemistry, but has a richer connotation of STEM domain knowledge that integrates subjects around issues of systems thinking, model-based reasoning, and quantitative reasoning. If we want students to coordinate essential principles, crosscutting concepts, and scientific practices, then we need more research on how to create a community of practice and reward structure that honors interdisciplinary teaching and learning.

The Introduction provides an excellent overview of learning progressions. I agree that the development of learning progressions is a promising way of moving from fact-based teaching to teaching for understanding. We are working on developing a learning progression on QR in the sciences. This progression is impacted by the interdisciplinary nature of QR, which we define as mathematics used within a real-world context. So the progression is examining student's development of reasoning, while many progressions focus on an enduring understanding or key concept within a discipline. Schwarz, et al. (2009) has also done work on reasoning with models. Development of learning progressions that address complex adaptive systems thinking, model-based reasoning, and quantitative reasoning will require us to meet with experts across multiple fields. It will be difficult and complicated work, in part because no single subject area claims these interdisciplinary concepts. We will face the challenge of grant funding to support the work and the restrictions of funding agency cycles on the longitudinal research that needs to be conducted. Three to five year funding cycles inhibit longitudinal studies of students' development of a conception over multiple grade spans.

Summary

There is a renewed emphasis on model-based reasoning in both the Next Generation Science Standards (Science and Engineering Practices: developing and using models, analyzing and interpreting data, using mathematics and computational thinking) (NGSS Lead States, 2013) and the Common Core State Standards—Mathematics (Mathematical Practices: model with mathematics) (NGA & CCSSO, 2010). There is a call to broaden students' view of science from the scientific process to the model building, refining, and testing process. Quantitative reasoning is essential to the process of model building, enabling students to move from qualitative accounts to quantitative accounts. Consider also the role that QR plays in understanding the hidden mechanisms (microscopic scale) that explain what students see in the system in which they live (macroscopic scale). These hidden mechanisms are often explained through the physical sciences, which are intensely quantitative in nature. Moving students from the macroscopic to microscopic scale immerses them in QR, as does having students generalize from the macroscopic scale to the landscape or global scale. Finally, science often models complex adaptive systems (CAS) such as the water cycle, carbon cycle, or biodiversity. QR supports both the modeling and systems thinking needed to understand CAS. So QR is extensively the Waterbury Summit.

Notes

1. This project is supported in part by a grant from the National Science Foundation: Culturally Relevant Ecology, Learning Progressions, and Environmental Literacy (DUE-0832173), which we refer to as *Pathways*.
2. For further information about the Waterbury Summit presentations, panels, and discussions go to the website at www.waterbury.psu.edu.

13

RESPONSE 2

Quantitative Reasoning: Capturing a Tension Between Structure and Variability

Rose Mary Zbiek

Mathematics as a STEM Field

Approximately 25 years ago in *On the Shoulders of Giants*, Steen (1990) articulated the richness of mathematics as he identified key elements of the field, including not only *quantity* but also *uncertainty*. I mention these two aspects of mathematics to keep quantity in perspective and to preface two observations about mathematics as a STEM field and about STEM learning, teaching, and curriculum.

A first critical observation is the "M" in STEM does not represent *mathematics* as much as it represents the *mathematical sciences*, including statistics. As a consequence, the "M" involves both quantity and uncertainty. Mathematics and statistics differ in terms of emphasis on deductive reasoning and on probabilistic reasoning, respectively. Mathematics offers useful ways to capture invariance and convey structure among quantities; statistics allows us to handle variability, as Peck, Gould, and Miller (2013) nicely argue. To reason with quantities in STEM settings requires identifying and understanding invariance as the basis of theory and theorems (Sinclair, Pimm, & Skelin, 2012). The role of context creates the need to balance the science of mathematics with the art of statistical work (Peters, 2010) to deal with the inevitable variance encountered in such settings. Although she references contexts outside mathematics, Johnson approaches quantitative reasoning from an essentially mathematical stance while Petrosino (this volume) embraces variation as he takes a statistical approach.

My second observation is that while one might argue that quantity is the mathematical area that most clearly overlaps with other STEM disciplines, mathematical modeling seems at least equally central to STEM integration. As Petrosino notes, NGSS (NGSS Lead States, 2013) cast model building as a way of developing new scientific knowledge. In Steen's (1990) work, modeling falls within uncertainty. Mathematical models and statistical models differ in terms of how they relate quantities and how they value invariance and variance. If we are to consider quantitative reasoning in the context of rich integration among STEM disciplines, we need to account for how quantity and uncertainty intertwine within context. Learning trajectories for quantitative reasoning across STEM disciplines would involve multiple views of quantity, including:

- The quantities that are the two (or more) variables of interest;
- A general relationship between two (or more) quantities and how they vary, as Johnson highlights;
- The particular structure of how the quantities are related, which would be the mathematical model;
- The relationships between quantities of true value and measurement error, to which Petrosino attend; and
- The essence of error as a quantity that is itself a composite of several quantities, such as measurement error and effect of other quantities.

Viewing quantities as they co-exist within a STEM problem situation suggests an intertwined perspective on quantitative reasoning in which these various quantities and relationships among them are seen in terms of structure and variation.

Broader Literature to Inform Trajectories

The task of developing learning trajectories that capture quantity and variation within the complexity of quantitative reasoning in STEM might be informed by a broader literature in mathematics education and statistics education—a literature with which the chapters by Johnson and Petrosino might implicitly resonate. For example, in statistics education literature, middle school students nontrivially develop a global view of data values as an aggregate distribution rather than treating them solely as individual cases (Ben-Zvi & Arcavi, 2001). In mathematics education, students do not immediately view a graph globally rather than in terms of particular points (Goldenberg, 1988). Johnson's description of students considering relationships between volume and height in terms of particular intervals might be an indicator of a trajectory from considering singular points to understanding aggregate effects.

A bifocal view of data seems to lie at the heart of understanding and representing center and spread. Peck and colleagues (2013) connect center and spread to structure and variability, respectively. Petrosino's work with measurement in a science setting suggests that learners might focus on neither structure nor variability but seek exact measurements, with a focus on a single value for each measuring event. These issues might relate to students' belief that a mathematics problem must have only one correct answer (Schoenfeld, 1989).

Mathematics education literature offers insights into other strategies that students might use in developing models. For example, given opportunities to identify quantities, measure, develop models, and interpret models, even teacher education students with substantial mathematics background (Zbiek, 1998) tend to foreclose on simplistic strategies when using a regression tool to develop a mathematical model and not to draw on characteristics of the functions they generate (e.g., domain, range, zeroes, monotonicity) when developing or evaluating a model. Simplistic strategies suggest that students might struggle with understanding modeling as a non-linear process of generating and testing possible models, and perhaps view it simply as a procedure to acquire. Such a procedural view of modeling seems to anticipate capturing structure but might ignore variability. Quantitative reasoning should inform how students transcend a procedural view.

Struggles of the types noted above, and those that Johnson notes related to multiplicative, algebraic, and covariational reasoning, seem important to address in describing, as does Petrosino, the development of learning progressions intended to support middle-school students' understandings related to concepts and practices of data modeling. Powerful learning progressions in quantitative reasoning will have value in the extent to which they resonate across areas of education, such as science education, mathematics education, statistics education, and learning science, and thus benefit from conceptions and strategies related to both structure and variability.

Mathematics in Classroom Tasks

In listening to the discourses of the Waterbury Summit, I am struck by the limited view of mathematics conveyed through the STEM activities in which students are engaged. Phrases heard and read emphasize the tool status of mathematics in the other areas, such as the need to "perform mathematical calculation(s)" and the possibilities to "use mathematics appropriately" in Advanced Placement Biology (Benson, this volume, pp. 119, 120).

Unfortunately, STEM integration in prekindergarten–grade 12 (and beyond) often happens in mainly strikingly superficial ways. It focuses on *borrowed concepts* that might be "quantities" of importance in science, technology, and engineering that are used in mathematics as numbers or measures and *appropriated procedures* that might be the mathematical tools that are applied for science, technology, and engineering purposes. Students are often told which quantities matter and are encouraged to apply known procedures to those quantities. One can imagine, for example, the bottle-filling situation found in Johnson's chapter implemented in this way. Such experiences obviously run contrary to the images suggested in the two chapters, and seem to privilege structure but not variability. Prescribed or limited use of mathematics in STEM settings might allow students to apply mathematics but do not fully engage students in quantitative reasoning while modeling.

Although tasks implemented in questionable ways seriously limit modeling experiences, such tasks could provide opportunities to develop aspects of quantitative reasoning and the tools of modeling—namely the mathematical concepts called upon in the structures that are being modeled. One might consider whether these types of classroom situations focus on developing a learning trajectory for a particular concept, such as Ellis and colleagues' exponential function learning trajectory, to which Johnson refers. Alternatively, the focus might be on a procedure, such as solving a linear equation or producing equivalent fractions. The goal of mathematics lessons featuring narrow mathematical opportunities likely target acquisition of *curricular mathematics* (Zbiek & Conner, 2006)—the mathematical content ostensibly most valued in school curriculum. Although the targeted concepts or processes might be mathematical tools that can be used in subsequent modeling, the learning focus in typical school curricula is not development of mathematical modeling practices.

Just as there is curricular mathematics, I suspect there also is curricular science—a collection of scientific principles and facts that, in the absence of model building, is the major focus of school science. In either a mathematics or science setting, a focus on conclusions that seemingly highlight students' understanding of structure rather than engage learners in processes that allow for variability averts opportunities for identifying quantities and minimizes experience in relating quantities and building interpretations based on quantity.

In short, I find it useful but insufficient to think about mathematics in general or quantity in particular as a tool for science, technology, and engineering. STEM areas provide settings for curricular mathematics to be learned but pursuing STEM fields suggest that modeling is also necessary. During the Summit, Petrosino referred to data modeling in terms of "the coordination of posing questions, generating measures and data representations, and making inference in light of uncertainty." Johnson summarizes Mayes, Peterson, and Bonilla's (2013) act of quantification as a process that is distinct from measuring and involves more than assigning a numerical value to a quantity. Both chapters describe quantitative activity in which measuring alone is useful but insufficient, interpretation is crucial, and the capture and representation of quantities of interest is critical. These process views of quantitative reasoning resonate with mathematical modeling as a practice, as in the Common Core State Standards for Mathematical Practices. They also lead to questions regarding how learning progressions might enrich classroom practice and enhance development of quantitative reasoning—reasoning of the kind that I described above as an intertwined perspective on quantitative reasoning that views quantities qualitatively and quantitatively and captures relationships among those quantities in terms of both structure and variation.

Conclusion

We need to address a key question that honors the "M" in STEM as a blend of mathematics and statistics in K-12 mathematics: How do understandings of quantity in terms of both structure and variability—including ways to represent structure and variability co-emerge for learners? Johnson provides evidence of how ideas related to quantity appear in the *CCSSM* (NGA & CCSSO, 2010). The development of learning progressions in quantitative reasoning is particularly important as *CCSSM* present contrasting views of modeling when one considers "model with mathematics" as a mathematical practice but notes particular standards that seemingly emphasize curricular mathematics; e.g., "Distinguish between situations that can be modeled with linear functions and with exponential functions" (NGA & CCSSO, 2010, p. 70). The chapters by Petrosino and by Johnson are useful in challenging ideas about how consideration of quantity and quantitative reasoning across science and mathematics can inform research, instruction, and assessment when we focus on tension between structure and variability.[1]

Note

1. I would like to thank Richard Lehrer for his thoughtful comments on this revised version.

Poster Presentation III

Fundamentals of Climate Science

Carla Zembal-Saul[1], Mark Merritt[1], Robert Crane[2]

[1] College of Education, Penn State University [2] College of Earth and Mineral Sciences, Penn State University

Scientific Understandings and Practices

- Energy and Energy Transfer
- Climate Systems Thinking
- Climate Literacy
- Constructing scientific explanations (CER)

Scientific Reasearch Context

About Rob...

- Director of the new Penn State Global Engagement Network, with the goal of establishing Penn State as a global university
- Climate scientist focusing on methodology, algorithms, and techniques for downscaling global climate models to regional-scale climate change
- Multidisciplinary work that seeks to apply these techniques to address specific social and environmental applications (e.g. water resources, malaria)
- Focused on Africa - Director of the Penn State Alliance for Education, Science, Engineering and Development in Africa (AESEDA)

Teaching Applications

- Exploring curriculum scenarios
- Analyzing video of classroom science teaching
- Thinking critically about classroom investigations/models (what do they represent and what misconceptions might they introduce)

Constructing Scientific Explanations

- Introduced the Claims-Evidence-Reasoning framework (McNeill, Lizotte, Krajcik, & Marx, 2006)
- Used throughout the course with class investigations and assessments
- Helps make thinking visible (formative assessment)

Climate Systems Thinking

- A course on climate is well suited for considering large complex systems.
- Models help us get at systems thinking by serving as a means to explain phenomena and predict outcomes.
- We have considered the following models that add together to tell the story of the earth's large scale climate patterns.

Earth's Energy Balance
↓
Differential Heating
↓
Convection
↓
Hadley Cell Circulation
↓
Coriolis Effect
↓
Atmospheric Circulation
↓
Wind Fields
↓
Transport of Heat and Moisture

Real Data

- First Hand Data Collection:
  - Differential heating; convection; greenhouse effect
- Large Scale Authentic Data Sets:
  - Mauna Loa; global maps of various climate parameters; ice cores, fossil pollen, African rainfall; IPCC Data

Broader Impacts

- Rob's reflections on science learning in higher education as a result of our collaboration:
  - Rethink general education science content courses for non-science majors
  - Limitations to content heavy, traditional disciplinary focused, general education college courses.
  - Even for science majors, general survey courses should not be the first courses taken.
- Parks and People South Africa
  - Study abroad program helps Penn State undergraduates consider interactions and tensions between human and natural systems.
  - Our collaboration provides opportunities for science education faculty and local teachers to work with South African teacher and students on science learning.

References

McNeill, K.L., Lizotte, D.J., Krajcik, J.,& Marx, R. W. (2006). Supporting students' construction of scientific explanations by fading scaffolds in instructional materials. *The Journal of the Learning Sciences, 15(2), 153-191.*

Theme III References

Association of American Colleges and Universities. (2010). *Quantitative Literacy VALUE Rubric.* Retrieved from http://aacu.org/value/rubrics/quantitative-literacy

Barab, S.A., & Squire, K. (2004). Design-based research: Putting a stake in the ground. *The Journal of the Learning Sciences, 13*(1), 1–14.

Behr, M., Harel, G., Post, T.R., & Lesh, R.A. (1994). Units of quantity: A conceptual basis common to additive and multiplicative structures. In G. Harel & J. Confrey (Eds.), *The Development of Multiplicative Reasoning in the Learning of Mathematics.* Albany: State University of New York Press.

Ben-Zvi, D., & Arcavi, A. (2001). Junior high school students' construction of global views of data and data representations. *Educational Studies in Mathematics, 45*(1–3), 35–65. http://doi.org/10.1023/A:1013809201228

Blanton, M., Levi, L., Crites, T., Dougherty, B., & Zbiek, R.M. (2011). *Developing Essential Understanding of Algebraic Thinking for Teaching Mathematics in Grades 3–5.* Reston, VA: National Council of Teachers of Mathematics.

Borko, H., Jacobs, J., & Koellner, K. (2010). Contemporary approaches to teacher professional development. In P.L. Peterson, E.L. Baker, & B. McGaw (Eds.), *International Encyclopedia of Education* (Vol. 7, pp. 548–56). Oxford, UK: Elsevier. Retrieved from http://public.eblib.com/choice/publicfullrecord.aspx?p=634860

Brown, A.L. (1992). Design experiments: Theoretical and methodological challenges in creating complex interventions in classroom settings. *Journal of the Learning Sciences, 2*(2), 141–78. http://doi.org/10.1207/s15327809jls0202_2

Carlson, M., Jacobs, S., Coe, E., Larsen, S., & Hsu, E. (2002). Applying covariational reasoning while modeling dynamic events: A framework and a study. *Journal for Research in Mathematics Education, 33*(5), 352–78.

Carraher, D.W., & Schliemann, A.D. (2007). Early algebra and algebraic reasoning. In F.K.J. Lester (Ed.), *Second Handbook of Research on Mathematics Teaching and Learning* (pp. 669–705). Charlotte, NC: Information Age Publishing.

Castillo-Garsow, C. (2013). Continuous quantitative reasoning. In R. Mayes & L. Hatfield (Eds.), *Quantitative Reasoning and Mathematical Modeling: A Driver for STEM Integrated Education and Teaching in Context* (Vol. 2, pp. 55–73). Laramie, WY: University of Wyoming College of Education. Retrieved from http://digitalcommons.georgiasouthern.edu/teaching-learning-facpubs/22

Castillo-Garsow, C., Johnson, H.L., & Moore, K.C. (2013). Chunky and smooth images of change. *For the Learning of Mathematics, 33*(3), 31–7.

Chinn, C.A., & Malhotra, B.A. (2002). Epistemologically authentic inquiry in schools: A theoretical framework for evaluating inquiry tasks. *Science Education, 86*(2), 175–218. http://doi.org/10.1002/sce.10001

Clements, D.H., & Sarama, J. (2004). Learning trajectories in mathematics education. *Mathematical Thinking and Learning, 6*(2), 81–9. http://doi.org/10.1207/s15327833mtl0602_1

Cobb, P., Confrey, J., diSessa, A., Lehrer, R., & Schauble, L. (2003). Design experiments in educational research. *Educational Researcher, 32*(1), 9–13. http://doi.org/10.3102/0013189X032001009

Confrey, J., & Smith, E. (1994). Exponential functions, rates of change, and the multiplicative unit. *Educational Studies in Mathematics, 26*(2–3), 135–64. http://doi.org/10.1007/BF01273661

Confrey, J., & Smith, E. (1995). Splitting, covariation, and their role in the development of exponential functions. *Journal for Research in Mathematics Education, 26*(1), 66–86.

Cooney, T.J., Beckmann, S., & Lloyd, G.M. (2010). *Developing Essential Understanding of Functions for Teaching Mathematics in Grades 9–12*. Reston, VA: National Council of Teachers of Mathematics.

Corcoran, T., Mosher, F.A., & Rogat, A. (2009). *Learning Progressions in Science: An Evidence-Based Approach to Reform. CPRE Research Report # RR-63*. Philadelphia, PA: Consortium for Policy Research in Education. Retrieved from http://eric.ed.gov/?id=ED506730

Daro, P., Mosher, F.A., & Corcoran, T. (2011). *Learning Trajectories in Mathematics: A Foundation for Standards, Curriculum, Assessment, and Instruction. CPRE Research Report # RR-68*. Philadelphia, PA: Consortium for Policy Research in Education. Retrieved from http://eric.ed.gov/?id=ED519792

Desimone, L.M. (2009). Improving impact studies of teachers' professional development: Toward better conceptualizations and measures. *Educational Researcher, 38*(3), 181–99. http://doi.org/10.3102/0013189X08331140

Dingman, S., & Madison, B. (2010). Quantitative reasoning in the contemporary world, 1: The course and its challenges: *Numeracy, 3*(2). http://doi.org/http://dx.doi.org/10.5038/1936–4660.3.2.4

Dougherty, B., Flores, A., Louis, E., & Sophian, C. (2010). *Developing Essential Understanding of Number and Operation*. Reston, VA: National Council of Teachers of Mathematics.

Duschl, R. (2008). Science education in three-part harmony: Balancing conceptual, epistemic, and social learning goals. *Review of Research in Education, 32*(1), 268–91. http://doi.org/10.3102/0091732X07309371

Ellis, A.B. (2011a). Algebra in the middle school: Developing functional relationships through quantitative reasoning. In J. Cai & E.J. Knuth (Eds.), *Early Algebraization a Global Dialogue from Multiple Perspectives* (pp. 215–38). Berlin; Heidelberg; New York: Springer. Retrieved from http://site.ebrary.com/id/10449575

Ellis, A.B. (2011b). Generalizing-promoting actions: How classroom collaborations can support students' mathematical generalizations. *Journal for Research in Mathematics Education, 42*(4), 308–45. http://doi.org/10.5038/1936-4660.3.2.4

Ellis, A.B., Ozgur, Z., Kulow, T., Dogan, M.F., Williams, C., & Amidon, J. (2013). An exponential growth learning trajectory. In *Proceedings of the 37th Conference of the International Group for the Psychology of Mathematics Education* (Vol. 2, pp. 273–80). Kiel, Germany: Psychology of Mathematics Education.

Goldenberg, E.P. (1988). Mathematics, metaphors, and human factors: Mathematical, technical, and pedagogical challenges in the educational use of graphical representation of functions. *The Journal of Mathematical Behavior*, 7(2), 135–73.

Gould, S.J. (1996). *Full House: The Spread of Excellence from Plato to Darwin*. New York: Harmony Books.

Greer, B. (1994). Extending the meaning of multiplication and division. In G. Harel & J. Confrey (Eds.), *The Development of Multiplicative Reasoning in the Learning of Mathematics* (pp. 61–85). Albany: State University of New York Press.

Jackiw, N. (2009). The Geometer's Sketchpad (Version 5). Emerville, CA: Key Curriculum Technologies.

Johnson, H.L. (2010). *Making sense of rate of change: Secondary students' reasoning about changing quantities* (Doctoral Dissertation). University Park, PA: The Pennsylvania State University.

Johnson, H.L. (2012). Reasoning about variation in the intensity of change in covarying quantities involved in rate of change. *The Journal of Mathematical Behavior, 31*(3), 313–30. http://doi.org/10.1016/j.jmathb.2012.01.001

Johnson, H.L. (2013). Designing covariation tasks to support students reasoning about quantities involved in rate of change. In *Task Design in Mathematics Education. Proceedings of ICMI Study 22* (pp. 213–22). Oxford, UK.

Johnson, H.L. (2014). A role of context in constructivist model building: What problem is the learner solving? *Constructivist Foundations, 9*(3), 339–41.

Johnson, H.L. (2015a). Secondary students' quantification of ratio and rate: A framework for reasoning about change in covarying quantities. *Mathematical Thinking and Learning, 17*(1), 64–90. http://doi.org/10.1080/10986065.2015.981946

Johnson, H.L. (2015b). Together yet separate: Students' associating amounts of change in quantities involved in rate of change. *Educational Studies in Mathematics, 89*(1), 89–110. http://doi.org/10.1007/s10649-014-9590-y

Klahr, D., Fay, A.L., and Dunbar, K. (1993). Heuristics for scientific experimentation: A developmental study. *Cognitive Psychology 24*(1); 111–146.

Koestler, C., Felton, M., Bieda, K., & Otten, S. (2013). *Connecting the NCTM Process Standards and the CCSSM Practices* (1st edition). Reston, VA: National Council of Teachers of Mathematics.

Konold, C., & Pollatsek, A. (2002). Data analysis as the search for signals in noisy processes. *Journal for Research in Mathematics Education, 33*(4), 259–89.

Kuhn, D., Black, J., Keselman, A., & Kaplan, D. (2000). The development of cognitive skills to support inquiry learning. *Cognition and Instruction, 18*(4), 495–523. http://doi.org/10.1207/S1532690XCI1804_3

Lehrer, R. (2013). A learning progression emerges in a trading zone of professional community and identity. In R. Mayes & L. Hatfield (Eds.), *Quantitative Reasoning and Mathematical Modeling: A Driver for STEM Integrated Education and Teaching in Context* (Vol. 2, pp. 173–86). Laramie, WY: University of Wyoming College of Education. Retrieved from http://digitalcommons.georgiasouthern.edu/teaching-learning-fac pubs/22

Lehrer, R., & Romberg, T. (1996). Exploring children's data modeling. *Cognition and Instruction, 14*(1), 69–108.

Lehrer, R., & Schauble, L. (2000). Modeling in mathematics and science. In R. Glaser (Ed.), *Advances in Instructional Psychology* (Vol. 15, pp. 101–59). Mahwah: Lawrence Erlbaum.

Lehrer, R., & Schauble, L. (2002). *Investigating Real Data in the Classroom: Expanding Children's Understanding of Math and Science.* New York: Teachers College Press.

Lehrer, R., & Schauble, L. (2004). Modeling natural variation through distribution. *American Educational Research Journal, 41*(3), 635–79. http://doi.org/10.3102/000283 12041003635

Lehrer, R., & Schauble, L. (2007). Contrasting emerging conceptions of distribution in contexts of error and natural variation. In M. Lovett & P. Shah (Eds.), *Thinking with Data* (pp. 149–76). New York: Lawrence Erlbaum Associates.

Lehrer, R., & Schauble, L. (2015). *Handbook of Child Psychology and Developmental Science.* (R.M. Lerner, L.S. Liben, & U. Mueller, Eds.) (7th ed., Vol. 2). Hoboken, NJ: Wiley. Retrieved from http://wiley.com/WileyCDA/WileyTitle/productCd-11181 36780.html

Lehrer, R., Schauble, L., & Petrosino, A. (2001). Reconsidering the role of experimentation in science education. In K.D. Crowley, C.D. Schunn, & T. Okada (Eds.), *Designing for Science Implications from Everyday, Classroom, and Professional Settings* (pp. 251–78). Mahwah, NJ: L. Erlbaum.

Lobato, J., & Siebert, D. (2002). Quantitative reasoning in a reconceived view of transfer. *The Journal of Mathematical Behavior, 21*(1), 87–116. http://doi.org/10.1016/S0732–3123(02)00105-0

Lobato, J., & Ellis, A.B. (2010). *Developing Essential Understanding of Ratios, Proportions, and Proportional Reasoning for Teaching Mathematics: Grades 6–8*. Reston, VA: National Council of Teachers of Mathematics.

Lobato, J., Hohensee, C., Rhodehamel, B., & Diamond, J. (2012). Using student reasoning to inform the development of conceptual learning goals: The case of quadratic functions. *Mathematical Thinking and Learning, 14*(2), 85–119. http://doi.org/10.1080/10986065.2012.656362

Louca, L.T., Zacharia, Z.C., & Constantinou, C.P. (2011). In Quest of productive modeling-based learning discourse in elementary school science. *Journal of Research in Science Teaching, 48*(8), 919–51. http://doi.org/10.1002/tea.20435

Madison, B., & Dingman, S. (2010). Quantitative reasoning in the contemporary world, 2: Focus questions for the numeracy community. *Numeracy, 3*(2). http://doi.org/10.5038/1936-4660.3.2.5

Mason, J. (2008). Making use of children's powers to produce algebraic thinking. In J.J. Kaput, D.W. Carraher, & M.L. Blanton (Eds.), *Algebra in the Early Grades* (pp. 57–94). New York: Lawrence Erlbaum Associates.

Mayes, R., Peterson, F., & Bonilla, R. (2013). Quantitative reasoning learning progressions for environmental science: Developing a framework. *Numeracy, 6*(1). http://doi.org/10.5038/1936-4660.6.1.4

Mayo, D.G. (1996). *Error and the Growth of Experimental Knowledge*. Chicago: University of Chicago Press. Retrieved from http://site.ebrary.com/id/10442169

Mayr, E. (2001). *What Evolution Is*. New York: Basic Books.

Medin, D.L. (1989). Concepts and conceptual structure. *American Psychologist, 44*(12), 1469–81. http://doi.org/10.1037/0003-066X.44.12.1469

Moore, K.C., & Carlson, M.P. (2012). Students' images of problem contexts when solving applied problems. *The Journal of Mathematical Behavior, 31*(1), 48–59. http://doi.org/10.1016/j.jmathb.2011.09.001

National Council of Teachers of Mathematics. (2000). *Principles and Standards for School Mathematics*. Reston, VA: National Council of Teachers of Mathematics.

National Council of Teachers of Mathematics. (2009). *Focus in High School Mathematics: Reasoning and Sense Making* (1st edition). Reston, VA: National Council of Teachers of Mathematics.

National Governors Association Center for Best Practices, & Council of Chief State School Officers. (2010). *Common Core State Standards for Mathematics*. Washington, DC: National Governors Association Center for Best Practices, Council of Chief State School Officers.

National Numeracy Network. (2011). What is Numeracy/QL/QR? Retrieved March 29, 2015, from http://serc.carleton.edu/nnn/resources/index.html

National Research Council. (1996). *National Science Education Standards Observe, Interact, Change, Learn*. Washington, DC: National Academies Press.

National Research Council. (2007). *Taking Science to School: Learning and Teaching Science in Grades K-8*. Washington, DC: National Academies Press.

National Research Council. (2009). *Learning Science in Informal Environments People, Places, and Pursuits*. Washington, DC: National Academies Press. Retrieved from http://site.ebrary.com/id/10315878

National Research Council. (2012). *A Framework for K-12 Science Education: Practices, Crosscutting Concepts, and Core Ideas*. Washington, DC: National Academies Press. Retrieved from http://www.nap.edu/catalog/13165/a-framework-for-k-12-science-education-practices-crosscutting-concepts

National Research Council. (2014). *STEM Integration in K-12 Education: Status, Prospects, and an Agenda for Research*. Washington, DC: National Academies Press. Retrieved from http://nap.edu/catalog/18612/stem-integration-in-k-12-education-status-prospects-and-an

Nesher, P. (1988). Multiplicative school word problems: Theoretical approaches and empirical findings. In J. Hiebert & M.J. Behr (Eds.), *Number Concepts and Operations in the Middle Grades* (pp. 19–40). Reston, VA: National Council of Teachers of Mathematics.

NGSS Lead States. (2013). *Next Generation Science Standards: For States, By States*. Retrieved from http://www.nap.edu/catalog/18290/next-generation-science-standards-for-states-by-states

Organization for Economic Cooperation and Development (OECD). (2000). Programme for International Student Assessment (PISA). Retrieved from http://www.oecd.org/pisa/home/

Peck, R., Gould, R., & Miller, S. (2013). *Developing Essential Understanding of Statistics for Teaching Mathematics in Grades 9–12* (1st edition). Reston, VA: National Council of Teachers of Mathematics.

Penuel, W.R., & Fishman, B.J. (2012). Large-scale science education intervention research we can use. *Journal of Research in Science Teaching, 49*(3), 281–304. http://doi.org/10.1002/tea.21001

Peters, S.A. (2010). Engaging with the art & science of statistics. *Mathematics Teacher, 103*(7), 496–503.

Petrosino, A.J., Lehrer, R., & Schauble, L. (2003). Structuring error and experimental variation as distribution in the fourth frade. *Mathematical Thinking and Learning, 5*(2–3), 131–56. http://doi.org/10.1080/10986065.2003.9679997

Piaget, J. (1970). *Genetic Epistemology*. New York: Columbia University Press.

Pickering, A. (1995). *The Mangle of Practice Time, Agency, and Science*. Chicago: University of Chicago Press. Retrieved from http://public.eblib.com/choice/publicfullrecord.aspx?p=648147

Pluta, W.J., Chinn, C.A., & Duncan, R.G. (2011). Learners' epistemic criteria for good scientific models. *Journal of Research in Science Teaching, 48*(5), 486–511. http://doi.org/10.1002/tea.20415

Porter, T.M. (1986). *The Rise of Statistical Thinking, 1820–1900* (Reprint edition). Princeton, NJ: Princeton University Press.

Robinson, K. (2013). *RSA Animate—Changing Education Paradigms*. Retrieved from https://www.youtube.com/watch?v=zDZFcDGpL4U&feature=youtube_gdata_player

Saldanha, L.A., & Thompson, P.W. (1998). Re-thinking co-variation from a quantitative perspective: Simultaneous continuous variation. In S.B. Berenson, K.R. Dawkins, M.L. Blanton, W.N. Coloumbe, J. Kolb, K. Norwood, & L. Stiff (Eds.), *Proceedings of the 20th Annual Meeting of the Psychology of Mathematics Education North American Chapter* (Vol. 1, pp. 298–304). Raleigh, NC: North Carolina State University.

Sarama, J., & Clements, D.H. (2009). *Early Childhood Mathematics Education Research: Learning Trajectories for Young Children*. New York; London: Routledge.

Schoenfeld, A.H. (1989). Explorations of students' mathematical beliefs and behavior. *Journal for Research in Mathematics Education, 20*(4), 338–55.

Schwartz, J.L. (1988). Intensive quantity and referent transforming arithmetic operations. In J. Hiebert & M.J. Behr (Eds.), *Number Concepts and Operations in the Middle Grades* (Vol. 2, pp. 41–52). Reston, VA: National Council of Teachers of Mathematics.

Schwarz, C.V., Reiser, B.J., Davis, E.A., Kenyon, L., Achér, A., Fortus, D., Shwartz, Y., Hug, B., & Krajcik, J. (2009). Developing a learning progression for scientific modeling: Making scientific modeling accessible and meaningful for learners. *Journal of Research in Science Teaching, 46*(6), 632–54. http://doi.org/10.1002/tea.20311

Shell Centre for Mathematical Education (University of Nottingham). (1985). *The Language of Functions and Graphs: An Examination Module for Secondary Schools*. Nottingham, UK: Shell Centre.

Simon, M. (1995). Reconstructing mathematics pedagogy from a constructivist perspective. *Journal for Research in Mathematics Education, 26*(2), 114–45. http://doi.org/10.2307/749205

Sinclair, N., Pimm, D., & Skelin, M. (2012). *Developing Essential Understanding of Geometry for Teaching Mathematics in Grades 6–8*. Reston, VA: National Council of Teachers of Mathematics.

Smith, C.L., Wiser, M., Anderson, C.W., & Krajcik, J. (2006). FOCUS ARTICLE: Implications of research on children's learning for standards and assessment: A proposed learning progression for matter and the atomic-molecular theory. *Measurement: Interdisciplinary Research and Perspectives, 4*(1–2), 1–98. http://doi.org/10.1080/15366367.2006.9678570

Smith, J.P., & Thompson, P.W. (2008). Quantitative reasoning and the development of algebraic reasoning. In J.J. Kaput, D.W. Carraher, & M.L. Blanton (Eds.), *Algebra in the early grades* (pp. 95–132). New York: Lawrence Erlbaum Associates.

Steen, L.A. (1990). *On the Shoulders of Giants: New Approaches to Numeracy*. Washington, DC: National Academy Press.

Steen, L.A. (2001). *Mathematics and Democracy: The Case for Quantitative Literacy*. Princeton, NJ: National Council on Education and the Disciplines.

Steen, L.A. (Ed.). (2004). *Achieving Quantitative Literacy: An Urgent Challenge for Higher Education*. Washington, DC: Mathematical Association of America.

Stefani, C., & Tsaparlis, G. (2009). Students' levels of explanations, models, and misconceptions in basic quantum chemistry: A phenomenographic study. *Journal of Research in Science Teaching, 46*(5), 520–36. http://doi.org/10.1002/tea.20279

Steffe, L.P. (1992). Schemes of action and operation involving composite units. *Learning and Individual Differences, 4*(3), 259–309. http://doi.org/10.1016/1041-6080(92)90005-Y

Steffe, L.P. (1994). Children's multiplying schemes. In G. Harel & J. Confrey (Eds.), *The Development of Multiplicative Reasoning in the Learning of Mathematics* (pp. 3–39). Albany: State University of New York Press.

Steffe, L.P., & Thompson, P.W. (2000). Teaching experiment methodology: Underlying principles and essential elements. In A.E. Kelly & R.A. Lesh (Eds.), *Handbook of Research Design in Mathematics and Science Education* (pp. 267–306). Mahwah, NJ: L. Erlbaum.

Stigler, S.M. (1986). *The History of Statistics: The Measurement of Uncertainty Before 1900*. Cambridge, MA: Harvard University Press.

Taylor, A., & Jones, G. (2009). Proportional reasoning ability and concepts of scale: Surface area to volume relationships in science. *International Journal of Science Education, 31*(9), 1231–47. http://doi.org/10.1080/09500690802017545

Thompson, P.W. (1993). Quantitative reasoning, complexity, and additive structures. *Educational Studies in Mathematics, 25*(3), 165–208. http://doi.org/10.1007/BF01273861

Thompson, P. W. (1994a). The development of the concept of speed and its relationship to concepts of rate. In G. Harel & J. Confrey (Eds.), *The Development of Multiplicative Reasoning in the Learning of Mathematics* (pp. 181–234). Albany: State University of New York Press.

Thompson, P. W. (1994b). Students, functions, and the undergraduate curriculum. In E. Dubinsky, A. H. Schoenfeld & J. Kaput (Eds.), CBMS issues in mathematics education. Washington, DC: Mathematical Association of America

Thompson, P. W. (2008). Conceptual analysis of mathematical ideas: Some spadework at the foundation of mathematics education. In O. Figueras, J.L. Cortina, S. Alatorre, T. Rojano, & A. Sepúlveda (Eds.), *Proceedings of the Joint Meeting of PME 32 and PME-NA 30* (Vol. 1, pp. 31–49). Mexico: Cinvestav-UMSNH.

Thompson, P. W. (2011). Quantitative reasoning and mathematical modeling. In S. A. Chamberlain & L. Hatfield (Eds.), *New Perspectives and Directions for Collaborative Research in Mathematics Education: Papers from a Planning Conference for Wisdom*[e] (Vol. 1, pp. 33–56). Laramie, WY: University of Wyoming College of Education.

Thompson, P. W. (2012). Quantitative reasoning and mathematical modeling: A driver for STEM integrated education and teaching in context. In R. M. Mayes & L. L. Hatfield, (Eds.), *Advances in Research on Quantitative Reasoning* (pp. 143–8). Laramie, WY: University of Wyoming.

Thompson, P. W., & Carlson, M. P. (in press). Variation, covariation and functions: Foundational ways of mathematical thinking. In J. Cai (Ed.), *Third Handbook of Research in Mathematics Education*. Reston, VA: National Council of Teachers of Mathematics.

Thompson, P. W., & Saldanha, L. A. (2003). Fractions and multiplicative reasoning. In J. Kilpatrick, W. G. Martin, & D. Schifter (Eds.), *A Research Companion to Principles and Standards for School Mathematics* (pp. 95–113). Reston, VA: National Council of Teachers of Mathematics.

Tzur, R., Johnson, H. L., McClintock, E., Kenney, R. H., Xin, Y. P., Si, L., Woodward, J., Hord, C., & Jin, X. (2013). Distinguishing schemes and tasks in children's development of multiplicative reasoning. *PNA*, 7(3), 85–101.

Tzur, R., Johnson, H.L., McClintock, E., & Risley, R. (2012). Culturally-mathematically relevant pedagogy (CMRP): Fostering urban English language learners' multiplicative reasoning. In L. R. Van Zoest, J.J. Lo, & J.L. Kratky (Eds.), *Proceedings of the 34th Annual Meeting of the North American Chapter of the International Group for the Psychology of Mathematics Education* (pp. 829–36). Kalamazoo, MI: Western Michigan University.

van Dyke, F., & Tomback, J. (2005). Collaborating to introduce algebra. *Mathematics Teaching in the Middle School, 10*(5), 236–42.

Von Glasersfeld, E. (1995). *Radical Constructivism a Way of Knowing and Learning*. London, UK: Falmer Press. Retrieved from http://site.ebrary.com/id/10057556

Weiner, J. (1994). *The Beak of the Finch: A Story of Evolution in Our Time*. New York: Vintage Books.

Windschitl, M., Thompson, J., & Braaten, M. (2008). Beyond the scientific method: Model-based inquiry as a new paradigm of preference for school science investigations. *Science Education, 92*(5), 941–67. http://doi.org/10.1002/sce.20259

Zbiek, R. M. (1998). Prospective teachers' use of computing tools to develop and validate functions as mathematical models. *Journal for Research in Mathematics Education, 29*(2), 184–201.

Zbiek, R. M., & Conner, A. (2006). Beyond motivation: Exploring mathematical modeling as a context for deepening students' understandings of curricular mathematics. *Educational Studies in Mathematics, 63*(1), 89–112. http://doi.org/10.1007/s10649-005-9002-4

THEME IV

Equity, Epistemic, and Ethical Outcomes

14

EDUCATIONAL AND ETHICAL DILEMMAS FOR STEM EDUCATION IN PENNSYLVANIA'S MARCELLUS SHALE GASFIELD COMMUNITIES[1]

Catharine Biddle and Kai A. Schafft

The Marcellus Shale natural gas extends a mile below the surface of two-thirds of mostly rural Pennsylvania, and extends north and south into parts of New York, West Virginia, Ohio, and Maryland. Long known to contain significant amounts of natural gas, technological developments in the last ten years have made it economically feasible to commercially extract shale gas. By the mid-2000s the Marcellus Shale formation was widely recognized as the largest reserve of natural gas in the United States (Engelder, 2009), and by 2007 unconventional natural gas extraction[2] in Pennsylvania had begun in earnest. For many Pennsylvanians, making sense of these changes to their local circumstance has been difficult within the context of a politically charged and on-going debate about the environmental and social effects of unconventional natural gas development.

Developed in the early 2000s in Texas' Barnett Shale region, unconventional gas extraction relies on a combination of technological innovations, including the use of horizontal drilling along underground shale layers and hydraulic fracturing involving a high volume, high pressure injection of slickwater, or a proprietary mix of water and chemicals (including biocides and lubricants, as well as a particulate matter such as silica). The slickwater injection into the well fractures the shale, while the particulate matter in the slickwater holds the shale fractures apart, releasing the natural gas into the well bore thereby allowing the natural gas trapped in the shale layer to move towards the surface to be extracted. While neither horizontal drilling nor hydraulic fracturing are new technologies, in their current usage they represent a highly industrialized activity.

A standard well drilling pad may be five acres in size and require 640 to 4,000 heavy equipment truck trips to build, drill, and hydraulically fracture (Ladlee & Jacquet, 2011; Moss, 2008). Unconventional drilling in the Marcellus Shale requires as much as four to ten million gallons of water for each well (Abdalla, 2010), in addition to added chemicals, including many identified as carcinogenic and/or that have been tied to respiratory, nervous system, and cardiovascular health risks (Finkel & Law, 2011). Although the chemical additives typically comprise less than 0.5 percent of the volume of the fluid, the total amount of fluid used is large enough that a three million gallon hydrofracture includes about 15,000 gallons of chemicals in the wastewater returning to the surface (Soeder & Kappel, 2009) which may also contain brines, heavy metals, and naturally occurring radioactive materials coming from the shale layer (Wilber, 2012). Pipelines also must be built, connecting wells to compressor stations used to transport the gas to consumer markets. Well pads, pipelines, and compression stations all involve significant changes to physical landscapes, as well as unprecedented demands on local roads and communities.

For small, rural communities that had experienced decades of economic decline with few prospects for revitalization, the new influx of people, money, and activity brought new opportunities and new revenues, especially for those with land to lease for drilling rights and those positioned either to take advantage of new employment within the industry or in supportive industries and businesses (Malin, 2013). As drilling proceeded, however, stories began to appear in the popular press and news media about the potential environmental consequences of unconventional gas extraction, as well as the actual tallies of accidents, spills, and well water contamination linked directly or indirectly to the industry (Adlesic, Gandour, Fox, Roma, & Fox, 2010; Pifer, 2011).

Local residents also saw other community-level impacts, including housing shortages and increased housing costs, dramatic increases in heavy truck traffic with attendant road damage and destruction, and newcomers from out of state securing the bulk of industry employment on drilling rigs, raising nagging questions about exactly how much of the new employment opportunities were actually going to local Pennsylvanians. Others noted new inequalities as some within local communities became "shaleionaires" from leasing revenues (Stahl, 2010), while others, especially low-income community residents, were priced out of the housing market in the face of rent prices that had tripled or quadrupled (Schafft & Biddle, 2014; Williamson & Kolb, 2011). Lastly, many residents raised questions about the longevity of the industry, raising the specter of the impacts of an economic "bust" following the initial gas boom. In sum, the experience of many local people in active gas extraction communities has involved an increasing ambivalence and a weighing of both opportunity and risk in the context of an unanticipated and unpredictable industry (Brasier et al. 2011; Schafft, Borlu, & Glenna, 2013).

Schools, in their role as key local institutions within communities, represent a useful lens through which to view and interpret rural community change and boomtown development (Schafft & Biddle, 2014; Schafft et al., 2013). Educators are uniquely positioned to influence youth career, residential, and educational aspirations, both through the relationships that they develop with the youth with whom they work, but also in the way they may interpret their roles in supporting workforce development through teaching and learning. Educators may increasingly be likely to interpret their workforce preparation role in this way within the context of a neoliberal federal policy environment that foregrounds the creation of human capital and global economic competitiveness as two of the most important goals of education (Schafft & Biddle, 2014; Theobald, 2009). Furthermore, the national focus in educational policy on preparing America's young people for careers in science, technology, engineering, and mathematical (STEM) careers could naturally lead many educators to conceive of educating youth about such opportunities in Marcellus-related industries as a professional responsibility.

This chapter explores the perspectives of youth and educators living in communities most profoundly affected by the Marcellus Shale gas development on the social, economic, and environmental changes that they have seen and the ways in which their schools and communities are responding. We look specifically at how STEM education and workforce development in the secondary school context are being shaped by perceptions of industry related community change, using interviews and focus group data gathered between 2011 and 2013 with educators, educational administrators, and high school students in school districts heavily impacted by shale gas development. Interviews and focus groups were conducted with a total of 88 educators (including educational leaders, teachers, and other relevant faculty, such as guidance counselors) and five focus groups were conducted with 36 high school students from Tioga, Bradford, and Lycoming Counties in northern Pennsylvania, and Washington and Greene Counties in southwest Pennsylvania (for a more detailed discussion of data and methods, see Schafft & Biddle, 2014).

STEM Education within the Context of Local Environmental and Community Change

When we asked educators about the extent to which they were adapting STEM curriculum in the context of Marcellus development, they responded by discussing opportunities that they had made available to students in order to support education around issues related to the workforce and industry development. One superintendent described a program, which had been started in the Southwestern area of Pennsylvania:

> We've incorporated curriculum for kids to benefit from the oil and gas industry. There's overflow on some jobs with one gas well. One of the things we want in the middle school for example, we've started talking about the process of the geology, the science behind it and what's all involved in it, and of course the break out, the many kind of jobs that are available with it all the way up through to high school.

Similarly, when asked the same question, youth focus group participants in the Northern and Southwest areas discussed off-handedly some activities that they did in their science classes related to Marcellus: Reports on the technology behind fracking, or having guest speakers come to their classes to talk about new job opportunities in natural gas development.

However, adaptations to curriculum were interpreted by educators almost exclusively through a workforce development lens with little mention of critical or place-based curricular integration. Furthermore, school districts did not seem to be rushing to integrate related content into the curriculum or to implement a renewed STEM focus as a result of the new Marcellus-related industry. Rather, school districts appeared to be leaving most of the gas industry-related educational and workforce development efforts to the Career and Technology Centers (CTCs), which, through the intermediate units[3], provide youth and adult vocational education and training.

We attribute the hesitance of administrators and teachers to incorporate Marcellus related content into the curriculum to both the uncertainty that they expressed regarding the future of the industry and the well-being of their communities as a whole, as well as their description of the lack of curricular freedom afforded by the current political context of curricular standardization and financial constraint. Throughout our discussions we heard numerous comments about the conflicting benefits, challenges and uncertainties associated with gas development, and the ambivalence towards the industry this tended to create for youth and educators.

From our focus group conversations with educators about STEM curricular offerings in response to Marcellus related community change (and from youths' account of their experiences of this programming), it was clear that, when Marcellus development was addressed at all, educators interpreted their role as helping students understand industry development using an "empirical" or value-free approach to the science of fracking, rather than exploring the meaning of that science within a local context. Youth in Bradford County told us about a guest speaker from one of the large gas companies developing the area who had visited all of the eighth and ninth grade Earth Science classes to talk about "how they get the gas" and "how they go through the ground." Students also talked about how careers in geology had been introduced as part of a "Career of the Month" initiative at their schools. Youth in Lycoming County mentioned having to write a paper on the shale development for their Earth Science class and that one of their teachers had, prior to teaching science, worked as a chemist for one of the gas companies.

This relatively uncritical approach to the science of fracking within the context of the STEM curriculum, however, stands in stark contrast to the remarks that both youth and educators made about the environmental effects of fracking within the rural landscape and communities in which they lived. For example, participants in every educator and youth focus group remarked on the rural, small-town character of their communities. "We like our rural community," one of the educators from Greene County told us, "We like the way we live." Another educator in the Northern Tier told us that it was an "ideal place to live" with the woods and the mountains but with a city only a few hours away. One youth described the "sense of peace" in his Southwestern community, and how "people are a lot nicer [here]." "It's a lot more comfortable than being in a city," we were told.

However, both youth and educators described the intense changes happening within their local landscapes related to Marcellus industrial development. Many of the youth described this change as a process of urbanization. However, rather than expressing excitement or anticipation about change coming to their rural communities and bringing with it urban qualities or amenities, these young people were far more circumspect. From their perspective, the "urbanization" they associated with industrial development was associated with the destruction of natural landscapes and wildlife habitats, the effects of noise and light pollution, the disruption of quiet community spaces, and increased social anonymity.

Educators often spoke about specific fears of environmental damage from gas industry activities, and in particular, water contamination. One administrator from Bradford County told us,

> there's been wells around my house for the last few years, but I always wonder if one day I go home, it might be today or tomorrow or a couple years, is my water all right? I think that's the biggest thing that I worry about. I don't really worry about traffic or the air quality. I don't really worry about any of that other stuff. I just worry about the water.

Fears such as these were common and several youth told us stories about specific instance of contamination. In one such instance though, a female student told us about babysitting for a local family whose house had, the student claimed, been affected by methane migration into the well water, as well as other contaminants:

STUDENT 1: (When) I first started babysitting for them, I had to learn, they had one of those gas lights that goes off if the levels get too high. They had to teach me if this goes off, we have an emergency bag, and where all their stuff was to get out cuz the house could . . .

MODERATOR: This is when you were babysitting?

STUDENT 1: Mm-hmm. It was a little scary. I was nervous, but, I mean, it never happened. We weren't allowed to use their water in their house. They had one of those filter—

STUDENT 2: Tank?

STUDENT 1: tanks, and the people had to come every so often and change their water. We weren't allowed to, only for washing dishes and stuff. When I was trying to give the little girl a bath, she would get burn marks on her arm sometimes from the water. It was just scary.

The curricular emphasis on geological science, as well as the process of fracking itself, seemed strangely at odds with youth and educator accounts of their perceptions of change to the rural landscape of the places in which they live and the feelings of loss which they express because of these changes. However, the emphasis on the science behind fracking could be read as an attempt to avoid the divisiveness of debates about the effects of fracking that educators and youth described as happening within the broader community. As one youth from Lycoming County told us about his community,

> there's a split between the people. There's "I don't wanna upset the environment" and there's people—not to offend anybody, but there's people in this town who are greedy and wanna get the money from the gas. Then there's the people that look at it the way everyone else here is talking. "Get out. You're ruining our place. There's no more trees. There's no more animals. It's urbanizing, and that's not what everyone wants here." Then again, there is people with a lot of land that want the money and want what comes with it.

Youth and educators described a social context surrounding the Marcellus development marked by huge financial opportunity unequally distributed across the local population, creating a "split community" as one administrator in the Northeast said,

> you have a group of people who think this is the best thing ever, and you have a group of people that think they're the spawn of hell coming in here to destroy the environment. It's very distinct.

Moreover, we were told that those who own land could become millionaires overnight, while low-income families and youth have been displaced by spiraling rent inflation due to the increased pressure put on the housing market by gas companies looking for housing for out-of-state workers. Educators described the pressure that this combination of haves and have-nots, as well as the on-going debate about environmental implications, had on their calculus for curricular programming. One teacher from Bradford County told us that, "someone had mentioned, like, okay, if you're gonna be telling the kids about the gas, you also need to have someone from like an environmental group come in and talk about—you know." Similarly, a principal told us that teachers themselves were very divided on these issues.

The context of environmental, economic, and social change associated with the industry seemed to lead educators to largely ignore addressing Marcellus related topics within the curriculum. When these topics were addressed, educators' perceived discussions of the science of fracking as the most safe, or controversy free, ways to address these topics, by focusing on cultivating young people's scientific literacy around fracking itself. However, the ambivalence amongst educators towards this integration is mirrored in their discussions of deciding how to address the low-skill, high-paying work opportunities that have become available through this development to youth.

One of the questions asked was whether the gas industry and the proliferation of new—and often lucrative—economic opportunities might encourage college-going, particularly within STEM fields such as geology and engineering, or whether it might discourage college-going given the opportunity costs of paying for post-secondary education while simultaneously foregoing current labor market opportunities. Many of the students with whom we spoke seemed uninterested in the opportunities available within the industry and expressed other post-secondary plans such as attending local colleges, entering the local workforce, or obtaining a post-secondary vocational-tech certificate. All of the students, however, knew of friends or family who had taken advantage of the expansion of opportunities in their areas associated with the development. In Lycoming County, students told us:

STUDENT 1: I know a lotta guys that are actually talkin' about it right now, who wanna just say "Forget school" after high school and just—

STUDENT 2: Yeah, they don't wanna go to college, they just wanna go to the gas lines.

STUDENT 3: Gas line, lotta money.

A student in Bradford County told us that he had already had offers to come work for some of the companies making $3,000 a week with only a high school diploma. An educator observed that typically these opportunities offered young people the chance to make "more than their mother or father a week." Within this context, it was clear that attending college, or even, potentially, finishing high school, could be considered a kind of opportunity cost when high-paying, low-skilled jobs were readily available. As one student from Washington County explained,

> I know last year we had a lot of graduates go straight to working on the wells . . . they didn't have to waste time at school. They just went right in and started making good money for being an 18-year old that just graduated.

We were also told stories of companies that potentially would hire students without a high school degree, or students who dropped out, worked for the gas companies and tried to work on their GED. Conversely, few students mentioned opportunities in the industry that could be obtained with a college degree.

Despite being aware of the lucrative nature of these opportunities, educators and youth also told us that these jobs could not truly be considered local.

The transient nature of the industry, which expanded and contracted on a month-to-month basis (with profound effect on the local economy), meant workers were constantly being moved around to areas where drilling would be the most profitable. An administrator in Bradford County made a similar observation in reference to what he perceived as a waning interest among youth in pursuing gas industry jobs:

> Two, three years ago, I think there was like this mentality that, 'Oh, I can make a lot of money. This is gonna be here forever. I've got it made. I'm 17, 18 years old. I could even quit school.' We had a lot of fear that that was gonna have an impact on our graduation percentages, our percentage of kids going on to college and that kind of thing. I think part of the dip, where everybody left the area, kind of woke everybody up. What I've been finding is the kids, when I'm talking to them, seem pretty savvy about what's going on here. The kids that were interested in that kind of stuff were kids that were thinking, 'This is an opportunity for me to stay in the area, have a good-paying job for 30 years.' I think what they're seeing is, 'Oh, a lot of these jobs, I'm gonna be like bouncing around the country. I'm not gonna be able to stay home, and the work hours are really difficult. It's hard work and I don't get to be with my family every night when I go home from work.'

This statement reflects many of the dilemmas that we heard educators express related to discussing industry related opportunities with young people. On the one hand, many educators felt that industry work represented high-paying opportunities, they were torn between endorsing a lifestyle that took young people outside of the community, as well as did little to encourage students to pursue post-secondary credentials that might better their long-term job security and financial future in the face of an uncertain and changing industry.

Last, educators also mentioned that an additional constraint preventing them from retooling STEM curriculum to be more responsive to local change were the institutional requirements of standardized curricula and high-stakes assessments. A superintendent told us that,

> we live and die by the test scores in the state of Pennsylvania. We have to try to make AYP. That's our biggest [priority]. We have to do instruction that's based on the standards, state standards. We have to ensure that those standards are in alignment with accountability . . . our standards and our curriculum are not based on this industry. The changes that we just recommended, the last two weeks, the change of composition, delivery of composition that had nothing to do with this industry.

This, ultimately, raises complicated questions about the purposes of education and the ways in which a concept like "accountability" is understood (Corbett, 2007; Schafft, 2010), particularly when accountability centers around bureaucratic mandates to the exclusion of ongoing dilemmas and complexities of community change faced by students, teachers, schools, and communities in areas undergoing boomtown development.

Conclusion

Youth and educator perspectives on how Marcellus Shale development has affected the places and the communities where they live paint a poignant picture of both ambivalence and constraint in constructing a school-based response to an uncertain industry. Redesigning the STEM curriculum to respond to the development is complicated by educator and youth perceptions of the social disruption and environmental destruction that has characterized the industrial development in their communities. Despite the fact that the development has brought with it unprecedented financial opportunity for some, educators and youth expressed reservations about the way in which the unequal distribution of this opportunity was changing the community itself. Furthermore, educators expressed ambivalence about preparing youth for work in an industry that could leave, through boom-bust economies, the community worse off in the long run (James & Aadland, 2011), and what the implications might be for students who decide to forego further formal education. And, both educators and youth expressed concerns about the exploitive nature of gas industry development and the risks of lasting environmental degradation, long after the main industry-related economic activity has ended.

The dilemmas facing educators who must consider how the school should respond to these changes is further circumscribed by the broader context of the global economic system and the way in which rural communities are positioned. Because of their role in human capital development, Pennsylvania educators are placed precariously between weighing the well-being of their rural communities and supporting their attempts at resilience and adaptation within a globally integrated world economy, or supporting a neoliberal ethos that not only prizes mobility and liquidity over durability, but further views rural areas as a periphery whose purpose is to provide resources, human and material, for an increasingly urbanized core (Corbett, 2007; Schafft, 2010). Given these fundamental conflicts about long-term community well-being and the reluctance of many youth to pursue industry-related STEM-based fields, educators must grapple with the ethical dilemma of what their roles are, not only as professional educators, but also community members, within this context.

One potential resolution to the constrained ability of educators to address and respond to local change could be using place-based instruction, a practice that has been tied to and lauded as a kind of rural pedagogy (Gruenewald, 2003). Place-based education has been looked to as a solution in the past for ways to address the inadequacies of traditional STEM instruction in drawing on the local literacies that rural students develop in the course of their daily lives (Avery & Kassam, 2011; Sobel, 2004). However, in the Marcellus shale case, simply including instruction about the empirical nature of the development—the technical specificities of drilling, the chemical composition of fracking fluid, and so forth—is not enough to address these deep-seated ambivalences, and thus calls into question the appropriateness and comprehensiveness of such an approach.

Merely integrating these concepts into the curriculum in an unproblematized way cannot do justice to the seriousness of what is at stake for youth and educators within these dilemmas. In this sense, this case raises important questions about how educators are called on to conceptualize the scientific literacy that they are meant to develop and enhance in their students. In what ways is this conceptualization sensitive to the complex relationship between human sustainability and ecological sustainability? Rather than treating science solely as a value-free, empiricist activity, our findings suggest that educators may be better served by integrating issues related to scientific ethics and the weighing of values and social ends in dilemmas presented within the practice and use of science: a fundamentally interdisciplinary and "systems-based approach" to STEM education (Duschl, Bismack, Greeno, & Gitomer, this volume) that incorporates analyses of not only physical systems, but how those physical phenomena, systems and processes may interact with human systems, and in turn what the implications are for social, environmental, and economic sustainability, local and global. In the end, these analyses involve questions not only for empirical physical and social sciences, but basic *ethical* questions about what constitutes a good society and how we might achieve it.

It is very unclear, within the current policy regime of common standards and high-stakes assessments, whether educators possess—or believe they possess—sufficient autonomy to incorporate such approaches into their development of student STEM literacy. The cost of not doing so, however, is to leave youth and educators to wrestle with these issues individually and largely outside the school context rather than engaging collectively and critically on topics that are controversial and locally relevant, and yet nonetheless are deeply embedded within and have critical implications for global environmental and social sustainability.

Notes

1. This chapter was written with support from the Penn State College of Education Waterbury Summit. Data used in this manuscript were gathered in studies funded by the Penn State Marcellus Center for Outreach and Research, The Penn State Children, Youth and Family Consortium, and the Center for Rural Pennsylvania. We want to acknowledge the assistance and contributions of Leland Glenna and Brandn Greene in gathering some of the data discussed in this chapter.
2. Unconventional gas reserves typically refer to gas found *within* geologic formations (rather than trapped and contained by them) such as shale and sandstone. Because the gas is trapped within the shale layer, unconventional gas extraction typically relies on the use of horizontal drilling and hydraulic fracturing (Wilber, 2012).
3. Intermediate units are regional organizations unique to Pennsylvania, which serve multiple school districts in providing educational support and services.

15

DEFINING A KNOWLEDGE BASE FOR REASONING IN SCIENCE

The Role of Procedural and Epistemic Knowledge

Jonathan Osborne

Introduction

The focus of many educational systems is rooted strongly in a belief that an understanding of science is so important that it should be a feature of every young person's education (American Association for the Advancement of Science, 1989; Confederacion de Sociedades Cientificas de España, 2011; Fensham, 1985; Millar & Osborne, 1998; National Research Council, 2012a; Sekretariat der Ständigen Konferenz der Kultusminister der Länder in der Bundesrepublik Deutschland (KMK), 2005; Taiwan Ministry of Education, 1999). Many of these documents and policy statements give pre-eminence to an education for citizenship. Likewise, the emphasis in the Programme for International Student Assessment (PISA) frameworks is on science for citizenship seeking to assess the competency of 15-year-old students to become informed critical consumers of scientific knowledge—a competency that all individuals are expected to need during their lifetimes (Organization for Economic Cooperation and Development [OECD], 2012). Drawing on the example of PISA, this chapter explores what kinds of knowledge bases are required to engage in such reasoning.

Scientific literacy in PISA 2015 is defined by the three competencies:

1. Explaining phenomena scientifically;
2. Evaluating and designing scientific inquiry; and
3. Interpreting data and evidence scientifically.

Clearly these competencies lie at the heart of what it means to reason scientifically and require knowledge of science or *content knowledge*. The second and third competencies, however, require more than a knowledge of what we know. Rather, they depend on an understanding of how scientific knowledge is established and the degree of confidence with which it is held. Historically, specific calls have been made for teaching about what has variously been called "the nature of science" (Lederman, 2007), "ideas about science" (Millar & Osborne, 1998) or "scientific practices" (National Research Council, 2012a). Within PISA, the 2006 framework operationalized this aspect of science using the term "knowledge about science." Recognizing and identifying the features that characterize scientific inquiry does, for instance, require a knowledge of the standard procedures of the diverse methods and practices used to establish scientific knowledge—which in this chapter will be called *procedural knowledge*. However, these competencies also require an understanding of the rationale for the common practices of scientific inquiry, the status of the knowledge claims that are generated, and the meaning of foundational terms such as theory, hypothesis, and data—essentially what might be called *epistemic knowledge*. Procedural and epistemic knowledge are necessary to identify questions that are amenable to scientific inquiry, to judge whether appropriate procedures have been used, to ensure that the claims are justified, and to distinguish scientific issues from matters of values or economic considerations. What then might be the constructs of procedural knowledge and epistemic knowledge? How might they be defined and how might they be demarcated from each other? This chapter offers an exploration of these constructs and a rationale for their importance. It will argue that such ideas help to clarify the forms of knowledge that should be an outcome of any K-12 science education.

What Knowledge Bases Are Needed for Scientific Reasoning?

Hill (2008) has argued that the societies that sustain their competitive edge will be "post-scientific" societies. In such societies, highly valued skills will be the ability to draw on a range of disciplinary knowledge, to think creatively and to evaluate new ideas in a critical, reflective and rational manner. Employers will require individuals who, while having a core understanding of scientific and technical principles, have the ability to communicate and synthesize knowledge in an original manner. Likewise, experts invited to a National Research Council (NRC) symposium in 2007 to explore the nature of the competencies required for the future workplace argued that the ability to solve problems creatively, sophisticated communication skills, self-management, and systems thinking would be needed (NRC, 2008). More recently, the NRC report *Education for Life and Work* (National Research Council, 2012b) stresses the development of three domains of competence—the cognitive, the intrapersonal, and the interpersonal. Central to the first of these is developing the ability to undertake the complex cognitive processes of critical thinking, non-routine problem solving, and constructing and evaluating evidence based arguments. Taken together, such arguments suggest that the valued competencies of the future will be the higher order reasoning abilities of evaluation, synthesis, and critique.

As Hill argues, if developing these competencies are to be a feature of science education, then it is important to "be certain that we emphasize what we want, for we shall surely get what we emphasize" (2008, p. 9). Yet, when it comes to the skill of reasoning scientifically, science education has suffered from a lack of clarity about "what we want." Despite the lip service that is paid to developing the ability to think critically, the picture that emerges from several studies is one where the prevailing classroom pedagogy in science focuses on transmission of a body of knowledge, is dominated by the use of closed, authoritative dialogue, and provides little opportunity to explore the domain of ideas in a critical and reflective manner (Lyons, 2006; Osborne & Collins, 2001; Tobias, 1990; Weiss, Pasley, Smith, Banilower, & Heck, 2003). Weiss et al. found that only "14 percent of lessons nationally having a climate of intellectual rigor, including constructive criticism and the challenging of ideas" (2003, p. 54), suggesting that a substantial gap exists between the rhetoric and the reality about the goals and possible value of any education in science. However, in approaching the task of identifying the knowledge base required for scientific reasoning in K-12 science education, I argue that two problems have confounded the field: first, is the disjuncture between domain-specific and domain-general conceptions of scientific reasoning; and second, a lack of clarity about the role of knowledge in scientific reasoning.

The first problem is a product of the many attempts to define scientific reasoning that historically have placed an emphasis on scientific reasoning as a *domain general* concept. One version of this perspective is seen among psychologists who commonly take a "nothing-special view" (Simon, 1966) arguing that general reasoning abilities account for the main characteristics of scientific reasoning, and are more *transferable*. Klahr and Simon (1999) acknowledge that domain-specific aspects are the basis for "strong methods" within the sciences, but claim that these only have limited value because they apply to well-defined contexts and problems. Instead, they contend that, when scientists move into new areas and meet ill-defined problems, general strategies or "weak methods" are the most important forms of initial reasoning. And it is these that should be taught to non-scientists—as "weak methods" can be deployed in new contexts external to science where the majority of students will be for their working lives.

The contrary argument, however, has been made both within psychology and science education. Perkins and Salomon made a seminal argument (Perkins & Salomon, 1989) that "general heuristics that fail to make contact with a rich domain-specific knowledge base are weak" and that "a domain-specific knowledge base without general heuristics . . . is brittle" (p. 24). Within science education, Passmore and Stewart (2002) have argued that "scientific practice is discipline specific" (p. 187) as do Duschl et al. (this volume) pointing to the specific role of models and quantitative reasoning. Sandoval (2003), too, concludes from his analysis of student explanations of Galapagos data "that epistemic and conceptual understanding are tightly interrelated" (p. 48).

I, too, wish to add to the argument that defining scientific reasoning as a domain general cognitive skill is both flawed and unsustainable. It is flawed because it misrepresents the nature of scientific reasoning as, while individuals in everyday situations may use reasoning akin to that of scientists, such forms of reasoning cannot be used to *define* what is distinct about scientific reasoning. Similarly, arguments for the value of teaching domain-general reasoning within science are unsustainable because general reasoning does not *have to* be taught in science. Other school subjects may be better at teaching, for example, moral reasoning, decision-making, and problem solving skills. A logical consequence then is that a conception for *domain specific* scientific reasoning is needed if we wish to claim that science education can foster higher order thinking skills required by contemporary society.

What, for instance, are the domain-specific entities used when reasoning within the domain of science? Essentially what is the role of knowledge in scientific thinking? The answer is the second critical problem that needs to be addressed in order to establish a better construct of scientific reasoning. Undoubtedly, a factor that has changed over the past six decades is the development in our understanding of what we mean by knowledge and reasoning—a development that can be illustrated by the changes that have occurred in Bloom's taxonomy of cognitive processes (1956) when compared to the revised version offered by Anderson and Krathwohl (2001). Bloom and his colleagues did not separate knowledge from knowing. Moreover, they did not include any form of "meta knowledge" (Krathwohl, 2002). Anderson et al., in contrast, split reasoning into two dimensions, a knowledge dimension and a cognitive process dimension, and use not one but four categories to describe their knowledge dimension: factual knowledge, conceptual knowledge, procedural knowledge, and metacognitive knowledge. Likewise, Li and Shavelson (2001) have used a similar framework splitting knowledge into declarative knowledge (knowing what), procedural knowledge (knowing how), schematic knowledge (knowing why), and strategic knowledge (knowing when, where, how knowledge applies)—a perspective which was influential in developing the NAEP 2009 framework for assessment in science (National Assessment Governing Board, 2008). The common point in these and other examples is that they alter the way we conceptualize the relationship between knowledge and reasoning. That is, what used to be explained as a generic "reasoning skill" linked only weakly to a body of undifferentiated knowledge is now dependent on a set of distinctive aspects of domain-specific knowledge of "what we know," "how we know," and "why it happens."

That reasoning within any domain requires domain-specific knowledge is also shown by Schunn and Anderson (1999) who point out how, lacking such knowledge, undergraduate subjects "did not use theories in designing their experiments" and "did not relate their results to their experiments," concluding that what the scientist has acquired "is not just a matter of general reasoning ability" (p. 368). Likewise, Koslowski (1996) in her critique of earlier work by Deanna Kuhn argued that in reasoning scientifically, "theory and data are both crucial, and theory and data are interdependent. Sound scientific reasoning involves bootstrapping: considerations of theory or mechanism constrain data, and data in turn, constrain, refine, and elaborate theory" (p. 86). More recently, Duschl (2008) has argued for the need to balance conceptual goals—that is a knowledge of the content with epistemic and social goals. However, to date, little has been done to define the nature of the domain-specific, procedural, and epistemic elements that might constitute the goals of any science education. For instance, given the epistemic importance of peer review or of double blind trials in establishing scientific claims, should some knowledge of these procedures be a basic element of all students' education in science?

Attempts to Define the Knowledge Base for Scientific Reasoning

More or less, ever since its inception as an element of the school curriculum, one of the arguments for the value of science education has been its value for training the "general faculties of mind" (Huxley, 1854). Then, as perhaps now (Roschelle, Bakia, Toyama, & Patton, 2011), this claim was contentious leading Faraday to ask "what is their [the natural sciences] effect on the mind? . . . or what does the mind learn by that training?" (Layton, 1973, p. 173). Similar sentiments were expressed nearly 50 years later by Layton who argued that "the question of content or process, subject matter or method of inquiry, is a recurring and still unresolved issue in the relatively short history of school science teaching" (p. 174). And, I would argue, still unresolved.

It is impossible, however, to review or discuss all the attempts that have been made to solve the challenge of operationalizing what we might wish to teach about scientific reasoning. Indeed, that would require a full history of science education and beyond, as the answers developed within science education also draw on developments in science studies and the learning sciences (as demonstrated by Duschl & Grandy, 2012, 2008). Rather what is presented here are some of the notable attempts to address the challenge and the ways in which they have succeeded or not. Importantly, the goal is not so much to criticize the failings of past approaches to teaching scientific reasoning but rather to learn from past mistakes—very much in the tradition of Santayana's sentiments that "those who forget the lessons of history are condemned to repeat the mistakes of the past."

Teaching Scientific Reasoning through Inquiry

A familiar and long-stranding approach to teaching scientific reasoning is to present learning as a process of inquiry. As long ago as 1759, Edmund Burke wrote that (Burke, 1909):

> I am convinced that the method of teaching which approaches most nearly to the method of investigation, is incomparably the best; . . . it tends to set the reader himself in the track of invention, and to direct him into those paths in which the author has made his own discoveries (p. 12).

In the teaching of science, Armstrong, the most well-known English proponent of this approach, argued for what he termed the "heuristic method," which "involves placing our students as far as possible in the attitude of the discoverer-using methods which involve their *finding out,* instead of being merely told about things" (Armstrong, 1902).

Joseph Schwab, who chaired the US development of the Biology Curriculum Studies in Science (BCSC) project, extended and elaborated the rationale for this approach to the teaching of scientific reasoning arguing that the teaching of science should be an "enquiry into enquiry[1]" (Schwab, 1962). The BSCS textbooks were committed to what Schwab called a "radical departure [from] conventional patterns" by embedding all content as an invitation to enquiry (Schwab, 1962, p. 39). An analysis of the textbooks would suggest that in print this objective was realized. Over 50 percent of the text was dedicated to helping students approach investigative problems—a feature which the most recent edition of the BSCS Human Approach textbook sustains (Chiappetta & Fillman, 2007).

Similar notions are articulated 30 years later in the National Science Education Standards (National Research Council, 1996) which argued that teaching science called for more than teaching science as a process and the specific skills of observing, experimenting, and inferring—a statement that suggests the authors were aware of the inherent problems of the process approach. Instead it offered a vision of science teaching where "inquiry is central to science learning" in which:

> students describe objects and events, ask questions, construct explanations, test those explanations against current scientific knowledge, and communicate

> their ideas to others. They identify their assumptions, use critical and logical thinking, and consider alternative explanations. In this way, students actively develop their understanding of science by combining scientific knowledge with reasoning and thinking skills (p. 23).

In one sense, such an argument is seductively attractive. It offers a conception of learning science that embodies the practice of science itself. But it is flawed in two ways. First is the flaw embodied in Ryle's distinction between "knowing how" and "knowing that" (Ryle, 1949). Learning science through a process of inquiry develops an extensive knowledge of the procedures required to produce new scientific knowledge e.g., "knowing how," but what it does not develop explicitly is a meta-level knowledge of the rules, procedures, and values that science uses and why these are required—what might be termed "epistemic knowledge." Thus a student may engage in a discussion over competing interpretations of a given data set, but such practice does not necessarily develop the knowledge and understanding that critique is essential to the successful construction of knowledge in science (Ford, 2008; Longino, 1990; Popper, 1962), or an understanding of the role and function of peer review. Rather, developing an explicit knowledge of the role of any practice requires a meta-level reflection on the practice itself that must be taught explicitly (Lederman, 2007; Monk & Osborne, 1997; Peters, 2009). Similar points can be made about building students' explicit knowledge of the procedural techniques that science uses. In short, simply engaging in a practice predominantly develops a body of tacit expertise (Polanyi, 1958). There is, for instance, little evidence that scientists hold sophisticated models of science. In contrast, there is a considerable body of evidence that school science leaves its students with naïve understandings of science because it fails to explicitly explore the nature of the discipline (Driver, Leach, Millar, & Scott, 1996; Lederman, 1992; Solomon, Scott, & Duveen, 1996).

Schwab did attempt to present a description of the epistemic and procedural constructs that govern scientific inquiry, but limited it to the "kind of conclusion, kind of data, and mode of verification" (p. 76), which are intrinsic to science. All sciences, Schwab argued, were reliant on a set of classificatory schema (taxonomic knowledge); measurement science which he described as "simple quickly describable patterns of enquiry"; causal science which seeks to show how one entity in a "system of mutually interacting and mutually determined parts acts as a concerted whole" (Schwab, 1962, p. 89); and analogical science which is the attempt to explain by inventing mechanisms not directly accessible to observation. The latter form of reasoning required the development of models and theories. Thus Schwab's work can—and should be seen as an attempt to define the knowledge bases required for scientific reasoning and address the lack of an explicit recognition. However, Schwab's work only managed to identify in general terms the epistemic constructs that are the foundation of scientific reasoning. Likewise, his work did not recognize that participating in inquiry required a body of procedural knowledge of specific constructs intrinsic to inquiry such as the notion of an independent/dependent variable, types of error, reliability and validity and more. Thus, in failing to identify sufficiently the forms and features of the knowledge on which scientific reasoning depended, his work lacked the insights necessary to build a deeper picture of what it means to reason scientifically leaving the second problem unresolved.

The second flaw in the argument for inquiry-based teaching of science is the conflation of the *doing* of science with the *learning* of science (Chinn & Malhotra, 2002; Roth, 1995) when the two are activities distinguished by their *differing* goals. In particular, that the goal of doing science is the discovery of new knowledge, whereas the goal of learning science is to comprehend the major features of the scientific landscape, to appreciate the intellectual achievement these ideas represent, and to understand how such knowledge has been derived. To suggest that inquiry—the major methodological tool of the scientist—can also be the major procedure for learning science is to make a category error akin to suggesting that learning to understand and appreciate music can only be achieved through playing an instrument—leaving no role for the education of the critic or music lover who has never played.

To his credit, this was not an error that Schwab made. Schwab did draw a distinction between presenting science as a product of a process of inquiry and using inquiry itself for learning and teaching—arguing that the inquiry curriculum was "by no means the only version nor necessarily the most desirable version in all schools for all students" (Schwab, 1962, p. 71). However, this was a distinction that others failed to recognize seeing "inquiry as central to science learning" (NRC, 1996). Retrospectively, the emphasis on teaching science through inquiry can be seen as an attempt to offer an account of what it means to do science that is more commensurate with contemporary scholarship. However, without a consideration of what forms of knowledge were required, and hence what kind of conceptual understanding such practice might seek to achieve, such efforts left the teaching of scientific reasoning akin to a ship lost at sea with no landmarks for guidance. For instance, Spillane et al.'s comment that "it is common for science teachers to report, for example, that they use inquiry in their classroom. Yet people can mean very different things by such a report" (Spillane, Reiser, & Reimer, 2002, p. 416). And as the recent NRC *Framework* commented "attempts to develop the idea that science should be taught through a process of inquiry have been hampered by the lack of a commonly accepted definition of its constituent elements' resulting in widely divergent objectives" (NRC, 2012a, p. 44). Part of the failure to develop a coherent or common understanding is, I would argue, due to a lack of clarity about what type of knowledge any given practice seeks to develop.

The Recognition of Procedural Knowledge

An initiative that led to the conception that there was more to "knowing science" than content knowledge came from the UK *Assessment for Performance Unit* (APU) project. This project, begun in the 1970s and lasting for 10 years, was an attempt to measure the performance of a representative sample of the educational systems in English, Mathematics, Science and Modern Foreign Languages akin to the US NAEP program. The group in charge of science developed an assessment framework (Murphy & Gott, 1984) that was "process oriented" and focused on students' performance on written and practical laboratory tasks. From the data of students engaging in empirical inquiry, the APU developed the notion of "procedural knowledge" which they distinguished from a "process skill" conceptualizing the relationship between the two as akin to that between "strategies and tactics" (Black, 1990)—a knowledge of common strategies being a body of explicit declarative knowledge whilst tactics were equivalent to the skills required to use such strategies judiciously.

Building on these findings and, in part as a response to Millar and Driver's critique of the process approach (Millar & Driver, 1987), Gott and Murphy (1987) argued against seeing investigation or inquiry simply as a "process skill." Instead, they sought to define inquiry as an "activity." Within such an "activity," they argued, students made use of both *conceptual* and *procedural* understanding. The latter was a knowledge and understanding of scientific procedures, or "strategies of scientific inquiry, "such as holding one factor constant and varying the other when controlling variables" (p. 13). This insight had emerged from an analysis of their results where they found that much of the variation in student performance on tasks could not be explained solely on the basis of an absence of appropriate conceptual knowledge e.g., the lack of a suitable model of the system being investigated. Rather, it was accounted for by "procedural failures," i.e., students not holding the necessary procedural knowledge. This finding led the APU to the conclusion that "the major influence on performance on this task is the availability to the child of certain relevant items of knowledge" and that "carrying out a scientific investigation, then, is primarily a display of understanding, and not of skill" (Gott & Murphy, 1987, p. 244).

Consequently, Gott and Murphy argued for the explicit teaching of procedural knowledge suggesting that "we must accept the need to develop an explicit underpinning [procedural] knowledge structure in the same way that we have developed such a structure for conceptual elements of the curriculum" (1987, p. 52). Developing this notion in a later paper, Gott and Mashiter (1991) explicitly presented *procedural knowledge* as the "missing element" that could form the basis of a theoretical response to Millar and Driver's (1987) question of "what develops" in their critique of the process approach. As such, it offered a new theoretical perspective that made the scientific reasoning required for empirical inquiry and critique knowledge-based rather than skill-based. This rationale was then developed further into a more complete framework, or "taxonomy," for procedural knowledge to be taught in school science (Duggan, Gott, Lubben, & Millar, 1996; Gott & Duggan, 1996) which the authors describe as "concepts of evidence" (Gott, Duggan, & Roberts, 2008).

Evidence of the importance of procedural knowledge in scientific reasoning has come from many research studies. One important study was the *Procedural and Conceptual Knowledge in Science* (PACKS) project that sought to identify what knowledge students actually apply when solving practical investigative tasks (Millar, Lubben, Gott, & Duggan, 1994). This research used written items to probe students' procedural knowledge and compared their performance with that of the investigative tasks. Correlation was high, indicating that what was tested in the written items accounted for 50 percent or more of the variance in practical tasks. The research also revealed "misconceptions" among students about scientific methodology. Many students expressed what Schauble, Klopfer, and Raghavan (1991) refer to as an "engineering framework" and interpreted the goal of the practical investigative task as an empirical test of the best solution rather than a means of testing a scientific hypothesis. Another important outcome was to identify levels of understanding from simple to advanced procedural knowledge (Lubben & Millar, 1996) thereby answering Millar and Driver's question of what develops. Thus the work of Gott and his colleagues identified a body of knowledge other than domain-specific conceptual knowledge required for scientific reasoning.

Such knowledge is essential for the 2nd competency in the PISA framework of evaluating and designing scientific inquiry (OECD, 2012). Evaluating any design requires questions to be asked about what variables have been identified, whether they have been appropriately controlled, how many measurements to take, how the potential for measurement error was minimized and more. Likewise, interpreting data and evidence is dependent on some knowledge of the data collection procedures. What, for instance, is a double blind trial and what is the role of the placebo? PISA defines this competency as the ability to:

TABLE 15.1 The Competency of Designing and Evaluating Scientific Inquiry (PISA, 2012)

Describe and appraise scientific investigations and propose ways of addressing questions scientifically demonstrating the ability to:

- Identify the question explored in a given scientific study;
- Distinguish questions that are possible to investigate scientifically;
- Propose a way of exploring a given question scientifically; and
- Evaluate ways of exploring a given question scientifically.

Describe and evaluate a range of ways that scientists use to ensure the reliability of data and the objectivity and generalizability of explanations.

Thus, to engage in critical deliberation about the validity and reliability of the data produced by any specific scientific inquiry, students need knowledge of:

- Concepts of measurement and of systematic observation;
- Ways of assessing and minimizing uncertainty;
- Mechanisms to ensure the replicability and accuracy of data;
- Concepts of variables;
- Common ways of abstracting and representing data using tables, graphs, and charts;
- Control of variables strategy and its role in experimental design and randomized controlled trials; and
- Appropriate designs for different scientific questions (e.g., experimental, field based or pattern seeking).

As such, this list represents an upper anchor for any progression in learning. What the steps and levels are for acquiring such knowledge and understanding is an important research topic for empirical investigation.

Although procedural knowledge offers an interesting alternative to the process skill paradigm, there remains one substantive problem; the claim that knowledge in scientific reasoning is *either* conceptual *or* procedural. Whilst such knowledge does address the goals of "knowing that" and "knowing how," "knowing why"—the epistemic goal—is not addressed. Duggan and Gott (2002) attempted to solve this problem by organizing concepts of evidence for scientific reasoning into "inner and outer layers," measuring and interpreting data concepts at *the core,* and understandings about science practices at the periphery. This approach, however, ignores the insights emerging in the past few decades where scientific knowledge is presented as the product of a set of both social and *epistemic* practices (Kuhn, 1962; Latour & Woolgar, 1986; Traweek, 1988). That is, students need to understand the features of scientific practice and their *epistemic* roles for procedural knowledge to have any meaning.

Scientific argumentation and epistemic knowledge

One feature emerging from the normative descriptions offered by philosophers (Kuhn, 1962; Lakatos & Musgrave, 1970; Nersessian, 1989; Popper, 1962) and from the ethnographies by sociologists (Latour & Woolgar, 1986; Pickering, 1983; Traweek, 1988) is a recognition that scientists rely on informal arguments of the type outlined by Toulmin (1958). In such arguments, claims are supported by data whose significance is established through warrants and underlying assumptions. As Quine (1951) pointed out, scientific theories do not exist in isolation but in competition with other theories. Hence, they must be argued for. Establishing the validity of any hypothesis requires both a deductive argument from premises to conclusion, as well as an inferential argument to the best explanation (an abductive argument). A notable example is in the paper by Watson and Crick (1953) proposing the structure for DNA where the authors begin not by arguing *for* their proposed solution but rather why two competing models were flawed. Likewise, Darwin's argument for the theory of evolution is both a hypothetico-deductive argument of the consequences of random mutation and natural selection and an inferential argument that it provides the best explanation for the diversity of species. Yet science education fails to make this kind of meta-level knowledge about the role and type of argument in science—what I wish to term epistemic knowledge explicit.

One significant attempt to remediate this omission was made by Driver, Newton, and Osborne (2000). Drawing on the ideas of science educators (Duschl, 1990), social epistemologists (Fuller, 1997), rhetoricians (Taylor, 1996), and science communicators (Gregory & Miller, 1998) who all argue for presenting science as a process in which scientific knowledge is socially constructed, and in which "discursive activity is central to the process of science" (p. 290), Driver et al. presented an influential paper making the case that, as argument was a core feature of science, it should likewise, be a central feature of science education. Arguing that "if students are to assimilate these norms [of argument], they also need the experience of rehearsing them for themselves" (p. 304), they advanced a case that engaging in argument had a role in developing conceptual understanding, enhancing investigative competence, and understanding the epistemology of science.

Another missing element is that of the role of explanatory models in science. Over the last 50 years, various scholars have identified that a key heuristic for scientific reasoning is building and using models of natural systems (Cartwright, 1983; Giere, 1984; Suppes, 1960). Models offer conceptual representations that enable deductive inferences about the behavior of systems or causal mechanisms. For instance, the wave model of light not only offers an explanation of how light can appear to bend round corners in terms of the concept of diffraction, but also predicts the behavior of light in thin films. Models and modelling are, therefore, "a signature of much research in the sciences" (Nersessian, 2008, p. 59), standard part of its practice and, hence, should be a core feature of science education—a point emphasized by Schauble (1996) who argued that "the goal of scientific reasoning is not primarily the formulation of inductive generalizations, but rather the construction of explanatory models" (p. 103). The philosopher Ron Giere, too, places the construction of models as a central feature of explaining scientific reasoning (Giere, 1984; Giere, Bickle, & Mauldin, 2005). Recognition within science education was marked in 2000 when the edited volume "Developing Models in Science Education" appeared (Gilbert & Boulter, 2000). Further contributions to developing model-based curricula can be found in Wells, Hestenes, and Swackhamer (1995), Van Driel and Verloop (1999), Grosslight, Unger, Jay, and Smith (1991) and Smith, Snir, and Grosslight (1992).

What this body of scholarship suggests is that the breadth and complexity of scientific reasoning requires the recognition that *epistemic knowledge* taught explicitly to students is a third element of a knowledge of science. Moreover, over the past two decades the field has been wrestling with how to incorporate such elements. The movement to address the nature of science specifically (Lederman, 1992;

Matthews, 1989) can be seen as one attempt to incorporate some aspects of epistemic knowledge. Likewise, Millar and Osborne's notion of "ideas-about-science" (1998) or the more generic term "how science works" are other formulations of what is a distinct body of knowledge *about science itself*. Coupled with empirical evidence suggesting that nature of science knowledge is a significant element of any science education (Osborne, Collins, Ratcliffe, Millar, & Duschl, 2003), such arguments have led to epistemic knowledge becoming a more prominent feature of contemporary curricula. Indeed, the 2007 version of the English national curriculum (Qualifications and Curriculum Authority, 2007) included a specific component entitled "How Science Works" while the *Framework* outlines eight scientific practices which should be an essential element of K-12 science (NRC, 2012a) though notably this document does not define any specific elements of epistemic knowledge.

The 2015 PISA assessment framework (OECD, 2012) does incorporate epistemic knowledge as consisting of two components—a) a knowledge of the constructs and defining features essential to the process of knowledge building in science, and b) the role of these constructs in justifying the knowledge produced by science. The features of the first component are a knowledge of:

- The nature of scientific observations, facts, hypotheses, models and theories;
- The purpose and goals of science (to produce explanations of the material world) as distinguished from technology (to produce an optimal solution to human need), what constitutes a scientific or technological question and appropriate data;
- The values of science e.g., a commitment to publication, objectivity and the elimination of bias; and
- The nature of reasoning used in science e.g., deductive, inductive, inference to the best explanation (abductive), analogical, model-base.

Whilst the features of the second component are a knowledge of:

- How scientific claims are supported by data and reasoning in science;
- The function of different forms of empirical inquiry in establishing knowledge, their goal (to test explanatory hypotheses or identify patterns) and their design (observation, controlled experiments, correlational studies);
- The function of a scientific hypothesis in enabling a testable prediction;
- How measurement error effects the degree of confidence in scientific knowledge;
- The use and role of physical, system, and abstract models and their limits;
- The role of collaboration and critique and how peer review helps to establish confidence in scientific claims; and
- The role of scientific knowledge, along with other forms of knowledge, in identifying and addressing societal and technological issues such as the prevention of disease, the supply of sufficient water, food, and energy and climate change.

Undoubtedly, such a list is partial and open to debate. Notably, though, the role of modelling is incorporated in this list. In addition, the lack of specificity in this list inevitably leaves it open to the critique of what exactly is meant. Does the failure to elaborate these constructs make them of little value? One response to this criticism is that the PISA framework states that only an understanding of what might be reasonably expected of an average 15 year old should be tested. While that does not answer the question, it does set limits. Moreover, I would argue that such a list performs the important first step of identifying and pointing towards what kind of knowledge is an essential requisite for scientific reasoning and is an important contribution to the solution of the second problem—the lack of an adequate recognition of the knowledge bases needed for scientific reasoning.

The latest vision of science elaborated in the *Framework* sees science as consisting of a set of social practices which take place in three "spaces of activity": a space for experimentation and investigation; a space for developing explanations, hypotheses, and models which enable prediction; and a space for critique and evaluation—elements which are captured by Figure 15.1 and which form the basis of the model in the chapter on practices (NRC, 2012a). The notable feature of this representation of science is that the role of modelling and the development of theoretical explanations are seen as being core features of science, as is argumentation from evidence. In that sense it goes a long way to remediating the common perception that the central practice of science is experimentation.

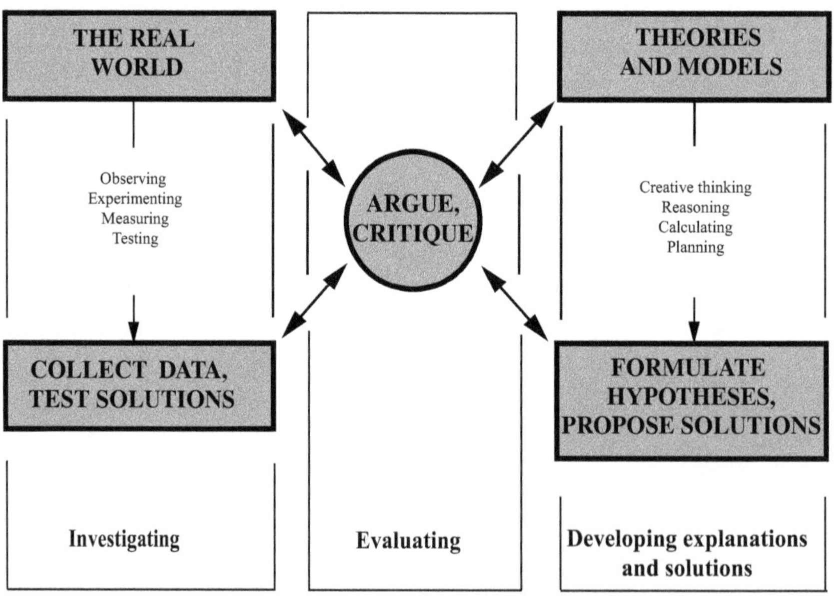

FIGURE 15.1 Scientific reasoning model; combining Klahr and Dunbar (1988) and Giere (2006)

Nevertheless, whilst the map is a representation of forms of scientific activity, it does not provide a vision of what it means to reason scientifically. To be a representation of scientific, as opposed to historical or other forms of reasoning, such activity is dependent on domain-specific disciplinary ideas, *procedural* and *epistemic* knowledge. While the *Framework* elaborates in great detail the first of these, the latter two are notable by their absence. Could sustained engagement in scientific inquiry and the practices of science develop the kind of knowledge of science that we might seek to attain as Duschl and Grandy (2012) argue? Such an argument is based on a view that knowledge is acquired by immersion in a practice. The first counter argument is that there is little evidence that the professional scientists develop a deeper philosophical or sociological understanding of science than the educated layperson. Lakatos memorably commented of scientists' that "most scientists tend to understand little more *about* science than fish about hydrodynamics" (Lakatos, 1970, p. 148). Second, the time available for compulsory science education is insufficient to develop epistemic and procedural knowledge without explicit reflection on practice and its purpose in attaining the goals of science. The goal of science education is not to develop a high level of proficiency with the practices of science, but rather, to use the engagement in a limited repertoires of scientific practices as a means of building as deep a knowledge and understanding of the content of science, the procedures that it commonly uses to achieve its goals, and the epistemic justification for their use as time permits. If so, then it is important that we should define the features of procedural and epistemic understanding that should be the outcome of any education in science and teach them *explicitly*. Asking students to engage in scientific practice must be seen as a means to an end. And, if so, it helps to be clear about what those ends might be and to have a model of science that has a degree of philosophical credibility (that is an accepted normative description) and a sociological and psychological credibility (in that it fits best with the empirical data). Only when armed with this kind of understanding can the teacher act as an effective guide to why any practice is an essential activity of science.

In summary, scientific reasoning may therefore be seen as requiring the use of three distinct types of knowledge: knowledge of scientific facts, theories, and concepts (content knowledge); procedural knowledge; and epistemic knowledge. It is, perhaps, self-evident that generating a scientific hypothesis requires content knowledge; less evident are the findings of research (Reiner & Gilbert, 2000) that suggest that epistemic knowledge or resources also plays a crucial role in its production. Experimentation, similarly, is not just a matter of knowing *how* to get reliable data—which is a procedural issue—but also *why* this is important—which is an epistemic issue. It is true that sometimes one phase of activity may be solved without involving all types of knowledge. Klahr (2000), for instance, has shown that his subjects generated correct hypotheses to a problem without considering any experimental data. But scientific reasoning is never dependent solely on *one* type of knowledge.

Conclusion and Implications for Practice

This chapter commenced by identifying two problems that have troubled science education for a long time and that, without resolution, will continue to cause confusion and, at worst, sustain poor pedagogic practice. The first problem identified was the lack of a definition of the domain-specific knowledge requirements for scientific reasoning. The appeal of a domain-general definition is that it would suggest that the capability developed by scientific reasoning is readily transferable to other areas. However, only if we can identify why any form of reasoning is unique to science can we defend the teaching of reasoning *in* science rather than in *any* subject. If, after all, reasoning is essentially domain-general, the argument that science offers some kind of unique training of the mind is flawed.

The second problem was the failure to identify the specific forms of knowledge needed for scientific reasoning. Finley (1983) saw the absence of domain-specific knowledge as the main challenge to Gagné's skills-based rationale for the teaching of science. The overemphasis on the domain-general at the expense of the domain-specific has inhibited the practice of science education itself. Nowhere has this emphasis been more evident than in psychologists' use of *content-lean tasks*, in which the aim has been to study reasoning without "interference" of the candidate's conceptual understanding—see for instance, D. Kuhn (1991).

The response to these two problems has been to provide a rationale for scientific reasoning which recognizes that it is a domain-specific activity that requires three categories of domain-specific knowledge: scientific conceptual knowledge, procedural knowledge, and epistemic knowledge. These forms of knowledge are seen as a requirement for the 3 competencies required to be scientifically literate within the new PISA framework for 2015, and the chapter has sought to elaborate a rationale for some of the features of procedural and epistemic knowledge. This is not to say that all cognitive skills are context-bound, but rather that the development of all reasoning must begin within a context—an argument first made by Perkins and Salomon (1989).

Engaging in scientific reasoning commonly requires at least two, if not all, three types of knowledge. The challenge for the field then is how such bodies of knowledge might be best taught. For instance, recent research (Allchin, 2011; Ford, 2008) challenges the common assumption that students first must learn the declarative features of such knowledge and then apply this in authentic situations. Ford (2008) conceptualizes the learning underlying experimental and epistemic understanding as a "grasp of practice," suggesting that they are learned by *enculturation* rather than explicit teaching. However, such an approach places too much emphasis on the acquisition of knowledge through practice and is challenged by evidence that suggests that such ideas need to be explicitly addressed (Lederman, 1992; Peters, 2009). School science does not have the luxury of time that is afforded the graduate student. Rather, engaging the student in any scientific practice should be seen as an opportunity to illuminate explicitly the nature and complexity of the knowledge and skill that is required of the professional scientist, just as the novice's attempts to play any musical instrument are used to build an understanding of what is necessary for a polished performance.

What then are the implications for science education? Only an account captured by Figure 15.1 or something similar demonstrates why it is theories and models that are the "crowning glory of science" and not empirical inquiry. In short, science is fundamentally about the ideas we have developed to explain the material world (Oakeshott, 1933). The phenomena are events to be explained but not the key feature of science itself. Second, it requires science educators to establish some kind of consensus about the features of epistemic and procedural knowledge that should be an essential element of the K-12 science curriculum.

The argument here has been that the failure to place scientific reasoning at the center of the science curriculum can be attributed to the lack of clarity of its role and function in learning science. To borrow from Kuhn's notion of *paradigms* within a discipline, this portrayal of the confusion and range of competing perspectives within science education resembles what Kuhn calls the period of *pre-science*, the period in which there is clearly a phenomenon of interest, worthy of study, but no coherence or agreement about its theoretical framing within the community. Present knowledge is now such that the time has come for the establishment of a period that is more akin to *normal science* where we can coalesce around a more unified vision of the nature of scientific reasoning and its role and function within any science education, and the knowledge bases required. This will require a consensus about its main characteristics and a common language within the community. Only then, can we offer students a less-distorted account of the scientific enterprise.

Note

1. Schwab chose to spell enquiry with an 'e'.

16

RESPONSE 1

Views from Above and Below: Access to Science Education

Nancy Brickhouse

The chapters by Osborne and by Biddle and Schafft (this volume) provide analytical lenses for viewing student opportunity to engage in science. Osborne's chapter provides a view "from 30,000 feet," articulating an intellectual framework for international standards and assessments, whereas the Biddle and Schafft chapter describes anthropologic work "on the ground," (i.e., the highly localized experience of contemporary boomtowns in the Pennsylvania's Marcellus Shale region of the United States and the challenges it raises for K-12 science educators).

From an equity perspective, these two chapters represent two very different frames for understanding educational opportunity. One frames educational opportunity in terms of access to high quality science teaching that is informed by well-articulated goals for science literacy. The other frames educational opportunity in terms of how local educational institutions help communities grapple with everyday experiences. I will first discuss each of the chapters individually in terms of implications for equitable access to science education and then conclude with some remarks about the white space between the two and how the ideas in these two chapters can somehow begin to reference one another. My argument will be that Osborne's view from above and Biddle and Schafft's view from below are both indispensible in achieving an accessible and equitable science education.

Jonathan Osborne's chapter represents a very important theoretical articulation of the conceptual, procedural, and epistemic knowledge needed in achieving scientific literacy. For many of us working in the realm of critical science literacy, these are highly valued components of scientific literacy because they are central to the evaluation of scientific knowledge and to the appreciation of scientific knowledge as a socio-cultural activity that advances through argumentation (Stanley & Brickhouse, 1994).

Osborne is right to focus his attention on the new Programme for International Student Assessment (PISA) and Next Generation Science Standards (NGSS). PISA is supported by an assessment used in countries across the globe to benchmark progress in achieving scientific literacy for 15-year-old students (OECD, 2012). While the NGSS have yet to be implemented, they have been adopted by 14 states in the US. The development of assessments that evaluate student performance in all three dimensions of the NGSS standards—core ideas, crosscutting concepts, and science and engineering practices—is underway (NRC, 2014). Internationally, educational systems that have had the most success with providing students access to science have done so by adopting a common set of high standards and aligning all other parts of the educational system to support student learning of these standards. For example, Darling-Hammond (2010) has documented how Finland, Korea, and Singapore created successful educational systems. While these countries invest wisely across the educational system (especially in teacher education), getting the standards right was important as a basis to support the development of effective curriculum development and teacher education programs.

What standards embody is the *intended or official curriculum*, i.e., "The curriculum set by officials at the national or system (school district, educational region, etc.) level that embodies the nation or system's goals and traditions" (Schmidt & Maier, 2009, p. 544). We know from the International Education Association (IEA) studies that there is considerable variability regarding how well the *intended curriculum* translates into the *implemented curriculum*, i.e., the instruction put in place in the classroom. For example, in Japan, the national standards drive both the content of instructional resources as well as the use of instructional time (Schmidt & Maier, 2009). Thus, the intended curriculum is synchronous with the implemented curriculum. In the US, alignment between standards, assessments, instructional materials, and instructional time has not been as strong. For example, based on research from the 1990s, we know that for 8th grade mathematics, the content of textbooks has had the greatest impact on teaching (Schmidt & Maier, 2009)—textbooks that are not strongly aligned to rigorous content standards. However, recent federal reforms begun by No Child Left Behind have stronger accountability measures for teaching common standards, causing changes in schools to align instructional time to tested content (Banilower et al. 2013). We do not yet have assessments that adequately evaluate growth in the full range of student knowledge and skills. The current system distorts reformers' intentions by providing strong incentives for schools to focus instruction on easily measured educational outcomes (Hamilton, Stecher, & Yuan, 2012).

Meadow's definition of a system (cited in Litzinger, this volume) is "an interconnected set of elements that is coherently organized in a way that achieves something." While we have a set of interconnected elements in the US educational system—teacher education programs in public, private, and for-profit institutions; curriculum developers; testing companies, federal, state, and local education agencies; schools; teachers; informal education providers—what is achieved is mediocre outcomes for most and inequities predicated on social class and race. An alignment between these elements to enable us to reach the goal of providing access and excellence to science education remains elusive due to inadequate systems of support that enable teachers and schools to implement high quality instruction in support of the standards (Cohen & Moffitt, 2009).

For those like Osborne who seek to improve science education for all by articulating appropriate goals and learning outcomes, success is highly dependent on aligning these goals with intellectual and material resources throughout the system to assure all students are given the opportunity to learn. Furthermore, the infrastructure must enable institutional learning to facilitate continuous improvement (Senge, 2000) and to facilitate that implementation of a science curriculum that aligns with the intended curriculum but does so in a way that is responsive to local cultures.

The challenge with achieving educational equity by focusing on the improvement of standards and assessments, however, is that schooling occurs in a *place*. The Biddle and Schafft chapter is an on-the-ground articulation of how science education intersects with the everyday concerns of communities. Situated in the Marcellus Shale region of Pennsylvania in the US, they articulate the concerns of a community profoundly affected by the new technologies that enable new natural gas drilling and produce new boomtowns in parts of the US

The local communities in which schools are embedded must not only embrace the educational aims of science education, they must also be able to articulate them in ways that speak to the needs and goals of particular communities. The world is not flat. The world is very, very bumpy—and the realities of a cultural clash between international and national standards and local communities are played across the US every day, most frequently in controversial areas such as climate change and in evolution by natural selection.

What Biddle and Schafft document here is not so much a cultural clash as an official curriculum that does not speak to the needs and goals of the local community. In spite of the fact that the community faces major challenges with the changes brought to their communities by the shale gas industries, the science curriculum, with the force of the standards and assessment regime, is being implemented without modification—in the same way that it would be implemented in urban and suburban communities with very different experiences with the shale gas industries. The changes necessitated by the new standards and those needed by the local community are essentially operating on parallel paths.

Biddle and Schafft express concern that schools are held accountable by distant government agencies whose interests are not the same as those of local communities. The standards that drive accountability are "place-less" in that they address learning outcomes that are the consensus of many people from across many different cultures and contexts, and few resources or incentives are provided to local schools to modulate the standards and develop curricula responsive to the needs to the community.

The answer to the dilemma posed by Biddle and Schafft is not a return to locally developed set of standards and assessments. Historically, this devolution of responsibility to local schools and districts has led to enormous inequity due to highly variable expectations and resources (Cohen & Moffitt, 2009). Children living in low-income communities are often stuck in schools where expectations for them are low and opportunities even lower. In many regions of the US, fundamental scientific theories such as evolution by natural selection and climate change are left out of the curriculum because the local community does not support them. Local responsibility is clearly an experiment that has been tried throughout the US and has failed to achieve the goal of access to a rigorous scientific literacy.

I suggest that a two-pronged approach is needed. One approach is to embrace the articulation of scientific literacy articulated by Osborne. If our standards and assessments were to require teaching not only the content and processes of science, but also the epistemic values of science, students would be far better positioned to grapple with local socio-scientific issues such as those laid out by Biddle and Schafft. A science education that engaged learners in understanding "the use and role of physical, system, and abstract models and their limits" and "the role of scientific knowledge, along with other forms of knowledge, in identifying and addressing societal and technological issues" (Osborne, this volume, p. 227). The articulation of these epistemic goals, as well as effective teaching strategies with professional development and assessments aligned with these goals, is an important first step.

This will not, however, be sufficient. The other approach must be support for local articulation of the broad, general goals of scientific literacy. NGSS includes a substantial amount of content on the use of natural resources and climate change. The role of these new technologies in enabling the wise use of natural resources, of assuring sufficient and adequate amounts of clean water, and the impact of these technologies on climate change could be the foundation of a vital science curriculum that also connects broad, general goals to local concerns.

Finally, Biddle and Schafft also remind us that the concerns of this community cannot be addressed solely by the science curriculum, as defined by NGSS or by PISA. The disciplinary definitions of the academy that have also been imposed on K-12 education are insufficient in preparing students to address the world's most serious challenges. Climate change, in particular, is not a disciplinary problem and will not be solved by science alone. It can only be addressed by influencing individual and collective human behavior and by developing a populace that understands how human cultures operate within economic systems. Only by including insights from the humanities and social sciences regarding the way that science operates in human and cultural contexts will we be able to make science education relevant to the world's most challenging problems. Both of these chapters are important steps in the right direction.

17
RESPONSE 2
The Values of Science Literacy

Nancy Tuana

Jonathan Osborne (this volume) focuses on the goals of science education, arguing that the emphasis of the PISA frameworks is "an education for citizenship" serving to ensure that students will become informed and critical consumers of science and technology. Biddle and Schaft (this volume), in their discussion of science education for those whose communities and lives are being transformed by the Marcellus Shale fracking, examine the practical, pedagogical, and ethical questions surrounding the purpose of education and their relationship with the communities they serve while working to elaborate on some of the ethical dilemmas faced by teachers and students in those communities. What I add to this dialogue is the importance, indeed, the urgency, of ensuring that science literacy includes an understanding of the ethical, as well as epistemic issues relevant to science and technology. I discuss the centrality of values to scientific practice and thereby argue that *science education that is to effectively serve as an education for citizenship must include ethical literacy.*

Osborne contends that an understanding of scientific reasoning is predicated upon understanding the goals of science. These goals are, for Osborne, reflected in the centrality of three questions:

1. What is the material world like? (the ontological question)
2. How can we explain what we observe? (the causal question)
3. How do we know or how can we be certain? (the epistemic question) (Osborne, 2011).

But this schema omits important dimensions of scientific reasoning and practice that involve values and have ethical significance. These include such questions as:

1. What are the relevant values? (values clarification)
2. What are the broader impacts of research? (the ethical question)
3. Why do we choose to focus research on certain topics and sideline or ignore others—known unknowns? (the coupled ethical-epistemic question)

In other words, Biddle and Schafft were quite right in noting the ethical issues relevant to science education that must be faced by science educators as they examine the complex needs of the communities that they serve. But I wish to take their insights further by pointing out that *ethical issues and value judgments are also a part of scientific reasoning and practice*. The realm of science literacy must thus encompass an understanding of the complex role of value judgments in scientific inquiry.

What Are the Relevant Values?

The practice of science involves values in at least three domains: epistemic values, aesthetic values, and ethical values. My focus in this short essay will be on epistemic and ethical values and their role in science. Too often it is mistakenly assumed that the inclusion of values will bias scientific deliberation. While it is true that scientific practice can be distorted by values as has been well demonstrated by scholars who have documented gender or race biases in science (Fausto-Sterling, 1985; Gould, 1981; Haraway, 1989), values also play an important role in scientific research. As documented by the research of Paul Thagard (2012), values legitimately affect scientific deliberations at all stages of research, including the discovery phase, assessments of findings, and practical applications. As Thagard explains, "the decisions that scientists and others need to make about what projects to pursue, what theories to accept, and what applications to enact will unavoidably have an emotional, value-laden aspect. Normatively, therefore, the best course is not to eliminate values and emotions, but to try to ensure that the best values are used in the most effective ways" (p. 300).

Take as an instance the self-reported values of one scientist, the Penn State geoscientist Klaus Keller, who is the Principle Investigator of the National Science Foundation (NSF) funded Sustainable Climate Risk Management Network (SCRiM, scrimhub.org). When asked to explain what motivates his scientific work in the SCRiM network, Keller listed four elements:

1. system understanding,
2. improved approximation to truth,

3. improving lives, and
4. empowerment through education.

What we see in this list is a balance between two types of values. System understanding and improved approximation to truth are *epistemic values*. Improving lives and empowerment through education are *ethical values*.

Predictive power, objectivity, robustness of evidence, and explanatory power are typical epistemic values embraced by the scientific community. However, there are frequently disagreements within the scientific community about how to prioritize them. A recent salient example concerns the inclusion of data about ice sheet melt in sea level rise projections. The question of how best to predict and communicate knowledge regarding future sea level rise has, for example, been one of the most controversial components of recent Intergovernmental Panel on Climate Change (IPCC) reports. As just one example, scientists involved in the Fourth Assessment Report (IPCC, 2007) had to decide whether data on ice sheet melt was sufficiently robust to include in the sea level rise projection models. Here scientists had to balance robustness with predictive power given that a decision not to include ice sheet data would result in the models being over-confident and very likely under-reporting future levels of sea level rise (cf. Rahmstorf, 2010). As we can see even with this brief sketch, decisions between epistemic values, in this case, epistemic robustness and predictive power, are not always clear-cut and scientists can legitimately disagree about how well a criterion is met or how to rank the importance of different epistemic values. As the IPCC reports are designed to provide information for policymakers to base decisions about adaptation, as well as mitigation strategies, clarity about value disagreements is crucial information for policymakers to understand potential tradeoffs.

There are often interactions between epistemic and ethical values. Given that there is always a degree of uncertainty governing the relationship between theories and evidence, Richard Rudner (1953), argues:

> The scientist must make the decision that the evidence is sufficiently strong or that the probability is sufficiently high to warrant the acceptance of the hypothesis. Obviously, our decision with regard to the evidence and how strong is "strong enough" is going to be a function of the importance, in the typically ethical sense, of making a mistake in accepting or rejecting the hypotheses . . . *How sure we need to be before we accept a hypothesis will depend on how serious a mistake would be* (p. 2).

How do we, as Thagard advocates, ensure that *the best values* are used in the most effective ways? An important first step is *values transparency*, in which the values, both epistemic and ethical, that are part of scientific discovery, explanation, and application are clear. Only in this way can these values be subject to the same careful and critical examination found in other dimensions of scientific practice. Thus science literacy requires values clarification and critical examination. Science education thus must include the ability to appreciate and respond to questions such as:

1. What epistemic and/or ethical values are embedded in this instance of scientific practice?
2. What do we need to know to judge which values are relevant?
3. Are there any values embedded that are not relevant and are potentially harmful or distorting?
4. What is the best way to rank the relevant values?
5. Are there any value disagreements within the scientific community and/or between the scientific community and the public relevant to this instance of scientific research and practice?

What Are the Broader Impacts of Research?

Ethical values in science often arise in the context of what the NSF labels the broader impacts of science, or the impacts of science on society. Scientists often have various motivations for the science that they practice. In addition to better understanding, scientists often select research topics that they believe will make a positive difference to society. One of the goals of many climate scientists, for example, is to contribute to human well-being through providing resources for nations and the global community to best prepare for the impacts of climate change.

The NSF's emphasis on the broader impacts of scientific research focuses on impacts on a) the educational and research infrastructure, b) the diversity of the fields themselves, as well as c) supporting projects with positive social impact. The latter is identified by the NSF as "connections between discoveries and their use in service to society" (NSF, 1997, p. 5). This is another domain in which both ethical and epistemic reasoning is relevant to the practice of science and thus is another reason education for science literacy must include ethical literacy. Included within this domain is a careful examination of such issues as:

- the effects of technological innovations on social ends such as health and well-being;
- whether pressing social and economic issues are likely to be addressed and if so, who benefits;
- the role of science in policy-making; and
- ethical concerns arising from the impact of society upon science, for example the impact of funding on research trajectories or the ways in which widespread societal biases can impact research trajectories.

As I previously argued (Tuana, 2010), literacy about the broader impacts of science includes a host of issues, such as:

- proper interpretation and application of the precautionary principle;
- furthering epistemically and ethically adequate societal understandings of issues of scientific uncertainty and risk;
- the role of scientists in the development of policy relevant research;
- the ethical responsibilities of scientists whose research reveals risks, for example, to human health or ecosystem sustainability.

What Is Known and What Is Ignored?

Science is constrained both by time and funding. Obviously not all questions can be explored and there are always complex trade-offs between what can be known at a particular time and what types of scientific research are in fact pursued. One additional question for the type of ethically informed science literacy I am advocating is the ability to weigh the pros and cons of different research trajectories. If we think of science education as "an education for citizenship," we need to ensure that students will be able to participate in the complex decisions about what knowledges are most worthy of resource allocations. This is a skill that is essential to determine how best to invest in research and it is one that scientists must reflect on in choosing their personal research trajectory. This is a domain in which the values of individual scientists, of the community of scientists, and of funders of scientific research once again come into play.

A topic is deemed worthy of research attention for a variety of reasons: it can be seen as a key to advancing an important body of knowledge; it can be seen as potentially lucrative; it might be judged to be essential to national security; etc. In addition, industry or other funding agencies may catalyze interest in a topic by funneling large sums of research dollars directed to a particular domain. We see this phenomenon, for example, with pharmaceutical funding.

Examining the role of interest on research extends far beyond the selection of the research topic and can impact, both negatively and positively, the trajectory of the research. For example, numerous studies have documented how interests, such as financial motivations, can actually distort what is ultimately known and what is ignored about a particular topic. The recent study of the negative impacts of pharmaceutical funding, *Bad Pharma: How Drug Companies Mislead Doctors and Harm Patients* (Goldacre, 2013), contends that the drug industry goes beyond simply financing clinical trials, but routinely withholds negative data and conducts trials on unrepresentative groups. Studies of the impact of the US tobacco industry on funding research, as well as lobbying government, are similarly revealing (Proctor, 1995) and show links between the tobacco industry's sponsored "manufacture of doubt" about the health risks of active and passive smoking and the industries who sponsored research to create doubt about environmental concerns, including anthropogenic climate change (Oreskes & Conway, 2010).

But what we know and do not know can also be distorted by unconscious bias. As argued in "The Speculum of Ignorance" (Tuana, 2006), the dearth of scientific knowledge in the arena of male contraceptive pills was the result of a series of gender biased beliefs including the belief that men would be less motivated than women to use such a contraceptive and that men would be less willing than women to accept the side effects of such a pill. As a result of those biases, little funding was made available for such research.

Conclusion

I offer an admittedly brief overview of three additional domains of knowledge relevant to science literacy that I would add to Osborne's list. In sum, if our goal for science education is to educate students in such a way that they can become critical consumers of science and technology, we cannot ignore the complex linkages between science literacy and ethical literacy.

Poster Presentation IV

Fossil Mammals and Biotic Response to Climate Change

Carla Zembal-Saul[1], Mark Merritt[1], Elizabeth Hufnagel[3], Russell Graham[2]
[1]College of Education, Penn State University [2]College of Earth and Mineral Sciences, Penn State University
[3]School of Education, University of Pittsburgh

Scientific Understandings and Practices

- Oranisms' responses to environmental change
- Ecological systems thinking
- Ethical and epistemic outcomes:
  - Constructing scientific explanations (CER)
  - Illuminating ethical dilemmas through emotions

Scientific Research Context

About Russ...

- Director of the Penn State Earth and Mineral Sciences Museum
- Vertebrate paleoecologist focusing on the dynamics of evolution, biogeographey, and extinction with respect to environmental change
- Cofounder of the Neotoma database, an online hub for data, research, education, and discussion about paleoenvironments
- Extensive experience in the excavation and analysis of Ice Age mammal sites and the taphonomy of fossil deposits
- Field work at cave sites in the mid-content of the United States, in particular, pit caves in the Black Hills of South Dakota

Teaching Applications

- Providing highly scaffolded opportunities for our students to work with small groups in local 1st and 5th grade classrooms
- Creating resources to be considered for the "for educators" section of the Neotoma database website

Constructing Scientific Explanations

- Introduced the Claims-Evidence-Reasoning framework (McNeill, Lizotte, Krajcik, & Marx, 2006)
- Used throughout the course with class investigations and assessments, and in students' teaching experience
- Helps make thinking visible (formative assessments)

Illuminating Ethical Dilemmas through Emotions

- Focus on students' emotions because they are inherently evaluative (Solomon, 2007; White, 2000).
- Students were asked their "topic emotions" (Brougnton et al., 2013) or emotions about related disciplinary ideas.
- Students' emotions and objects of their emotions provided insights into how students relate climate to their lives including:
  - how they perceived themselves as part of the problem,
  - if they desired to help address the problem.
  - whether or not they knew what to do about the problem, and
  - how they understood climate change as a scientific theory.
- Findings prompted real-time curricular and pedagogical course revisions.
- While valuable to the design and teaching of the course, Beth's dissertation research also closely examines students' emotions to highlight ways in which students respond to learning about climate change.

Real Data

- Students used screen wash concentrate from Russ' research sites to sort, identify, and solve problems related to paleoecological reconstructions.
- Students used the Neotoma Database to map large scale changes in biotic communities and ecosystems in response to past global climate changes.
- Students participated in investigations common to K-8 classrooms where their evidence was constructed from data generated first-hand.

Broader Impacts

- Input from science teacher educators in developing the educational components of the Neotoma database website
- Students' use of the Neotoma database provided feedback for improving the usability and efficiency of the database for others
- Science teacher educators and practicing K-12 teachers visiting and working at Russ' active cave research sites
- Students contributing to Russ' active research project by doing an initial sort of the screen wash concentrate as part of class investigations

References

Broughton, S., Sinatra, G., & Nussbaum, E (2013). "Pluto has been a planet my whole life!" Emotions, attitudes, and conceptual change in elememtary students' learning about Pluto's reclassification. *Research in Science Education, 43*, 529–550.

McNeill, K. L., Lizotte, D. J.,Krajcik, J., & Marx, R. W. (2006). Supporting students' construction of scientific explanations by fading scaffolds in instructional materials. *The Journal of the Learning Science, 15*(2), 153–191.

Soloman, R. C. (2007). *Ture to our feelings: what our emotions are really telling us*. New York, NY: Oxford University Press.

White, G. M. (2000). Representing Emotional Meaning: Category, Metaphor, Schema, Discourse. In M. Lewis & J. M. Haviland-Jones (Eds.), *Handbook of Emotions* (2nd ed., pp. 30–44). New York, New York: Guilford Press.

Theme IV References

Abdalla, C.W. (2010). Water withdrawals for development of Marcellus Shale Gas in Pennsylvania. University Park, PA: Penn State College of Agricultural Sciences, Cooperative Extension.

Adlesic, T., Gandour, M., Fox, J., Roma, D., & Fox, J. (2010). *Gasland 2010.* United States: New Video Group. Retrieved from https://youtube.com/watch?v=6mp4ELXKv-w&feature=youtube_gdata_player

Allchin, D. (2011). Evaluating knowledge of the nature of (whole) science. *Science Education, 95*(3), 518–42. http://doi.org/10.1002/sce.20432

American Association for the Advancement of Science. (1989). *Science for all Americans: A Project 2061 report on literacy goals in science, mathematics and technology*. Washington, DC: AAAS.

Anderson, L.W., & Krathwohl, D.R. (Eds.). (2001). *A Taxonomy for Learning, Teaching, and Assessing: A Revision of Bloom's Taxonomy of Educational Objectives.* New York: Longman.

Armstrong, H.E. (1902). The heuristic method of teaching. *School Science and Mathematics, 1*(8), 395–401.

Avery, L.M., & Kassam, K.-A. (2011). "Phronesis": Children's local rural knowledge of science and engineering. *Journal of Research in Rural Education, 26*(2), 1–18.

Banilower, E.R., Smith, P.S., Weiss, I.R., Malzahn, K.M., Campbell, K.M., & Weis, A.M. (2013). *2012 NSSME Report of the 2012 National Survey of Science and Mathematics Education.* Chapel Hill, NC: Horizon Research, Inc. Retrieved from http://horizon-research.com/2012nssme/research-products/reports/technical-report/

Black, P. (1990). Science—The past and the future. *School Science Review, 72*(258), 13–28.

Bloom, B.S. (1956). *Taxonomy of Educational Objectives; The Classification of Educational Goals, Handbook 1, Cognitive Domain.* New York: Longmans.

Brasier, K.J., Filteau, M.R., McLaughlin, D.K., Jacquet, J., Stedman, R.C., Kelsey, T.W., & Goetz, S.J. (2011). Residents' perceptions of community and environmental impacts from development of natural gas in the Marcellus Shale: A comparison of Pennsylvania and New York case studies. *Journal of Rural Social Sciences, 26*(1), 32–61.

Burke, E. (1909). *On Taste. On the Sublime and Beautiful. Reflections on the French revolution. A Letter to a Noble Lord.* New York: Collier.

Cartwright, N. (1983). *How the Laws of Physics Lie.* Oxford, UK; New York: Clarendon Press.

Chiappetta, E.L., & Fillman, D.A. (2007). Analysis of five high school biology textbooks used in the United States for inclusion of the nature of science. *International Journal of Science Education, 29*(15), 1847–68. http://doi.org/10.1080/09500690601159407

Chinn, C.A., & Malhotra, B.A. (2002). Epistemologically authentic inquiry in schools: A theoretical framework for evaluating inquiry tasks. *Science Education, 86*(2), 175–218. http://doi.org/10.1002/sce.10001

Cohen, D.K., & Moffitt, S.L. (2009). *The Ordeal of Equality: Did Federal Regulation Fix the Schools?.* Cambridge, MA: Harvard University Press.

Confederacion de Sociedades Cientificas de España. (2011). *Informe ENCIENDE. Enseñanza de las Ciencias en la Didáctica Escolar para edades tempranas en España.* Madrid, Spain: Confederacion de Sociedades Cientificas de España.

Corbett, M.J. (2007). *Learning to Leave: The Irony of Schooling in a Coastal Community.* Halifax, Nova Scotia: Fernwood Publications.

Darling-Hammond, L. (2010). *The Flat World and Education: How America's Commitment to Equity Will Determine Our Future.* New York: Teachers College Press.

Driver, R., Leach, J., Millar, R., & Scott, P. (1996). *Young People's Images of Science.* Buckingham: Open University Press.

Driver, R., Newton, P., & Osborne, J. (2000). Establishing the norms of scientific argumentation in classrooms. *Science Education, 84*(3), 287–312. http://doi.org/10.1002/(SICI)1098-237X(200005)84:3<287::AID-SCE1>3.0.CO;2-A

Duggan, S., & Gott, R. (2002). What sort of science education do we really need? *International Journal of Science Education, 24*(7), 661–79. http://doi.org/10.1080/09500690110110133

Duggan, S., Gott, R., Lubben, F., & Millar, R. (1996). Evidence in science education. In M. Hughes (Ed.), *Teaching and Learning in Changing Times* (pp. 149–68). Oxford, UK: Blackwell Publishers.

Duschl, R. (2008). Science education in three-part harmony: Balancing conceptual, epistemic, and social learning goals. *Review of Research in Education, 32*(1), 268–91. http://doi.org/10.3102/0091732X07309371

Duschl, R.A. (1990). *Restructuring Science Education: The Importance of Theories and Their Development*. New York: Teachers College Press.

Duschl, R.A., & Grandy, R. (2012). Two views about explicitly teaching nature of science. *Science & Education, 22*(9), 2109–39. http://doi.org/10.1007/s11191-012-9539-4

Duschl, R.A., & Grandy, R.E. (Eds.). (2008). *Teaching Scientific Inquiry: Recommendations for Research and Implementation*. Rotterdam, Netherlands: Sense Publishers.

Engelder, T. (2009, August 18). Marcellus. *Fort Worth Basin Oil & Gas Magazine*.

Fausto-Sterling, A. (1985). *Myths of Gender: Biological Theories about Women and Men*. New York: Basic Books.

Fensham, P.J. (1985). Science for all: A reflective essay. *Journal of Curriculum Studies, 17*(4), 415–35. http://doi.org/10.1080/0022027850170407

Finkel, M.L., & Law, A. (2011). The rush to drill for natural gas: A public health cautionary tale. *American Journal of Public Health, 101*(5), 784–85. http://doi.org/10.2105/AJPH.2010.300089

Finley, F.N. (1983). Science processes. *Journal of Research in Science Teaching, 20*(1), 47–54. http://doi.org/10.1002/tea.3660200105

Ford, M. (2008). Disciplinary authority and accountability in scientific practice and learning. *Science Education, 92*(3), 404–23. http://doi.org/10.1002/sce.20263

Fuller, S. (1997). *Science*. Minneapolis: University of Minnesota Press.

Giere, R.N. (1984). *Understanding Scientific Reasoning*. New York: Holt, Rinehart, and Winston.

Giere, R.N. (2006). *Understanding Scientific Reasoning*. Belmont, CA: Thomson/Wadsworth.

Giere, R.N., Bickle, J., & Mauldin, R. (2005). *Understanding Scientific Reasoning* (5 Edition). Belmont, CA: Cengage Learning.

Gilbert, J., & Boulter, C.J. (2000). *Developing Models in Science Education*. Dordrecht, Netherlands: Kluwer Academic Publishers.

Goldacre, B. (2013). *Bad Pharma: How Drug Companies Mislead Doctors and Harm Patients*. New York: Faber and Faber.

Gott, R., & Murphy, P. (1987). *APU Science Reports for Teachers 9: Students Aged 13 to 15—National STEM Centre*. Department of Education and Science. Retrieved from http://nationalstemcentre.org.uk/elibrary/collection/729/apu-science-reports-for-teachers-students-aged-13-to-15

Gott, R., & Mashiter, J. (1991). Practical work in science—A task-based approach. In B.E. Woolnough (Ed.), *Practical Science: The Role and Reality of Practical Work in School Science* (pp. 53–66). Milton Keynes, UK: Open University Press.

Gott, R., & Duggan, S. (1996). Investigative work in school science. *International Journal of Science Education, 18*(7), 791–806.

Gott, R., Duggan, S., & Roberts, R. (2008). *Concepts of Evidence.* Durham, UK: University of Durham.

Gould, S.J. (1981). *The Mismeasure of Man.* New York: Norton.

Gregory, J., & Miller, S. (1998). *Science in Public Communication, Culture, and Credibility.* New York: Basic Books. Retrieved from http://public.eblib.com/choice/publicfullrecord.aspx?p=625033

Grosslight, L., Unger, C., Jay, E., & Smith, C.L. (1991). Understanding models and their use in science: Conceptions of middle and high school students and experts. *Journal of Research in Science Teaching, 28*(9), 799–822. http://doi.org/10.1002/tea.3660280907

Gruenewald, D.A. (2003). The best of both worlds: A critical pedagogy of place. *Educational Researcher, 32*(4), 3–12.

Hamilton, L.S., Stecher, B.M., & Yuan, K. (2012). Standards-based accountability in the United States: Lessons learned and future directions. *Education Inquiry, 3*(2), 149–70. http://doi.org/10.3402/edui.v3i2.22025

Haraway, D.J. (1989). *Primate Visions: Gender, Race, and Nature in the World of Modern Science.* New York: Routledge.

Hill, C.T. (2008). The post-scientific society. *Issues in Science and Technology, 24*(1), 78–84.

Huxley, T.H. (1854). *On the Educational Value of the Natural History Sciences.* London: J. Van Voorst.

Intergovernmental Panel on Climate Change. (2007). *Fourth Assessment Report Climate Change 2007.* Cambridge, UK: Cambridge University Press.

James, A., & Aadland, D. (2011). The curse of natural resources: An empirical investigation of US counties. *Resource and Energy Economics, 33*(2), 440–53. http://doi.org/10.1016/j.reseneeco.2010.05.006

Klahr, D. (2000). *Exploring Science the Cognition and Development of Discovery Processes.* Cambridge, MA: MIT Press.

Klahr, D., & Simon H. A. (1999). Studies of scientific discovery: Complementary approaches and convergent findings. *Psychological Bulletin, 125*(524–43).

Koslowski, B. (1996). *Theory and Evidence the Development of Scientific Reasoning.* Cambridge, MA: MIT Press.

Krathwohl, D.R. (2002). A revision of bloom's taxonomy: An overview. *Theory Into Practice, 41*(4), 212–18. http://doi.org/10.1207/s15430421tip4104_2

Kuhn, D. (1991). *The Skills of Argument.* Cambridge, UK: Cambridge University Press.

Kuhn, T.S. (1962). *The Structure of Scientific Revolutions.* Chicago: University of Chicago Press.

Ladlee, J., & Jacquet, J. (2011). *The Implications of Multi-Well Pads in the Marcellus Shale* (No. 43) (pp. 1–4). Ithaca, NY: Cornell University Community and Regional Development Institute.

Lakatos, I. (1970). Falsification and the methodology of scientific research programmes. In I. Lakatos & A. Musgrave (Eds.), *Criticism and the Growth of Knowledge* (pp. 91–196). Cambridge, UK: Cambridge University Press.

Lakatos, I., & Musgrave, A. (Eds.). (1970). *Criticism and the Growth of Knowledge.* Cambridge, UK: Cambridge University Press.

Latour, B., & Woolgar, S. (1986). *Laboratory Life: The Construction of Scientific Facts* (2nd ed.). Princeton, NJ: Princeton University Press.

Layton, D. (1973). *Science for the People: The Origins of the School Science Curriculum in England.* New York: Science History Publications.

Lederman, N. G. (1992). Students' and teachers' conceptions of the nature of science: A review of the research. *Journal of Research in Science Teaching, 29*(4), 331–59. http://doi.org/10.1002/tea.3660290404

Lederman, N. G. (2007). Nature of science: Past, present and future. In S. K. Abell & N. G. Lederman (Eds.), *Handbook of Research on Science Education* (pp. 831–879). Mahwah, NJ: Lawrence Erlbaum Associates.

Li, M., & Shavelson, R. J. (2001). Examining the links between science achievement and assessment. In *Annual Meeting of the American Educational Research Association*. Seattle, WA.

Longino, H. E. (1990). *Science as Social Knowledge: Values and Objectivity in Scientific Inquiry*. Princeton, NJ: Princeton University Press.

Lubben, F., & Millar, R. (1996). Children's ideas about the reliability of experimental data. *International Journal of Science Education, 18*(8), 955–68. http://doi.org/10.1080/0950069960180807

Lyons, T. (2006). Different countries, same science classes: Students' experiences of school science in their own words. *International Journal of Science Education, 28*(6), 591–613. http://doi.org/10.1080/09500690500339621

Malin, S. (2013). There's no real choice but to sign: Neoliberalization and normalization of hydraulic fracturing on Pennsylvania farmland. *Journal of Environmental Studies and Sciences, 4*(1), 17–27. http://doi.org/10.1007/s13412-013-0115-2

Matthews, M. R. (1989). A role for history and philosophy in science teaching. *Interchange, 20*(2), 3–15. http://doi.org/10.1007/BF01807043

Millar, R., & Driver, R. (1987). Beyond processes. *Studies in Science Education, 14*(1), 33–62. http://doi.org/10.1080/03057268708559938

Millar, R., & Osborne, J. (1998). *Beyond 2000: Science Education for the Future: A Report with Ten Recommendations*. London: King's College London.

Millar, R., Lubben, F., Gott, R., & Duggan, S. (1994). Investigating in the school science laboratory: Conceptual and procedural knowledge and their influence on performance. *Research Papers in Education, 9*(2), 207–48. http://doi.org/10.1080/0267152940090205

Monk, M., & Osborne, J. (1997). Placing the history and philosophy of science on the curriculum: A model for the development of pedagogy. *Science Education, 81*(4), 405–24. http://doi.org/10.1002/(SICI)1098-237X(199707)81:4<405::AID-SCE3>3.0.CO;2-G

Moss, K. (2008). *Potential Development of the Natural Gas Resources in the Marcellus Shale: New York, Pennsylvania, West Virginia and Ohio*. Denver, CO: National Park Service, US Department of the Interior, Geologic Resources Division.

Murphy, P., & Gott, R. (1984). *APU Science Report for Teachers 2: Science Assessment Framework at Age 13 and 15—National STEM Centre*. London, UK: Department of Education and Science. Retrieved from http://nationalstemcentre.org.uk/elibrary/resource/3102/apu-science-report-for-teachers-2-science-assessment-framework-at-age-13-and-15

National Assessment Governing Board. (2008). *Science framework for the 2009 National Assessment of Educational Progress*. Washington, DC: US Department of Education.

National Research Council. (1996). *National Science Education Standards*. Washington, DC: National Academies Press.

National Research Council. (2008). *Research on future skill demands: A workshop summary*. Washington, DC: National Academies Press.

National Research Council. (2012a). *A Framework for K-12 Science Education: Practices, Crosscutting Concepts, and Core Ideas*. Washington, DC: National Academies Press. Retrieved from http://www.nap.edu/catalog/13165/a-framework-for-k-12-science-education-practices-crosscutting-concepts

National Research Council. (2012b). *Education for life and work: Developing transferable knowledge and skills in the 21st century*. Washington, DC: National Academies Press.

National Research Council. (2014). *Developing assessments for the next generation science standards*. Washington, DC: National Academies Press.

National Science Foundation. (1997). *GPRA Strategic Plan, FY 1997-FY 2003*. Retrieved from http://www.nsf.gov/od/gpraplan/gpraplan.htm

Nersessian, N. (2008). Inquiry: How science works; Model-based reasoning in scientific practice. In R.A. Duschl & R.E. Grandy (Eds.), *Teaching Scientific Inquiry: Recommendations for Research and Implementation* (pp. 57–79). Rotterdam, Netherlands: Sense Publishers.

Nersessian, N.J. (1989). Conceptual change in science and in science education. *Synthese, 80*(1), 163–83. http://doi.org/10.1007/BF00869953

Oakeshott, M.J. (1933). *Experience and its Modes*. Cambridge, UK: The University Press.

Oreskes, N., & Conway, E.M. (2010). *Merchants of Doubt: How a Handful of Scientists Obscured the Truth on Issues from Tobacco Smoke to Global Warming*. New York: Bloomsbury Press.

Organization for Economic Cooperation and Development (OECD). (2012). The PISA 2015 Assessment Framework: Key competencies in reading, mathematics and science. Retrieved from http://www.oecd.org/pisa/pisaproducts/pisa2015draftframeworks.htm

Osborne, J. (2011). Science teaching methods: A rationale for practices. *School Science Review, 93*(343), 93–103.

Osborne, J., & Collins, S. (2001). Pupils' views of the role and value of the science curriculum: A focus-group study. *International Journal of Science Education, 23*(5), 441–67. http://doi.org/10.1080/09500690010006518

Osborne, J., Collins, S., Ratcliffe, M., Millar, R., & Duschl, R.A. (2003). What "ideas-about-science" should be taught in school science? A Delphi study of the expert community. *Journal of Research in Science Teaching, 40*(7), 692–720. http://doi.org/10.1002/tea.10105

Passmore, C., & Stewart, J. (2002). A modeling approach to teaching evolutionary biology in high schools. *Journal of Research in Science Teaching, 39*(3), 185–204. http://doi.org/10.1002/tea.10020

Perkins, D., & Salomon, G. (1989). Are cognitive skills context-bound? *Educational Researcher, 18*(1), 16–25. http://doi.org/10.3102/0013189X018001016

Peters, E.E. (2009). Developing content knowledge in students through explicit teaching of the nature of science: Influences of goal setting and self-monitoring. *Science & Education, 21*(6), 881–98. http://doi.org/10.1007/s11191-009-9219-1

Pickering, A. (1983). *Constructing Quarks*. Edinburgh, UK: Edinburgh University Press.

Pifer, R.H. (2011). What a short, strange trip it's been: Moving forward after five years of Marcellus Shale development. *University of Pittsburgh Law Review, 72*(615), 615–60.

Polanyi, M. (1958). *Personal Knowledge: Towards a Post-critical Philosophy*. Chicago: University of Chicago Press.

Popper, K.R. (1962). *Conjectures and Refutations: The Growth of Scientific Knowledge*. New York: Basic Books.

Proctor, R. (1995). *Cancer Wars: How Politics Shapes What We Know and Don't Know About Cancer*. New York: Basic Books.

Qualifications and Curriculum Authority. (2007). *Programme of Study: Science KS4*. London: Qualifications and Curriculum Authority.

Quine, W.V. (1951). Main trends in recent philosophy: Two dogmas of empiricism. *Philosophical Review, 60*(1), 20–43.

Rahmstorf, S. (2010). A new view on sea level rise. *Nature Reports Climate Change*, (1004), 44–45. http://doi.org/10.1038/climate.2010.29

Reiner, M., & Gilbert, J. (2000). Epistemological resources for thought experimentation in science learning. *International Journal of Science Education, 22*(5), 489–506. http://doi.org/10.1080/095006900289741

Roschelle, J., Bakia, M., Toyama, Y., & Patton, C. (2011). Eight issues for learning scientists about education and the economy. *Journal of the Learning Sciences, 20*(1), 3–49. http://doi.org/10.1080/10508406.2011.528318

Roth, W. M. (1995). *Authentic School Science: Knowing and Learning in Open-inquiry Science Laboratories*. Dordrecht, Netherlands: Kluwer Academic Publishers.

Rudner, R. (1953). The scientist qua scientist makes value judgments. *Philosophy of Science, 20*(1), 1–6.

Ryle, G. (1949). *The Concept of Mind*. London: Hutchinson.

Sandoval, W. A. (2003). Conceptual and epistemic aspects of students' scientific explanations. *Journal of the Learning Sciences, 12*(1), 5–51.

Schafft, K. A. (2010). Economics, community, and rural education: Rethinking the nature of accountability in the twenty-first century. In K. A. Schafft & A. Y. Jackson (Eds.), *Rural Education for the Twenty-first Century: Identity, Place, and Community in a Globalizing World* (pp. 275–90). University Park, PA: Pennsylvania State University Press.

Schafft, K. A., & Biddle, C. (2014). School and community impacts of hydraulic fracturing within Pennsylvania's Marcellus Shale region, and the dilemmas of educational leadership in gasfield boomtowns. *Peabody Journal of Education, 89*(5), 670–82. http://doi.org/10.1080/0161956X.2014.956567

Schafft, K. A., Borlu, Y., & Glenna, L. (2013). The relationship between Marcellus Shale gas development in Pennsylvania and local perceptions of risk and opportunity. *Rural Sociology, 78*(2), 143–66. http://doi.org/10.1111/ruso.12004

Schauble, L. (1996). The development of scientific reasoning in knowledge-rich contexts. *Developmental Psychology, 32*(1), 102–19.

Schauble, L., Klopfer, L. E., & Raghavan, K. (1991). Students' transition from an engineering model to a science model of experimentation. *Journal of Research in Science Teaching, 28*(9), 859–82. http://doi.org/10.1002/tea.3660280910

Schmidt, W. H., & Maier, A. (2009). Opportunity to learn. In G. Sykes, B. L. Schneider, & D. N. Plank (Eds.), *Handbook of Education Policy Research* (pp. 541–59). New York: Routledge.

Schunn, C. D., & Anderson, J. R. (1999). The generality/specificity of expertise in scientific reasoning. *Cognitive Science, 23*(3), 337–70. http://doi.org/10.1016/S0364-0213(99)00006-3

Schwab, J. J. (1962). The concept of the structure of the discipline. *Educational Record, 43*(3), 197–205.

Sekretariat der Ständigen Konferenz der Kultusminister der Länder in der Bundesrepublik Deutschland (KMK). (2005). *Bildungsstandards im Fach Biologie für den Mittleren Schulabschluss (Jahrgangsstufe 10)*. Berlin: Sekretariat der Ständigen Konferenz der Kultusminister der Länder in der Bundesrepublik Deutschland (KMK).

Senge, P.M. (2000). *Schools that Learn: A Fifth Discipline Fieldbook for Educators, Parents, and Everyone who Cares about Education*. New York: Doubleday.

Simon, H. A. (1966). Scientific discovery and psychology of problem solving. In R. G. Colodny (Ed.), *Mind and Cosmos: Essays in Contemporary Science and Philosophy* (pp. 22–40). Pittsburgh, PA: University of Pittsburgh Press.

Smith, C., Snir, J., & Grosslight, L. (1992). Using conceptual models to facilitate conceptual change: The case of weight-density differentiation. *Cognition and Instruction, 9*(3), 221–283. http://doi.org/10.1207/s1532690xci0903_3

Sobel, D. (2004). *Place-based Education: Connecting Classrooms & Communities* (1st ed.). Great Barrington, MA: The Orion Society.

Soeder, D.J., & Kappel, W.M. (2009). Water resources and natural gas production from the Marcellus Shale (pp. 1–6). Reston, VA: US Department of the Interior, US Geological Survey.

Solomon, J., Scott, L., & Duveen, J. (1996). Large-scale exploration of pupils' understanding of the nature of science. *Science Education, 80*(5), 493–508. http://doi.org/10.1002/(SICI)1098-237X(199609)80:5<493::AID-SCE1>3.0.CO;2–6

Spillane, J.P., Reiser, B.J., & Reimer, T. (2002). Policy implementation and cognition: Reframing and refocusing implementation research. *Review of Educational Research, 72*(3), 387–431. http://doi.org/10.3102/00346543072003387

Stahl, L. (2010). *Meet the "shaleionaires"* (Vol. 43). New York: CBS. Retrieved from http://www.metacafe.com/watch/cb-7uw5fb4BqGyR/extra_meet_the_shaleionaires_season_43_episode_42/

Stanley, W.B., & Brickhouse, N.W. (1994). Multiculturalism, universalism, and science education. *Science Education, 78*(4), 387–98. http://doi.org/10.1002/sce.3730780405

Suppes, P. (1960). A comparison of the meaning and uses of models in mathematics and the empirical sciences. *Synthese, 12*(2–3), 287–301. http://doi.org/10.1007/BF00485107

Taiwan Ministry of Education. (1999). *Curriculum outlines for "Nature Science and Living Technology."* Taipei, Taiwan: Taiwan Ministry of Education.

Taylor, C.A. (1996). *Defining Science: A Rhetoric of Demarcation*. Madison, WI: University of Wisconsin Press.

Thagard, P. (2012). *The Cognitive Science of Science Explanation, Discovery, and Conceptual Change*. Cambridge, MA: MIT Press.

Theobald, P. (2009). *Education Now: How Rethinking America's Past Can Change its Future*. Boulder, CO: Paradigm Publishers.

Tobias, S. (1990). *They're Not Dumb, They're Different: Stalking the Second Tier*. Tucson, AZ: Research Corp.

Toulmin, S. (1958). *The Uses of Argument*. Cambridge, UK: Cambridge University Press.

Traweek, S. (1988). *Beamtimes and Lifetimes: The World of High Energy Physicists*. Cambridge, MA: Harvard University Press.

Tuana, N. (2006). The speculum of ignorance. *Hypatia: A Journal of Feminist Philosophy, 21*(3), 1–19.

Tuana, N. (2010). Leading with ethics, aiming for policy: New opportunities for philosophy of science. *Synthese, 177*(3), 471–92. http://doi.org/10.1007/s11229-010-9793-4

Van Driel, J.H., & Verloop, N. (1999). Teachers' knowledge of models and modelling in science. *International Journal of Science Education, 21*(11), 1141–53. http://doi.org/10.1080/095006999290110

Watson, J.D., & Crick, F.H.C. (1953). A structure for deoxyribose nucleic acid. *Nature, 171*, 737–38.

Weiss, I.R., Pasley, J.D., Smith, P.S., Banilower, E.R., & Heck, J.D. (2003). *A Study of K–12 Mathematics and Science Education in the United States*. Chapel Hill, NC: Horizon Research, Inc.

Wells, M., Hestenes, D., & Swackhamer, G. (1995). A modeling method for high school physics instruction. *American Journal of Physics, 63*(7), 606.

Wilber, T. (2012). *Under the Surface: Fracking, Fortunes and the Fate of the Marcellus Shale*. Ithaca, NY: Cornell University Press.

Williamson, J., & Kolb, B. (2011). *Marcellus Shale Initiative Publications Database*. Williamsport, PA: Lycoming College Center for the Study of Community and the Economy. Retrieved from http://www.marcellus.psu.edu/resources/PDFs/housingreport.pdf

THEME V

STEM Communication and Policy Outreach

18

WHY PEOPLE CARE ABOUT CHICKENS AND OTHER LESSONS ABOUT RHETORIC, PUBLIC SCIENCE, AND INFORMAL LEARNING ENVIRONMENTS

Stacey Pigg, William Hart-Davidson, Jeff Grabill, and Kirsten Ellenbogen

When, where, and how do people learn science? In response to this question, the National Academy of Sciences report "Learning Science in Informal Environments" (National Research Council [NRC], 2009) stressed the importance of everyday experiences, designed spaces like museums and science centers, non-school science education programs, and science media. The report built on an array of scholarship attuned to science learning as a lifelong, often self-motivated endeavor. The findings are not surprising. In all cases, we spend more of our lives learning outside of classrooms and other formal learning institutions than we do inside them (Gerber, Cavallo, & Marek, 2001). The situation is analogous when we think about when, where, and why people engage public science. Often the scholarly literature focuses on deliberation in related normative forums, yet most of us engage science issues in ways (and in places) less structured and more connected to circumstances of daily life (Barron, 2006; Falk, Storksdieck, & Dierking, 2007). Indeed, in these less structured forums, what we do would not often be considered "deliberation" at all by scholars. This is particularly true for learning and engagement online, which can be easily understood as too messy to be useful (Grabill & Pigg, 2012).

We made a number of claims in that first paragraph, and in what follows, we intend to be more precise in developing our thinking. Our chapter dwells in the informal and the digital and focuses as well on learning as important to science communication. Specifically, we argue that scientific communication in an age of social media might benefit from focusing more on creating engaging and *useful* learning environments than on transmitting scientific information. Online learning environments can tap into interests of public participants and facilitate engagement with both scientific concepts and scientific ways of thinking and making knowledge. With their ability to act as both collective memory and epistemic medium, online writing environments can involve participants in ongoing inquiry-based learning. This is a fundamentally different way of thinking about the role writing can play in mediating public encounters with science—as a tool for knowledge creation, rather than reception.

In this chapter we provide a descriptive example of this approach to science communication through findings from a recent study of the Science Museum of Minnesota's *Science Buzz*, a popular blog that uses writing to create space for conversation about current science. Our larger study provides us with a way to characterize the nature of the discourse on *Science Buzz* and to discuss indicators of learning in this forum. However, we spend much of our chapter discussing one key blog entry entitled "The Chicken and the Egg," which has generated hundreds of public comments over several years. This thread, known as the "Chicken Thread," demonstrates a model of knowledge building driven by the interest and identities of participants who leverage experiences caring for chickens to engage science. To be more precise, the participants are most often discussing chicken *eggs* and exploring the conditions of incubation. A text analysis of this thread to isolate important terms reveals that two words, in particular, stand out as used far, far more often than others (by an exponent of nearly 3, in fact): 1) "egg" and 2) "if." When we read the threads in both systematic ways as researchers and as interested observers following the conversation, we are reminded that science is driven by quite practical questions like "If the eggs get too cold, will they still hatch?"

But is this how we think of science in policy discussions about enhancing STEM learning? Do we understand the forms of reasoning central to "doing science" to be grounded in deeply practical questions and answers? The answer, we suggest, is not nearly often enough. With science itself driven by questions of application and practical import and embedded in complex political, ethical, and linguistic webs, it is only natural that online learning and engagement environments should open conversation to practical scientific questions that matter deeply to citizens. In response, we forward this conversational dynamic of shared concern, multiple stakeholder engagement through practical questioning and answering, and persistence through writing as a model for twenty-first century online science learning.

Complexities of Public Science Conversation, Engagement, and Learning

Public understanding of science scholarship has long called for engagement efforts that place scientists and non-scientists in conversation about matters of concern (see Irwin & Michael, 2003). How these dialogues are to be structured, however, and to what ends remains a matter of debate. For example, Bickerstaff,

Lorenzoni, Jones, and Pidgeon (2010) identified tensions in Royal Society National Forum dialogues facilitating public discussions between scientists and non-scientists. They argued that within the National Forum project, "choices about problem framing and purpose, roles and identities, as well as the detail of the day's structure, were predefined by the Society" (p. 482–3). With responsibility for framing solely in the institution's hands, members of the public were given "very little capacity to either shape the trajectory of debate or, subsequently, the ways in which the outputs were (institutionally) interpreted and deployed" (p. 489).

Bickerstaff et al. (2010) suggested that argument framing, or shaping the parameters of discussion, often becomes a latent influence controlled by *someone* or *something* in public engagement discourse. When tightly controlled framing undercuts the public's capacity to participate, it recalls the perceived hierarchies between scientists and members of the public that have been the focus of public understanding of science scholarship for twenty years now. Wynne (1992), for example, argued that the social gulf between stakeholder groups led to the difficulty for scientists and non-scientists to learn from one another's expertise. As Wynne pointed out, with different "social identities" come different understandings of argument and how it functions. Discussing post-Chernobyl interactions between scientific authorities and Cumbrian sheep farmers, Wynne showed how "[e]ach side only recognized even as possible evidence, claims expressed within its cultural style" (p. 296). Thus, Wynne concluded that significant differences between scientists and non-scientists are generally "social conventions" that distribute "relative authority in ways which may be inappropriate" (p. 297). Renegotiating these hierarchies is key to productive conversations between scientists and non-scientists. Gross (1994) connected this conversation to rhetoric studies through his focus on the contextual model's "interaction between science and its publics," whereby public understanding becomes "the joint creation of scientific and local knowledge" (p. 6).

In spite of this work, assumptions like those foundational to a "deficit model" of public science knowledge have not disappeared. As Bauer, Allum, and Miller (2007) have described, survey-based measures of scientific literacy in the 1980s measured public knowledge by "quiz-like textbook items," and low scores on individual items in these inventories were often reported as "indicators of public ignorance and reasons for moral panic" (p. 81). Indeed, with the current similar perceived crises related to high-stakes standardized test performance across educational sites and levels, science learning contends not only with scientific institutional hierarchies and tensions but also educational ones. The problem with this scenario is not only its negative positioning of publics, but also its narrow view of science, which becomes reduced to a mass collection of facts divorced from contexts of production and use.

Against this backdrop, however, recent approaches to science learning emphasize the need for developing habits of mind that lead to inquiry and ongoing interaction with science over time. These approaches reframe science learning not as passive reception of established facts but rather an active process of interacting with—not merely aligning with—scientific ways of knowing. Recognizing that science language use in general (Lemke, 1990) and scientific argumentation in particular (Bricker & Bell, 2008; Duschl & Osborne, 2002) are central to science learning, Goldman, Duschl, Ellenbogen, Williams, and Tzou (2003) focused on the role online discourse can play in creating "thinking spaces" that support peer argumentation and scaffold dialogues in which learners reason through science thinking together. According to Goldman et al., online systems for science dialogue mediate thinking space differently for at least two reasons: they restructure spatial/temporal expectations of participation, and they make insights concrete and visualized: part of a collective memory as a "written record of the dialogue" that is "available for inspection and reflection at any time" (p. 3).

We build on this scholarship as we call for increased attention to the possibilities new web-based social writing platforms offer for facilitating science learning and engagement through interactive communication. Outside of classrooms, it is hard to imagine individuals encountering the scaffolded systems Goldman and her colleagues describe. However, less structured networked writing systems like blogs and social networking sites are already an active part of many individuals' social lives (see Lenhart, Purcell, Smith, & Zickuhr, 2010). These online writing environments support characteristics of exchange that have proven successful in classroom implementations. For example, social software generally create collective memory through written archives that can be revisited and extend participation possibilities by making conversations available to individuals who would not be able to attend formal dialogues due to space/time constraints. In addition, with their logic built on a flattened participatory rather than top-down transmission model, online networked writing systems hold promise for building new forms of social community (Wellman & Guilia, 1999; Zappen, 2005). Although power dynamics clearly do not disappear online (Nakamura, 2010), social writing spaces can potentially support written interaction among individuals with different forms of expertise, an important component of both science engagement and learning.

Science Buzz: Public Science Meets Online Participation

Before moving to our extended example of online science learning and engagement, we should first provide context about the particular online science forum we are discussing, as well as the methods used to systematically analyze it. *Science Buzz* is a nationally recognized blog maintained by the Science Museum of Minnesota (SMM). *Science Buzz* engages a broad audience about relevant science topics ranging from health to earth science to the latest findings from chemistry and physics. The blog postings on *Buzz* are written by a group of moderators who compose short writings intended to garner interest, promote feedback, and sustain ongoing argumentation. Broader conversations develop within the comments attached to these initial postings, and we focus primarily within these conversational comments to discuss public learning and engagement.

We take our analysis in this chapter from "Take Two: A Study of the Co-Creation of Knowledge on Museum Web 2.0 Sites," a multi-year study of the impact of web technologies on museum learning and practice. *Take Two* explored whether and how blogs and other web-based technologies support outcomes that are important to museums and other institutions of informal learning. Supported by the Institute of Museum and Library Services (IMLS), *Take Two* brought together a multi-institutional team of investigators focused on the following questions:

- What is the nature of the community that interacts through *Science Buzz*?
- What is the nature of the online interaction?
- Do these online interactions support knowledge building for this user community?
- Do online interactions support inquiry, learning, and change within the museum—i.e., what is the impact on museum practice?

This chapter reports on the results of the discourse analysis, which focused on participant interaction on *Science Buzz* and enabled us to identify indicators of learning.

Because many Internet-based interactions are written, discourse analysis provides a systematic method for describing, characterizing, and understanding what is happening in online interactions. One of the initial tasks of the study was to build an analytical tool that could accurately describe the discourse on *Science Buzz*. Taking an approach that emphasizes rhetoric in detail (Johnstone & Eisenhart, 2008), our analytical tool focused on how participants build argument, identity, and community, as well as how they explore new ideas. Our data analysis was structured in two phases. In the first phase, we coded the *Science Buzz* sample at the T-unit level focusing on indicators of argument, identity, community building, and the invention of new ideas.[1] In the second phase, we used a more interpretive analytic that built on the coding results of the first phase but was explicitly structured by the National Academies report "Learning Science in Informal Environments: People, Places, and Pursuits."

Our results suggest that the activity and community that form *Science Buzz* are highly complex and variable. Some topics and threads attract a large number of participants with diverse levels of knowledge and perspective and who participate from locations both inside and outside the physical museum. Furthermore, argumentative strategies such as making claims are the most common discursive moves on the *Buzz* blog. These reasoning strategies are the focus of many prior discourse-based studies of concepts like "learning" in group environments. Yet our analysis of the *Buzz* blog also shows a range of strategies that we associate with constructing individual identities (more common) and group identities (less common). These discourse strategies are often ignored in studies of learning or knowledge construction, but they are clearly integral to any understanding of online spaces made available by museums to the public.

Our analysis also shows *Science Buzz* to be a learning environment. When we offer the claim that *Buzz* is an informal learning environment, we are doing so in a precise fashion: We are not claiming that there is learning per se, which is extremely difficult to measure and impossible given the context and methods of this study. Rather, we are claiming that there are *indicators* of learning in evidence in meaningful representations, patterns, and densities. In particular, our discourse analysis identified practices that are aligned with categories of informal learning identified by the National Academies as important, such as developing interest in science, engaging in scientific reasoning, and identifying with the scientific enterprise.

For example, as we look across our sample, we see considerable evidence that participants are interested in science and that they are compelled to participate in conversations about science for a number of reasons ranging from entertainment to intellectual engagement to fear and anxiety. We also see indicators of scientific reasoning in each thread, including a number of argument moves (e.g., use of evidence). In addition, there is evidence of inquiry (asking questions and engaging in a process of answering them). We see numerous examples of participants applying ideas, concluding with and from evidence and thinking with others in conversation. In some threads, we also see sophisticated reasoning practices and people thinking together to work through ideas. Finally, identity work is persistent and complex. In some threads, participants are authorizing "expert" and/or "expert-like" identities. Sometimes these dynamics lead those who identify most strongly with the scientific enterprise to dominate threads and ignore questions from those less knowledgeable or comfortable expressing expertise. However, this is not always the case. Performing identities also allows people to draw from experience in productive ways and call for answers to questions of concern. This is particularly important in threads where people are motivated to discuss scientific issues because of their relationship to problems that directly affect them.

Science Buzz is thus substantial in terms of the amount of science discourse it has generated, as well as with the extent to which it aligns with what we know about how people learn about science in informal environments. As we have suggested, the various threads on the site differ considerably in how these key dynamics of interest, reasoning, and identification develop through conversation. The remainder of this chapter explores the intersections of informal learning, science communication, and web technologies in detail by paying close attention to one exceptional blog conversation (in terms of participation) that has been a part of *Science Buzz* for over five years. Within this exchange, we see high levels of interest and argument—participants thinking together in writing—and through discourse moves such as these, they construct an engaging learning environment.

Caring about and for Chickens: The Chicken and the Egg

Begun in Fall 2006, "The Chicken and the Egg" (see www.sciencebuzz.org/blog/chicken-and-egg) has received 850 public comments at the date of this draft (with more added every day). Key contributors to the thread include a museum blogger and staff facilitator, a poultry scientist from the University of Minnesota, and a number of public posters (some self-identified and others remaining anonymous). Our analysis revealed that "The Chicken and the Egg" contained high amounts of argument. Nearly half of the T-units in the Chicken Thread (48 percent) were assigned to our code, "Building an Argument." This is not surprising compared to our larger sample. Where the Chicken Thread most stands out compared to other *Science Buzz* conversations is in the intersection of this argumentation with discursive moves that build community connectedness. Around eighteen percent of T-units (17.71 percent) in the Chicken Thread were assigned the code, "Building Community Identity," compared to an average of 11.5 percent across our total sample. According to our analytical scheme, this meant that the Chicken Thread contained many discursive acts in which participants explicitly requested that others explain or respond, articulated shared experiences or roles, or connected their contributions concretely to ideas or people mentioned previously in the thread. This initial analysis helped us understand that the reasoning patterns of the Chicken Thread extended beyond a collection of disconnected claims. Participants were (and are) talking to and thinking with one another.

The co-presence of engagement and argumentation moves is a substantial finding in the context of our larger study, which did not always show evidence of blog commenters creating community or engaging one another. Our closer narrative analysis of the Chicken Thread helped us to understand how these moves functioned in this particular case (and perhaps gives clues to why these particular moves were not present in others). To draw on the language of the "Learning Science in Informal Environments" report, a key to the discursive patterns of the thread is high individual interest. Participants care about chickens. Perhaps even more importantly, they care *about* chickens because they care *for* chickens. As stewards, their public interest in conversing and learning more about poultry and poultry science is less driven by a sense of "wonder" than of "application" (Fahnestock, 1986, p. 279). When individuals are this deeply invested in a topic or problem, an online writing environment like *Science*

Buzz has the potential to become an online learning environment, sustaining engagements that facilitate learning and communication. Within the Chicken Thread, we see this play out as forms of reasoning connected to application and action that prompt continually more nuanced accounts about science. This inquiry model is similar to the dynamics of scientific rhetoric in more official genres and discourse communities. Our stance here further supports the substantive role for rhetoric outlined by Gross as a means to reconstruct both the facts and the methods of science in ways that mesh with matters of importance to the general public. In order to further describe this dynamic, we first characterize the Chicken Thread's discourse in more detail before focusing on two short samples from the conversation that illustrate themes we have introduced.

"The Chicken and the Egg" and the Discourse of Engagement and Learning

Like all "threads" on *Science Buzz*, "The Chicken and the Egg" begins with a short blog post created by museum staff to facilitate a conversation. In this case, the initial post positions the museum staff as an interested, inquisitive member of the public who has a question relevant to science issues (lines 1–27).[2] She opens simply by describing how she has "noticed a lot of dead fledglings and raided nests lately" and along with one of them "an unhatched baby bird." This act of noticing, the writer says, led her to a question about everyday experience: "Are grocery store eggs fertilized?" (line 3) and to a set of follow-up questions: "Why do you never seem to see a grocery store egg with a baby chicken in it? And, if fertilized eggs are the norm, why would the chicken 'spend the energy required to produce unfertilized eggs?'" (line 5).

Her initial post proceeds to describe her next steps: after conducting an informal web search, the staff member found relevant evidence from scientists to answer her questions. She cites the Howard Hughes Medical Institute to describe how chicken eggs form and develop and provides a simple discussion of the major differences between fertilized and non-fertilized chicken eggs (e.g., if kept at the right temperature, an egg that has been fertilized will "differentiate to form a chick, which will hatch after 21 days" whereas a nonfertilized will never differentiate and, thus, yolk material remains). We don't see baby chickens in grocery store eggs, she continues, because eggs to be sold in large commercial operations generally come from chickens that never had contact with a rooster. Importantly, the poster claims that she not only found reputable sources to help her answer her question, but also learned that "lots of other people had the same question" (line 8). She furthermore applies what she has learned to revise prior thinking. For example, she mentions how her reading helped her rethink an idea she had been taught as a child: that a "blood spot" means an egg is fertilized. Now she knows that blood spots are just broken blood vessels.

On the *Science Buzz* blog, the initial post is important. It serves a facilitative role that the museum staffer continues in subsequent engagement with thread participants. In this case, however, the initial post also models a process of learning science:

1) The post foregrounds the interest of the writer and her identification with science.
2) The post initiates an inquiry (if not scientific) by asking questions.
3) The post engages—it facilitates—by opening up the learning process to others rather than providing the answer to the original question.

Rather than simply reporting science facts, the poster models an inquiry process and way of thinking and wondering that is central to active learning (and by extension, we suggest, to doing science as well). What makes the Chicken Thread remarkable is that the comments that follow enact a similar process of learning through inquiry, but do so through argumentation that involves multiple participants. The blog comments are largely driven by what appear to be genuine questions from participants (mostly non-scientists) who are interested and/or invested in caring for chickens. Commenters in the Chicken Thread often direct questions to the facilitator, the scientific expert, or to a "you" with an unclear referent; however, people with a range of social positions offer answers based on different forms of reasoning and evidence. Active questioning and responding in the Chicken Thread connect individuals' interests to broader scientific conversations and create shared objects of concern that can be addressed from different points of view.

To provide an example, Figure 18.1 represents one early question and several of its responses. In this example, a public participant writes for information about why her chicken has stopped laying eggs. As Figure 18.1 illustrates, both institutional representatives (museum staff) and other members of the public address the problem of chicken fertility raised by this poster. Notably, however, none of their answers simply settles the issue but rather brings new evidence and possibilities to bear on the question of concern by speaking of the general issue in a newly situated way.

Questions like this one about chicken fertility persist temporally within the thread discourse, as both staff and public posters add to the building collection of responses and information. They also recur and are revisited when new posters enter the conversation. In repeatedly addressing these questions, citations of online authorities become layered with descriptions of concrete experience. At times, a particular kind of evidence might more or less useful to answer a question. For example, as we describe later, there are moments in which the hypothetical response of a scientist is of little use in helping an interested citizen make a contextualized decision about future action. In the same way, a public poster's description of caring for chickens might be irrelevant or mismatched to a more theoretical question. What is consistent, however, is that multiple forms of

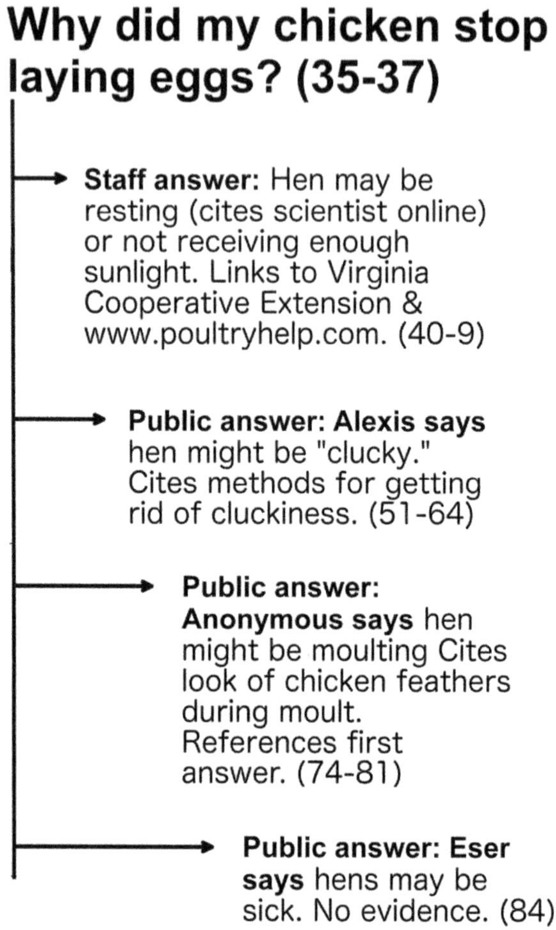

FIGURE 18.1 Sample early question and subsequent responses

response and evidence can be sustained through written engagement. Much like in the initial posting, the point of this exchange is not merely to transmit scientific concepts but to enact an inquiry cycle that does not end with one answer to one question. Science learning, instead, emerges as a sustained process of engagement by asking, answering, and then asking again. This process continually tests the limits of shared understanding by considering related cases of a similar phenomenon. Technologies matter here as well, of course, as a blogging platform (in this case) is also a participant in creating a thinking space in which ideas persist in spite of distances of time and space, dialoguing individuals who might not otherwise have access to one another.

Our description to this point has been concerned with making a narrow but hopefully meaningful argument, and so we have left out other issues of importance. For instance, the dynamic we have described is not flat. That is, hierarchies between experts and non-experts exist. Indeed, it is notable that much of the thread follows an "ask the expert" logic wherein one member of the public poses a question and an expert in the thread answers that question, often by linking to relevant content on the web. From the poster who asks the scientist whether it is possible to find yolk material on commercial chicken meat (line 1053) to the poster who asks for advice about whether it is possible to keep chickens in her neighborhood (line 1069), some questions do not launch broader inquiry. In other moments of the thread, scientific experts draw on markers of status in order to defend knowledge during moments of community negotiation, citing authority as a way to add weight to evidence built from formal scientific procedures. Clearly, power dynamics from outside *Science Buzz* influence production and interpretation of discourse and are part of how the blog functions as an environment. Dynamics that we associate with power-knowledge relationships are operational in *Buzz*, just as they are meaningful in any scientific communication process. However, it is not always the case that power/knowledge asymmetries lead to shutting down conversation and learning. Based on our work with *Science Buzz*, we suggest that such dynamics can also enable public learning. Discursive markers of authority, for instance, are used to model inquiry processes and ways of arguing, add nuance to conversations, or refine ways of asking or answering questions.

Importantly, in the Chicken Thread, science facilitators have the opportunity to model practices important to scientific thinking because public participants inquire in ways that are both useful and personally applicable. Having said this, we want to close with two examples that move beyond the basic inquiry dynamic that we have described so far and toward more subtle engagement practices characterized by participants testing the applicability of scientific knowledge and norms. We trace, first, how scientists and non-scientists differently support the persistent discussion of a question raised by a public poster, adding nuances and applications as commenters situate the question within varied experiences and knowledge bases. We follow this by focusing on a moment of community negotiation in which public posters engage science by questioning the "truth" of a forum facilitator's claim.

Asking toward Action: Why Public Posters Reason about the Difference Between Fertilized and Non-Fertilized Eggs

Within the Chicken Thread, one recurring question—how to tell whether an egg is fertilized—becomes a site from which to observe that public engagement in scientific reasoning (an important indicator of learning) is often driven by questions of future action[3], even when that action is unarticulated. In the post that launches this conversation, a commenter named Tom asks, "Can you clarify the fertilized egg issue . . . if you have a rooster in the hen house, how can you tell which eggs are fertilized and which are not?" (line 197). Addressing an unclear "you" here, he follows his question by discussing why candling—a common method for looking through the shell of an egg to see its contents—would not be an appropriate method to answer his question. Candling is only useful after an egg has been kept at a proper incubation temperature for around 10 days. For Tom, who thinks it would be most economical to eat unfertilized eggs and hatch chickens from fertilized ones, incubating an unfertilized egg would render it inedible because of high incubator temperatures.

The museum staff member who wrote the blog's initial post is the first to take up Tom's question, claiming there is every reason to believe all eggs laid by a hen that has had previous contact with a rooster are fertilized. Elaborating this claim, the staff member assumes that a moral/value imperative had motivated Tom's question, which she addresses in her closing thoughts: "I don't think there's any reason you shouldn't eat a freshly-laid, fertile egg if that's why you're keeping the chickens. After all, if you eat the adults . . . " (line 208). Tom responds to this assumption by pointing out that, in fact, a different imperative had prompted his thinking: "My question is, how do you avoid the waste. I'm not worried about eating fertilized eggs" (line 217).

We understand this as an important moment in the Chicken Thread for illustrating how the public interest that motivates engagement in scientific reasoning often emerges from practical, contextualized problems that require action. While the National Academies report (NRC, 2009) lists participation in scientific reasoning and developing an interest in science as two separate strands of learning, it is clear that the two are closely connected in this case. The science facilitator misreads the particular motive for Tom's question but intuitively realizes that his form of interest and form of reasoning are linked. Tom's response to the staff member is further notable because he does not simply accept the general answer the facilitator provides; he persists by refining his question building on the context that she helps to draw out. Following up, then, Tom continues by practicing a key discursive move of this conversation: he restates his original question, now situating it more clearly within the ethic of care for chickens that motivated it. He suggests that there "must be a more accurate way" of telling the difference between a fertilized and non-fertilized egg to avoid "destroy[ing] a potential chick that could have strengthened your chicken supply" while avoiding the waste of incubating eggs that are not fertilized.

Tom's refined question now invites a new kind of response. At this point, two anonymous posters suggest (again) that Tom candle the eggs to tell whether they are fertilized or not. These answers to the question miss the context Tom built earlier for why candling would not be an effective method for determining fertilization in his particular case. However, while their answers do not help Tom very much, they do prompt additional participation that adds to the ongoing knowledge building within the thread (in this instance, on the problem of fertilization identification). Specifically, their responses prompt a poultry scientist to enter the conversation. The scientist echoes Tom's sentiment that candling will not reveal whether an egg is fertilized until an egg has been incubated (something that wouldn't happen with a egg to be eaten). The role that the scientist plays at this particular point is to continue the conversation. She does not put the issue to rest by offering a definitive answer (something that does happen by those playing the role of "poultry scientist" at other moments in the discourse). Instead, she continues the process of knowledge accumulation that had begun through Tom's question, the museum blogger's response, and the ongoing refining and reiteration.

This practice of knowledge-building continues when a public poster who calls herself "a nonny mouse," echoes the museum staff member's original claim but adds new information, including suggestions on how to hatch chickens and when candling *would* be an appropriate method to determine whether an egg is fertilized. The echoing of this point by another poster who calls herself "I have chickens 2" still does not sufficiently answer the question for purposes of the thread. Almost immediately, a poster named Mel asks a question strikingly similar to Tom's, but differently contextualized: "ok so my neighbor gave me and my friends these three eggs" but "i really want to know if the egg's he gave us are fertilized." Here again, the scientist responds by pointing out that candling will not reveal whether an egg is fertilized under all or even most conditions. The scientist doesn't stop here, though. She uses the opportunity of the open question to build additional knowledge into the conversation. Specifically, she shares a common method for candling, describing the process and linking to descriptions online. The scientist further uses this point to add nuance (via her insertion of hypotheticals) that continue to have resonance both for Tom's original question and for the much earlier question about chicken infertility: "If there are too many roosters for the number of hens, however, you can get a case where the roosters spend more time fighting with each other instead of mating with the hens." Following the scientist, another public poster inserts himself to address the museum staff member's focus on the moral discussion about eating fertilized and non-fertilized eggs. Here, this poster who calls himself "Tim the Enchanter" says, "It's actually a vegetarian issue. Some vegetarians will eat non-fertilized eggs, but fertile eggs are off-limits. And of course, in that case, you wouldn't eat the adults either" (line 341).

We could continue tracing the intricacies of this fertilization conversation for many more pages. The conversation never exactly comes to a halt. For the sake of space, however, we will simply share a few more places in which the question recurs:

- line 707 ("Is there any other way to tell if the egg has been fertilized, other than candling?");
- line 724–5 ("I do not know if the eggs are fertile. Will she continue to lay on the eggs for many weeks if they do not hatch and if so how long should I leave it before removing the eggs?");

- line 812 ("we separated the hens and roosters for a brief time due to a large amount of feathers missing from hens back. Probably a week has gone by and one of the hens is sitting on her eggs, are these eggs fertile?");
- line 975 ("how can you tell if a egg is fertile?"); and
- and line 1156–7 ("I have 10 bantam chicken eggs taken from the farm where I work. They are all in an incubator but how do I know if they are fertilized?").

Seeing this recurring question, we might easily ask, *Why can't they settle on an answer? Why must the question be repeated? Is nobody listening or learning?* What we hope our reading shows is that this iterative questioning process is a means of engagement that demonstrates participant interest, facilitates argument, and therefore helps to construct precisely the sort of environment in which learning is possible. What is notable about the recurring questions around incubation is not only the way rhetoric is used to bring evidence to claims, but also the way rhetoric becomes a medium for a broader construction of understanding. Gross (1994) gives this phenomenon a name: a forum of reconstruction. The "re" in "reconstruction" is especially important for Gross, as the forum functions to forge linkages between scientific knowledge and local knowledge, something that must be done over and over again to establish the truth value of scientific facts relative to what members of the lay public know from experience. We see this played out quite vividly in the Chicken Thread in a number of ways as those who keep chickens raise questions and relate anecdotes about specific conditions and circumstances pertaining to fertilization.

The dynamics of the forum of reconstruction also reveal an important role for the Scientist vis-à-vis the public: that of the facilitator. This role is predicted by Gross' discussion of the way scientific truths in public forums are mediated by the identity of experts, rising and falling in credibility in proportion to the trust accorded to the expert. We would expect to see our poultry expert reiterating scientific facts, but doing so in relation to the subtleties of each new formulation of public understanding of the phenomenon in question. And that is precisely what we see. Yet the discourse practices of public posters are equally important. Tom's initial question initiates multiple points of connection from which others build, and his move to reiterate a question based on a clearer sense of this particular case or context draws new detail into the conversation. Motivated by the issue of how best to care for chickens, Tom's questions also test the boundaries of the applicability of scientific knowledge in everyday life. Questions of chicken stewardship are not strictly scientific. They are practical. They are rich with ethical, economic, and even emotional issues. As the conversation demonstrates, science tells us some things that can inform the conversation. But it will never give a single, definitive answer to such a probabilistic question as: how should I best care for chickens? Thus, public science learning requires a complex synthesis of multiple forms of evidence and knowledge.

This communicative complexity, we suggest, would not be reached through a public engagement model that seeks simply to inform members of the public about science concepts. Nor would it likely emerge as part of dialogues in which scientists over-determine argumentative frameworks, including questions or topics of concern. Acts of questioning launch an exchange of different forms of expertise from descriptions of how one should care for chickens, to lay knowledge about how many roosters is too many roosters, to scientists' takes on the best methods for candling. Posters revisit topics of previous interest that were raised in responses to questions (like when Tim picks back up the moral implications of eating fertilized eggs) and layers of new commentary emerge, even when thread participants ask questions that have already presumably been answered.

When Reasoning Styles Conflict: Why Public Posters Care If Refrigerated Eggs Will Hatch

Our analysis further suggests that conflicts may occur when learning facilitators assume a reasoning process aligned with "wondering" about a topic when public interlocutors are actually reasoning toward decisions or action (Fahnestock, 1986, p. 334). This tension arises during a conversation between several thread participants in response to one poster's plea for advice after her child's beloved pet chicken, Sally, disappeared: "Tonight, I went to put the chickens in for the night and Sally was not around. We have three eggs from the last three days (they are in the ref.) and I was wondering if they would hatch? Please let me know" (line 87). This question was meaningful not only because it involved kids and pets—touchy subjects—but also because it was a recurring question: one that had been raised and addressed previously in the conversation. Just previously, a poster had asked, "Are unfertilized eggs that have been stored in a refrigerator considered dead or alive?" (line 87). The museum blogger at that time responded, first, by suggesting that unfertilized eggs are not living (line 92) and then by citing an expert website that says it is possible to hatch a chicken from a fertilized egg that has been in the refrigerator (under specific circumstances—when the refrigeration happens before incubation and not after).

While this answer is technically correct, the new way of posing the question involving Sally the pet chicken leads several public posters to express dissatisfaction with the staff member's original response. One writes, "If ref. is short for refrigerator and they've been in there for three days the chances of them hatching is 1 to 1,000 so don't even try to do anything about it" (line 114) and another takes issue with the initial poster's claim that an egg in a refrigerator could hatch. He writes, "An egg cannot get below 40 degrees or it is no longer viable. You cannot put a fertilized egg in the refrigerator and expect it to hatch when you put it in the incubator. People need to do research before answering somebodies question" (line 119).

The conflict that arises in this short exchange recalls the dissonance between different stakeholders' forms of reasoning and interest illustrated in early studies like Wynne's. The staff member is attempting to model a process of scientific practice by emphasizing the importance of supporting claims through the evidence of citation and peer review that comes from a record of previous research. However, public posters' reasoning toward action leads them to either misunderstand the facilitator's thinking process or to reject it as inappropriate to address the kinds of interests they have in science.

Naturally, the staff member defends her answer to the question by referencing her citation of evidence: "I cited my source—a reputable expert. And my answer was reviewed, after posting, by local poultry expert [name removed]. So I stand by it" (line 125). However, two public posters next jump in to continue questioning what constitutes an appropriate response in this case. "I have chickens 2" (also an active participant in the first thread sample we discussed) further supports the anonymous posters, claiming "you really need to do some research before answering anyone's questions, they really want the truth and really want to know what to do, how to do it and all the how to's" (line 132). Following up, "I have chickens 2" elaborates this claim, notably drawing on the idea that it is "really like common sense" that eggs from the icebox would become too cold to hatch, strengthening his/her claim through his/her authority as someone with much experience caring for chickens (line 133). Important in both these responses are appeals to practical knowledge of common sense and the need for reflective action when caring for chickens.

The museum staff member, confronted by this continual questioning, suggests a return to analyze the context of the original question. She thus reiterates the initial context for inquiry: "The poster wanted to know if it was possible that eggs kept in the refrigerator for three days, that had never been incubated, could still hatch. And my answer was, it's possible" (line 143). In this discussion, the staff member reveals that her reasoning process was suited to the original question about what was possible, as opposed to the following question about practical action. She follows up by explaining that she never suggested that hatching eggs from a refrigerator was a "recommended practice," but instead just a possibility. Interestingly, a member of the public then steps in to defend the museum blogger's claim, providing evidence from her own experience: "We hatched our pet turkey Charlie from a group of four eggs that were in the refrigerator for about 24 hours. Only one of the eggs hatched so the chances are slim but possible!" (line 153). Finally, the scientist enters the argument at this point to echo the sentiments of the previous two posters: while refrigerating eggs is not recommended practice because "[m]ost refrigerators store feed at 45F, which is too cool for successful hatching of fertile eggs," it is true that "[c]hicks are extremely resilient and anything is possible" (lines 164).

In this exchange, we see a dynamic common in many public science communication contexts: public challenges to scientific expertise. Such challenges can be (and are) understood in many different ways, most of them negative (e.g., uninformed, emotional, NIMBYism). But these participants do not appear to be merely "flaming," and instead offer reasonable challenges to a view of scientific reasoning motivated by generalized knowledge as opposed to practical action. As with the previous vignette, the questions and challenges are not merely scientific, and this is precisely one of the points we wish to make. Public communication of science is almost never simply about science. It is always also about policy, ethics, politics, experience, and so on. Efforts to isolate the technical or scientific issues and the risks associated with them often falter because of their narrowness. New technologies provide mechanisms by which participants can help frame conversations with representatives of scientific institutions. Furthermore, such technologies can help support conversations in which disagreement can happen productively and in ways that encourage participants to reexamine assumptions. In this vein, it is important not only for public posters to revisit prior scientific knowledge but also for science facilitators to develop an emerging sense of what is driving public interest in science and shaping the forms of reasoning that are possible. There are significant cultural issues at stake as well, and the primary way that we have characterized these cultural issues is our focus on "learning" rather than "communication" as perhaps a more productive way to think about structuring public engagement around science.

As we close our discussion, then, we ask: How might we best use the resources of the Internet in the interest of advancing scientific knowledge among the general public? Our investigation suggests that there is still value in creating what Gross (1994) calls "forums of reconstruction" that explicitly encourage experts and members of the lay public to make links between scientific facts and local knowledge. This can happen around issues that not only stir controversy but also coincide with routine matters about which people care deeply. Like caring for chickens. Or, to offer another example, caring for a loved one with a newly diagnosed chronic illness. The need to draw on scientific evidence to guide practice and decision-making in such situations creates tremendous motive for engagement. But the means to engage—the forum for reconstruction—may not be readily apparent or easily accessible for most people. Conceiving of online communicative spaces as environments for active learning and engagement is one possible first step toward reaching out to the significant number of individuals who have interests or concerns like those of the Chicken Thread's participants.

Notes

1. We computed simple agreement for each code combination for a sub-sample of threads. By code combination, we mean that we expected agreement not at the level of the category "building an argument," for instance, but rather at the level of the discourse move within the category ("building an argument, claim"). Agreement was well above 90 percent on most codes.

2. We include transcript line numbers in our detailed analysis of the blog comments in order to help readers orient to how the conversation is unfolding. These line numbers are not available on the *Buzz* blog. However, we have posted a segmented version of the thread in an online Appendix as a reference for readers. At this time, that version of the thread is available at the following online location: http://tinyurl.com/cjnca49.
3. Fahnestock's (1986) application of rhetorical stasis theory to public science provides useful vocabulary for accounting for misunderstandings that can arise between public posters and science facilitators during interactive online conversations. Drawing on a classical rhetorical invention method, Fahnestock describes how stasis "defines and orders the kinds of questions that can be at issue," including questions of conjecture, definition, quality, and finally of action (p. 290).

19

NEW ENVIRONMENTS FOR PROFESSIONAL DEVELOPMENT

Situating Science Learning and Teaching in a Framework and NGSS World

Jean Moon

Introduction

"Among the many challenges facing the transformation of PreK-16 science teaching and learning is that of coordinating conceptual, epistemological and social competencies and learning goals in disciplinary domains" (Duschl, Bismack, Greeno, & Gitomer, this volume, p. 1). This statement takes us to the heart of the *Framework* (National Research Council, 2012a) and the *Next Generation Science Education Standards [NGSS]* (NGSS Lead States, 2013). Together they specify new multi-dimensional targets for students, teachers, preservice and in-service faculty and informal science educators in the contexts of curriculum, professional development, and assessment. Combined they create a whole system of change in how science learning and teaching is enacted (NRC, 2015).

This chapter examines one piece of that system, professional development (PD), and focuses on supporting teachers as they discover how epistemic, knowledge-building practices, such as modeling, argumentation, and explanation, are essential in learning science content. Moon, Passmore, Reiser, and Michaels (2013), Reiser (2013), and Wilson (2013) make the case about the central importance of knowledge-building practices and their role in helping students learn core disciplinary concepts through active engagement in doing science.

The challenges then are: What does it really mean to say we are involving students in using epistemic practices to investigate phenomena or to respond to an engineering challenge? What do these practices look like in a classroom or out-of-school learning environment? What are the indicators that might provide some reality to what the three dimensions for science teaching and learning look like when practiced by students? And, in turn, what new demands do such learning goals place on teaching practice in science? What are promising ways to move forward in supporting science educators to rethink their practice in line with the reform visions put forward in the *Framework* and *NGSS*?

This chapter examines three aspects of the current realities influencing the call for "re-tooling" science teaching, with particular attention given to the central role of epistemic practices and the importance of teachers developing a deep knowledge of those practices. After some comment on the state of current and future professional development in science, we turn to the role cyber-enabled learning environments might play in developing new models of science-based PD. We then explore one cyber-enabled professional learning environment in science education—Next Generation Science Exemplar Learning System or NGSX—as an example of a designed professional learning environment that weaves together a specific core disciplinary idea with epistemic practices. Throughout, a number of questions are addressed frequently coupled with this new vision for science education:

- What qualities of professional learning environments are needed in order to support science educators as they learn, experience, and try out scientific and engineering practices (knowledge-building practices) in the context of disciplinary core ideas?
- What are the larger realities in STEM as part of our 21st century world that are influencing models of teacher learning in science?
- Why are intentional and continuous connections to classrooms in emergent models of professional development so central in helping teachers make critical and lasting changes in their teaching practice?
- How can these shifts in professional learning begin to occur at scale across states in a reasonable time frame and in a cost-efficient way?

The State of Professional Learning for Teachers of Science

At the moment, the state of PD for educators is a smorgasbord, and a disappointing one at that. A lack of sustained attention to research-based findings on what makes effective PD has contributed to the emergence of what Wilson (2013) has characterized as "a carnival of PD options" in the U.S. These options include the abundant "one-shot" one- or two-day workshops focused on topics that teachers can choose to add to their practice without necessarily fundamentally rethinking that practice. By contrast, engaging in the knowledge-building practices identified in the *Framework* will require new models of PD that weave together content knowledge with how that knowledge is learned and by extension taught. Simply put, the new PD models are the critical means to realizing the end goals envisioned by *NGSS*.

Teachers must themselves acquire knowledge-building strategies and new pedagogical practices, as well as receive ongoing support and practice in transferring those practices to the classroom or other learning environments (Alozie, Moje, & Krajcik, 2010; Jiménez-Aleixandre, Rodríguez, & Duschl, 2000; Moon, Singer, & Reiser, 2012; Windschitl, Thompson, & Braaten, 2008). Lampert's (2010) call for PD models in which teachers "learn in, from, and for practice" remains true. Moreover, characterizing the new work of science classrooms as engaging key intellectual or epistemic practices locates the work not in scripted routines, isolated definitions, or recipe-oriented labs, but in classrooms working as a community to co-construct knowledge collaboratively and progressively over time, using knowledge-building practices such as modeling to develop robust content understandings (Passmore & Svoboda, 2012).

For example, successfully orchestrating public reasoning in a science classroom and building an overall culture of learning through productive discussion is improvisational work, calling for a new level of skill in listening and observing students (Michaels, O'Connor, & Resnick, 2008). The PD that is needed to prepare teachers for this kind of unscripted, complex pedagogy requires guided practice with colleagues, as well as support with resources that are accessible and continuous. This kind of PD can be connected productively to one's classroom through high quality video cases of students and teachers engaged in making shifts in their teaching practice (Moon et al., 2013, 2012; Reiser, 2013; Wilson, 2013). The development of these models of PD are particularly critical for reforms, such as NGSS, that call for change not only in what is taught, but also with how it is taught (Reiser, 2013).

Learning and Knowledge in the 21st Century

In her seminal *Science* article (2013), Suzanne Wilson calls for a more complex view of teacher learning. Based on a corpus of research on professional learning, she sees PD as a more dynamic and iterative experience, connecting teachers' experiences in classrooms with formal opportunities for collective reflection on shared problems of practice, sensemaking, and trying out high leverage epistemic practices such as argumentation, making one's reasoning with evidence public, and modeling (Windschitl, Thompson, Braaten, & Stroupe, 2012). Wilson's call for a more complex view of teaching and learning is well founded. Simply put, we are now located in a time where how we learn, how we teach, where we learn, what we learn, and what we use as tools for learning has made a profound shift with the dawn of this century—and why such a shift?

First, a rapidly accumulating corpus of knowledge on learning in and out of formal educational environments, particularly over the last two decades, is supporting new pedagogical approaches, especially those mediated through the use of technology-based hardware and software. Second, emerging cyber-enabled learning and working environments are providing models rich in functionality and support for learning and working. Third, the rapid rise of interdisciplinary and transdisciplinary domains of study reflect epistemic practices as critical tools in the work of these new discipline arenas. Arguably the confluence of these events is driving a seismic shift in education in general, as well as in science education.

Education has to play catch-up with the realities of learning and working in the 21st century in which knowledge and the relationship of the learner and worker to knowledge is changing (Bereiter & Scardamalia, 2012). This message has clear implications for teachers, learners, researchers, and policy makers, helping to situate what is occurring in science learning and teaching within a larger context of recurring themes across disciplines, educational levels, professions, and learning environments.

The emphasis on learning practices and competencies embedded in new content standards is well aligned with messages contained in a wide variety of national reports, not just the *Framework, NGSS,* and the *Common Core State Standards* (*CCSS*) (National Governors Association Center for Best Practices & Council of Chief State School Officers, 2010). *A New Biology for the 21st Century* (NRC, 2009) makes the case that the kinds of complex problems to be solved in the 21st century in the areas of health, water, air quality, the environment, climate, food, quality of life, and energy require the integration of core concepts and practices across scientific disciplines and sub-disciplines, aided by an ever increasing corpus of sophisticated technologies. *Vision and Change in Undergraduate Biology Education* (American Association for the Advancement of Science, 2011) stresses the role of core competencies, much like those contained in the *Framework*, as essential to begin preparing biologists to work successfully in problem-solving or "figuring out" contexts. *The Engineer of 2020* (NRC, 2004) embraces cross-disciplinary fertilization and interdisciplinary engagement with non-engineering disciplines.

Further, the competencies for pre-medical and medical students outlined in *Scientific Foundations for Future Physicians* (Association of American Medical Colleges and Howard Hughes Medical Institute, 2009) are largely interdisciplinary, drawing heavily on a range of science disciplines and mathematics. Similarly an emphasis on using knowledge was identified in the NRC's study on "deeper learning," *Education for Life and Work* (NRC, 2012b). The study committee defined deeper learning as the process through which a person becomes capable of taking what was learned in one situation and applying it to new situations—in other words, learning for "transfer."

While knowledge of core subject matter in science must be a primary outcome of schooling, the ability to generate new knowledge, to solve complex problems or create new products in a small group or team format, and to transfer knowledge to new and novel situations is becoming increasingly significant across subject matter and educational levels. The shift is away from an emphasis on merely knowing to an emphasis on using knowledge. This is the larger societal, educational, and STEM context in which science teaching and learning in an

NGSS world and the professional learning for teachers essential for supporting it are now deeply situated.

What can be learned from reviewing these reports and others like them is that connecting core disciplinary ideas with epistemic practices is not the sole province of K-12 education (NRC, 2012a; Singer, 2011). The theme of knowledge-building with shared practices in the pursuit of solving problems or "figuring out" the affordances and constraints in applying knowledge is pervasive in literatures on K-16 education and workforce preparation. These disruptive influences to the Industrial Age model of education are critical to understanding the role of the *Framework*, *NGSS*, and the field's promising response to each in designing and enacting new models of preservice and in-service teacher learning for science education.

Cyber-Enabled Learning Environments

In considering new PD models for science educators it is critical to remember another new reality. We visualize, learn, have access to, and share information in vastly different ways today than we did even a decade ago. With advances in network technologies and tools, the Internet is evolving as an important community-based learning environment for teachers. These virtual learning settings can be a place where active participation allows teachers to "try things out" with colleagues in meaningful contexts that are close to classrooms via high quality video and student work. Through technology, teachers can interact in an environment that allows them to connect continuously with image-rich resources, tools, and tasks. Simply put, a web-based environment is unique in its capacity to provide learners with a highly functional, social, relational, and image- and media-rich environment.

The Internet's capacity to upload, store, and link text documents, pictures, videos, and other hyper-linked resources provides a unique learning environment that is a promising PD environment, given the changes at hand. The accessibility of the web, regardless of geographic location, supports learning as a continuous activity, rather than an episodic event with superficial end points. The power and reach of a networked world helps to reinforce the integration of professional learning with professional practice, regardless of location, district resources, or expertise (Moon et al., 2012).

In her *Science* article Wilson (2013) put forward two grand and mutually supporting challenges to teacher PD, the first linked to an exploration of cyber-enabled learning.

1. Harness new technologies and social media to make high-quality science PD available to all teachers.
2. Identify the kind of PD that will best prepare teachers to meet the challenges of the Next Generation Science Standards.

Wilson appears to be spot-on with these two grand challenges, yet at the same time it is important to acknowledge the difficult work ahead. First, we do not understand the particular ways in which "cyber-enabled learning environments" or the use of social media-based strategies can support the kinds of changes in teaching practice (reasoning-oriented, strategic, content-linked, and improvisational) called for in the *NGSS*. Programs of research need to be conceptualized, supported, and enacted.

Second, and closely linked to the first, we lack reliable, valid, and sharable measurement tools to know if our designs work (i.e., what features in cyber-enabled environments work well and why). Without a well-researched, useable, and shareable suite of measurement tools for assessing complex learning in PD environments and their transfer to the classroom, we will not be able to make rapid or coherent progress in the field (Michaels, 2013). Prototypes are needed.

One such prototype of a designed cyber-enabled PD environment is the Next Generation Science Exemplar System (NGSX). Begun in 2012 and supported by the National Science Foundation (Award #1251611), NGSX is the result of my collaboration with professors Sarah Michaels of Clark University, Cindy Passmore of the University of California—Davis, and Brian Reiser of Northwestern University. We are seeking to develop a PD model consistent with both research-based principles of effective PD and the goals of the *Framework* and *NGSS*, and to do so in a highly functional and resource rich cyber-enabled learning environment.

Next Generation Science Exemplar System (NGSX)

Model Overview

NGSX is designed to help science educators apply the three dimensions of science teaching and learning introduced in the *Framework* and *NGSS* to their own practice. This is a hybrid or blended model that couples a face-to-face learning environment organized into study groups with a web-based system of learning. This web-based system poses tasks for each session and provides content and pedagogical expertise through video TED-like talks by scientists, classroom video cases, as well as supportive tools, tasks, and resources, including scaffolding tools to guide participants in their learning. It is asynchronous as well. After each face-to-face session participants have continuous access to the website to revisit videos, look at uploaded reflections from group activities, drawings, and group representations, and complete "On Your Own Activities" between sessions.[1]

NGSX is not exactly like the more common "hybrid" learning environments (enhanced with Moodle or asynchronous as well as social media functions). There is no external PD provider, discussion-board moderator, or course instructor.

Rather, NGSX is designed to locate expertise and knowledge building in the participants themselves. Each study group has a dedicated facilitator, prepared for this role through a five unit NGSX Facilitator Pathway focused on building certain kinds of skills to support an NGSX study group to work with tasks, tools, videos, images, and texts embedded in each subject-matter pathway. The NGSX Facilitator Pathway also pays particular attention to preparing facilitators to guide productive discussions such as a consensus discussion, as well as productive talk among the study group participants. Facilitators selected to complete the NGSX Facilitation Pathway are generally district or state science leaders or school-based teacher leaders, preservice teacher education faculty, as well as informal science educators. This selection process happens at the local or state level.

In situating expertise in the learning system itself we are drawing from the literature on "communities of learners" and "learning ecologies," examining the role that participation, tool use, and learner roles contribute to the development of teacher identity, agency, and skill (Lave & Wenger, 1991; Michaels, Sohmer, O'Connor, & Resnick, 2009; Sawyer, 2011). In addition to not making this learning environment dependent upon importing external expertise, the NGSX system and all the learning pathways associated with the system, are shaped by the following design principles:

Embed PD in subject matter. Teachers' knowledge of how to support student learning draws on general ideas (e.g., building on students 'prior conceptions'), but critically depends on understanding how those general ideas are revealed when connected to specific subject matter issues (e.g., the nature of matter) and the challenges students face in making sense of this subject matter (Putnam & Borko, 2000). Moreover, supporting student learning depends upon a teacher's ability to understand how to design and enact lessons based on investigations of natural phenomena and helping students develop, test, and refine their explanatory ideas while continuing to motivate ongoing questions about phenomena to help further refine their models.

Organize PD around sensemaking of classroom cases. Study group participants analyze video cases from K-12 classrooms to investigate how students using epistemic practices in studying science phenomena and teacher pedagogical strategies work together. These classroom cases are not examples of perfected teaching. They are exemplars to think with, to discuss, and to consider: How do *students* engage in science practices to build the explanatory idea, piece by piece, over time? How do *teachers* support practices to build these ideas? How do the *tasks* situate students' engagement in practices in making sense of phenomena? What is known about motivating students to explore, in a sustained way, natural phenomenon at a deep level?

Focus on high leverage practices. Video cases also involve the practices most challenging and central to the *NGSS*—engaging in epistemic practices of argumentation and explanation to develop and use explanatory models. These cases should make the three dimensions of the *NGSS* (disciplinary core ideas, practices, and crosscutting concepts) realistic, recognizable, and relevant to science classrooms, demonstrating how practices and disciplinary core ideas work together.

Organize the work and learning of study groups to advance participant learning and apply reforms to their own practice. PD tasks support teachers in working with their peers to apply what they learn about classroom discourse and practices to their own classrooms. Video cases and study group tasks model examples of how iterative coherence (i.e., disciplinary core ideas increasing in sophistication over time) occurs.

Fundamental to these design assumptions is the principle of NGSX as a functioning system of highly integrated and mutually reinforcing component pieces. To "pull out" a particular unit, to make that unit a stand-alone module breaks down the integrity of the system and undermines the power of a systemic, coherent approach to professional development based on a particular set of research-based design principles.

We are simultaneously exploring and researching the affordances of focusing on each of these design principles in influencing participant growth in understanding three-dimensional learning that is at the heart of the *Framework* and *NGSS*. Importantly, we are also exploring how the functionality of a web-platform supports and makes these design principles accessible and interactive.

Design of the NGSX Exemplar System

Professional development occurs in the context of a learning pathway that consists of nine units of instructional engagement, each unit lasts approximately three to four hours. Each unit is divided into eight to fifteen inter-connected, mutually supporting steps. Units are intended to be flexible, allowing a study group to move through a pathway at their own pace. A typical study group is variable, ranging between eighteen and twenty-two teachers.

The initial learning pathway unit is titled *Argumentation, Explanation, and Modeling the Behavior of Matter.* This pathway focused on physical science is guided by a question that is linked to the *NGSS* student performance expectations, "How do particles combine to form the variety of matter one observes?"

How Do Participants Engage with NGSX?

Let's imagine ourselves as NGSX participants. In doing so, visualize yourself seated around a table with a group of K-12 teaching colleagues from the district or region where you teach, as well as a group of educators from a local science and technology center who have joined you. Everyone has a laptop computer or a tablet of some kind like an iPad. In front of the room is a large screen where the NGSX website is projected.

The NGSX facilitator generally has her laptop connected to a projector so that the entire group can see the NGSX website in addition to viewing it on their own individual screens. Participants are asked to bring a laptop or tablet, as there frequently are times in the unit when participants, in a large or small group format, are asked to record comments, make observations, participate in a poll, upload pictures or break into small groups to watch a video, perhaps making notes along the way. The NGSX web platform serves as the center of the PD.

Once everyone is logged into the website, usernames and password functioning, participants are greeted via video by a unit designer(s) talking with participants. Participants are provided with an overview of the learning goals for that particular unit. As the participants move through the unit they receive content explanations through TED-like talks by science and learning experts. They are able to watch science demonstrations, video cases, text documents, as well as instructions for science activities that are generally done in small groups.

Throughout, the NGSX facilitator is guiding the process as participants move through each step in a unit, as well as helping to make conceptual connections to previous steps in a unit. The facilitator helps to lead with whole group discussions, manage gallery walks productively, and gather data on participants' work

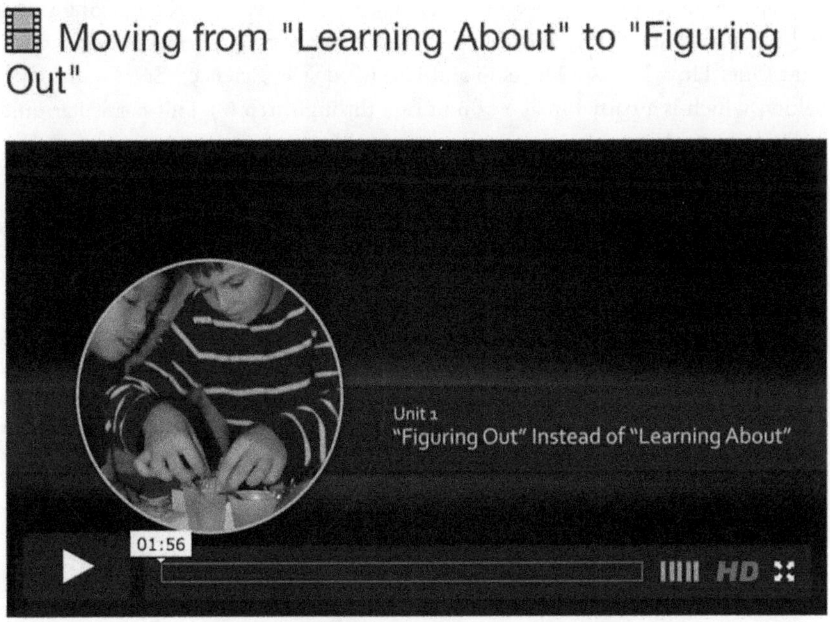

FIGURE 19.1 Screenshot of an embedded video case within NGSX

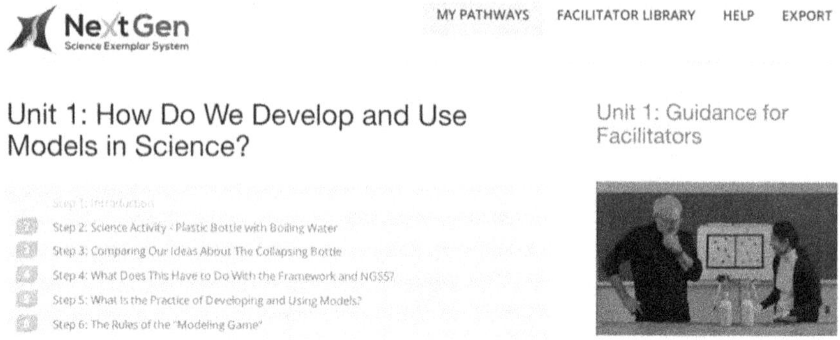

FIGURE 19.2 Part of Unit One: How do we develop and use models in science?

as she visits small group work. Figure 19.1 is taken from the NGSX and reflects how an embedded video case appears on a participant's screen.

If you want to log into the NGSX website at some point after the group PD session to review a video, complete the "on your own" assignment, or re-read one of the resource essays you will be able to do so, as long as you have your user name and password and are an NGSX participant. As a participant the PD learning pathway is accessible to you 24/7. Some study group participants use Skype, *gotomeeting*, or another communications software program to be in contact with study group colleagues between PD sessions.

After an initial practice session with navigating the website, recognizing icons and what they stand for, and doing a practice activity, participants begin with Unit One: How Do We Develop and Use Models in Science. (See Figure 19.2 below, which is a partial image of unit one through step 6.) This particular unit is divided into 15 steps.

The initial phase for piloting the prototype-learning pathway on matter occurred in nine states (Arkansas, Illinois, Michigan, Minnesota, Oregon, Washington, Wisconsin, California and Vermont), beginning in 2013. Since then NGSX has been a part of PD programs in Connecticut, Michigan, Denver Public Schools, and again in Vermont.

Summary

The *Framework* and *NGSS* have given science education a new roadmap for teacher and student learning, as well as a conceptual map that aligns with conceptual work happening across the STEM disciplines or subject areas. Taking advantage of this rich conceptual time in science education by connecting it to complementary work on epistemic practices in STEM education and work place sectors is the right way forward. It would allow the STEM community to take not only a promising step, but also a coherent one in leveraging our understanding of the role of these practices in knowledge-building and knowledge application.

Critical next moves to enact this roadmap will be determined by the bold and creative work to be taken up by the field, particularly how the field will organize and collaborate in a time that is inviting exploration. Where science education is concerned, mutually reinforcing, forward-thinking and well-connected programs of research to guide the design and enactment of teacher learning, for practicing teachers as well as preservice students and aligned with the *Framework* and *NGSS* will be essential to the success of the *NGSS* and of science educators.

There is an ongoing program of research for NGSX and the results to date have been positive in terms of participants' understanding and use of epistemic practices, as well as changes in beliefs governing how students learn the ideas of science. Building a research agenda focused on new models of teacher learning, especially cyber-enabled or technology-mediated-learning, is a task that awaits multiple communities of scholars and researchers.

While we do know some important things about what works to create effective teacher learning, as well as the role of technology-mediated PD contexts, our knowledge is limited and needs to be expanded. What can be drawn for this existing corpus of research to inform the design of teacher learning environments to develop pedagogical strategies to help students use epistemic, knowledge-building practices such as model building, argumentation, and planning investigations? What kinds of research are well suited to providing a knowledge base on these complex, unscripted pedagogical strategies? What types of research are needed to help the field flesh-out and understand the work of knowledge-building practices? Unless well-conceived research programs are undertaken in the short term, we run the very real risk of these practices being viewed as yet another list of things one might do or not.

The research and scholarship agenda necessary to support the shifts in teaching practice called for in the *NGSS* and the *Common Core State Standards* is daunting, yet it is exciting and professionally challenging. But how do we move forward on all of this as a community? What professional learning projects, like the NGSX, are underway at least in a prototype manner? What is being learned? How do we stay connected to programs of research being contemplated and that may benefit from work already underway? How can we "unsilo" this work? Should professional meetings designate a section of their annual meeting to focus on next generation research in STEM subjects and PD? Is this a potential way to find leverage points to strengthen and quicken work in the field? Does the challenge before us in systematically changing teaching practices meet the standard of one of the top ten challenges over the coming decade?

Note

1. These post-session asynchronous activities are designed to reinforce major ideas covered in a current unit or an upcoming unit as well as provide a continuous connection to the NCSX learning environment.

20

RESPONSE 1

School-System Contexts for Professional Development

Edward J. Fuller

Introduction

There have been long-standing efforts to employ teacher professional development (PD) to build the capacity of in-service teachers to deliver high-quality curricula in a pedagogically effective and engaging manner (Sun, Penuel, Frank, Gallagher, & Youngs, 2013). Unfortunately, however, such efforts have not been particularly effective in achieving such goals (Sun et al., 2013). More recent efforts have called for the need for STEM teachers to attend to the communication practices of K-12 teachers and other science educators in an effort to develop the ability of individuals to learn and do science (Duschl et al., this volume). Yet, as is the case with PD, such efforts have fallen short of widespread adoption across science education (Duschl & Osborne, 2002).

In this response, I review two chapters that address these critically important topics within the field of STEM education. Jean Moon's chapter proposes cyber PD as a new and potentially more effective manner to build the capacity of teachers to improve teaching and learning. The other chapter, by Stacey Pigg, William Hart-Davidson, Jeff Grabill, and Kirsten Ellenbogen, examines the cyber-based discursive and argumentative practices of scientists and community members on a museum's blog about everyday issues of science.

I divide this response into three sections. First, I provide a brief review of Moon's chapter and provide some suggestions for ensuring the effectiveness of teacher PD in general and cyber PD, in particular. Second, I review Pigg et al.'s chapter and provide some suggestions for moving the important research forward by connecting to broader areas of work within the STEM education field. I conclude my comments in section three in which I provide some additional suggestions about specific areas of needed research within the broader STEM research agenda.

New Environments for Professional Development: Situating Science Learning and Teaching in a Post-Framework and NGSS World

Moon's chapter (this volume) examines the need for teacher PD to enact changes in curriculum and pedagogy. More specifically, Moon addresses two major topics: the state of current and future professional development in science; and the role of cyber-enabled learning environments in the context of new models of science-based professional development. In an effort to provide concrete details about how such PD might actually look in practice, she also explores a current project that is working to create and implement a cyber-based PD effort.

Moon makes a cogent argument that teachers will need to adopt and implement new pedagogical practices in order to implement the *Next Generation Science Standards (NGSS)* (NGSS Lead States, 2013). This is particularly true, according to Moon, because the forms of teaching required to implement the *NGSS* cannot be accomplished through scripted lessons or, I would argue, computer-assisted instruction. Further, she notes that developing a culture of learning that includes the ability to reason and converse with others will require improvements in the abilities of teachers to guide classroom conversations about complex information.

Given the sheer number of teachers who will need to acquire these high-level skills, Moon argues the field will need to build new forms of PD and enact a wave of new and better research on the types of PD that are effective in different contexts and for different types of teachers. As one potential solution, Moon describes a new cyber-based model of PD that overcomes one barrier to effective PD—the need for sustained collaboration among teachers who may not work in the same school or even the same district (Sun et al., 2013).

Moon is correct in that teachers will need sustained, high-quality PD for the *NGSS* to be implemented well in classrooms (Allen & Penuel, 2014; Banilower, Heck, & Weiss, 2007; Sun et al., 2013). This is particularly true, because the *NGSS* are standards, not a distinct curriculum, thus teachers will need assistance in developing the specific curricular items, as well as the instructional and assessment strategies employed in teaching. High-quality PD is particularly important for the approximately 29 percent of U.S. teachers in rural and town districts (Fuller, 2015) that have less access to quality face-to-face PD than teachers in other geographic locales (Darling-Hammond, Wei, Andree, Richardson, & Orphanos, 2010).

Critiques and Questions

Moon makes a convincing argument that cyber-PD overcomes three of the primary challenges regarding effective PD in schools: (1) the need for sustained assistance to teachers as they learn to implement new standards, (2) the need for PD to address subject-specific content and skills, and (3) the need to provide participants active learning experiences (Sun et al. 2013).

Other challenges to effective PD, however, remained unexplored. Of course, no chapter can address every issue and Moon could certainly not directly attend to how cyber-PD would address the numerous challenges to the effective implementation of teacher PD documented in the literature (e.g., Coburn, 2005). Many of the other challenges concern larger systemic issues such as inequitable funding, teacher evaluation policies, testing systems, among others, that could never be addressed by developers or providers of PD. Such challenges must be addressed in the larger policy arena.

Yet other challenges reside within the school setting and could potentially be addressed by providers of PD. For example, one school-level challenge to effective PD is the manner in which school leaders influence the adoption of new curricula and instructional strategies (Coburn, 2005; Youngs & King, 2002; Yuen, Law, & Wong, 2003). Research suggests school leaders influence the sensemaking processes of teachers (Coburn, 2005), create conditions within a school that support the new efforts (Coburn, 2005), and allocate resources (e.g., money, time, etc.) that allow for effective implementation (Youngs & King, 2002). In sum, school leaders can substantially influence the effectiveness of PD efforts—including cyber PD—to improve teaching and learning (Neumerski, 2013; Wilson, 2013).

Given the important role school leaders play in the success of teacher development (Neumerski, 2013; Wilson, 2013), my suggestion to Moon would be to think strategically about how efforts to provide online PD might include school leaders in an effort to increase the odds that the teacher PD will be successful. In particular, I raise the following questions as a means for Moon and others to think about how teachers and school leaders might be involved in PD efforts to ensure greater leader support and overall instructional coherence:

- What does effective school leader PD that is supportive of teacher PD entail?
- Should the teacher PD involve school leaders or should separate PD be provided to school leaders?
- Can school leader PD that supports teacher PD be provided effectively through an online option?

Attempting to answer such questions while concomitantly examining effective cyber PD that supports effective adoption of the *NGSS* will increase the odds that the teacher PD is effective and sustainable.

Why People Care about Chickens and Other Lessons about Science Communication, Rhetoric, and Informal Learning Environments

The Pigg et al. chapter (this volume) reports on a real-world example of an informal science initiative based at a museum to examine the discourses between scientists and the public on a particular issue located on the museum's *Science*

Buzz blog. In particular, Pigg et al. seek to use the discourse examples to answer four questions:

- What is the nature of the community that interacts through *Science Buzz*?
- What is the nature of the online interaction?
- Do these online interactions support knowledge building for this user community?
- Do online interactions support inquiry, learning, and change within the museum (i.e., what is the impact on museum practice)? (this volume, pp. 257–8)

The discourse analysis focused on how the blog participants built arguments, constructed a community identity, and explored new ideas. Pigg et al. found that almost 50 percent of the postings were related to building an argument—clearly a crucial practice within science education, as well as in education overall. Further, about 18 percent of the postings were related to building a community identity.

Pigg et al.'s chapter provides an excellent description of how a blog writer or any educator can model the act of "doing science" rather than simply focusing on facts and findings. Indeed, as Pigg et al. describe in great detail, some (but not all) of the blog thread follows the initial modeling of the original post by engaging in the practice of questioning and searching for answers. Further, much of the discussion remained steeped in the hierarchy of "expert scientist" and "non-expert public." Yet, Pigg et al. contend that this "power/knowledge" asymmetry did not shut down conversation and learning; rather, communication and learning continued.

Ultimately, one of the major findings from the study relevant to the issue of science education and communication is how important the role of facilitator can be for the science educator. This is true not only for the blog author, but also for teachers in classrooms, scientists engaging with the public, and teachers interacting with members of their school community. Indeed, I would argue—and Pigg et al. imply as well—that the modeling of the behavior of "doing science" through the cycle of questioning, seeking answers, and refining the questions is far more important to science communication and education than providing definitive scientific answers.

Another finding is that the perspectives and assumptions of the scientist and the public may differ and that these differences can cause confusion and even conflict. In my own experience, this is a common occurrence in conversations between experts and non-experts across all fields of education. As Pigg et al. note, part of the conflict arises because of different views and perspectives of what actually constitutes evidence. The "science expert" relies on authoritative figures and peer-reviewed evidence while the "public non-experts" rely on personal experience or hearsay from other non-experts.

This chapter is extremely important in that it underscores the need for far more research that investigates the interactions of scientist/educators and those in the public sphere. Such interactions are key to not only improving the scientific literacy of the populace, but also to engender greater support for the types of teaching and learning called for by advocates of the *NGSS*.

Critiques and Questions

I have two suggestions for Pigg et al. and others as such work continues into the future. First, when engaged in similar types of studies one might ground their analyses in existing research about argumentation in general and science argumentation in particular. For example, authors might employ an argumentation framework such as the one proposed by Driver, Newton, and Osborne (2000) or use the perspectives employed by Duschl and Osborne (2002). This would allow for a more systematic examination of the issues and make a stronger connection to similar research in other areas of research such as teacher PD and pre-service teacher education.

Second, the author could broaden the connections of the study to different areas within STEM education. The problems regarding argumentation and discourse are not unique to informal science education, but cut across multiple areas within the broader field of STEM education. For example, the very important issues raised so well by Pigg et al. are also important in the areas of teacher education, the preparation of academics, teacher-student interactions, and the interactions of teachers with the public, as ambassadors of the field. Making such connections might better situate such studies within the larger domain of argumentation in STEM education.

An Agenda for Research

To conclude my response, I propose a few specific areas to include in a broader research agenda that would help address some of the immediate and long-term challenges in ensuring widespread and equitable adoption of efforts such as those identified in the *NGSS*.

Cyber-Based Education

One common thread across the two chapters is the use of cyber-based strategies to improve education. Given the increasing use of cyber-based educational efforts, STEM researchers should integrate online, cyber-based education efforts in both formal and informal education settings into the existing research agenda. Ideally, researchers could use experimental designs to compare the efficacy of traditional face-to-face and cyber-based education efforts such as the type of PD described by Moon. Researchers should also make use of the rich documentations of discursive and argumentative practices that can easily be captured in cyber conversations such as those detailed by Pigg et al. and discussed by Gitomer (this volume). Given that an increasing proportion of conversations occur in the cyber world, efforts such as those by Pigg et al. are increasingly relevant and important to the field of STEM education.

Factors Influencing Teacher PD

We need a far more robust understanding of the factors that facilitate and impede the ability of teachers to learn from others through formal PD (Dede, Ketelhut, Whitehouse, Breit, & McCloskey, 2008), instructional leadership (Neumerski, 2013), and informal "spillover" teacher-to-teacher collaboration (Supovitz & Turner, 2000). While Borko (2004) has provided a comprehensive map of the terrain for PD and teacher learning at the student, teacher, and group levels, work in these more micro arenas needs to be integrated with quality investigations of how school and district leaders influence quality professional development within a maze of varying local, state, and national policies that influence teaching and learning. For example, we need more and better studies of how school leaders interpret and modify policies and how those actions influence PD. We also need more and better studies about the relationship between state policies such as high-stakes testing, school accountability, and teacher evaluations and how they impact PD efforts to improve teaching and learning. Without examining the effects of these more macro issues, we cannot accurately determine the effectiveness of strategies such as cyber PD targeted at influencing efforts in classrooms.

Patterns of Communication in STEM

I believe researchers in the STEM field—and the field of education in general—should closely examine how students, teachers, and other adults communicate, argue, and make decisions within STEM formal and informal education settings. The field needs far more studies such as the Pigg et al. chapter (this volume), Petrosino chapter (this volume), and *Taking Science to School* (NRC, 2007), among others, that have started to document the conversations and argumentations that occur in classrooms and within other formal and informal education spheres.

More importantly, we need investments that support longitudinal studies of the impact of engagement with the discourses, argumentation, and reasoning called for by Duschl (Duschl, 2008), Duschl & Ellenbogen (2009), and Petrosino (this volume), among many others. Specifically, we need longitudinal studies that examine whether engagement in such practices develops the ability of participants to employ the discourse and reasoning that occurs in scientific communities (Duschl & Ellenbogen, 2009). Further, such studies should examine the impact of such practices in a variety of formal and informal STEM education settings. Such longitudinal studies could determine if engagement in these practices influenced the abilities of participants to communicate and reason in ways that reflect the communication and reasoning that occurs within STEM fields.

Final Thoughts

As mentioned previously, one commonality of these two chapters by Moon and Pigg et al. is the cyber-based nature of education. As education increasingly becomes cyber-based (Dede et al., 2008), the STEM research field should increasingly engage in the type of quality research on cyber efforts exhibited by these two authors. Indeed, I argue that the overall STEM research agenda should be infused throughout with issues related to cyber-based education. Moreover, I encourage the field to build upon the exciting and encouraging work of Moon and Pigg et al. These authors are providing outstanding insights into STEM education and their work should be expanded upon as efforts to ensure the goals of the *NGSS* are achieved throughout formal and informal STEM education spaces.

21

RESPONSE 2

Technology-supported Communication in Science: Conjectures on Expertise and Evaluation

Drew H. Gitomer

Practices associated with communication in science can be supported by technology applications, as nicely articulated in the work being done by Jean Moon and Stacey Pigg, William Hart-Davidson, Jeff Grabill, and Kirsten Ellenbogen. These are examples of using technology to *do better things* rather than using technology to *do conventional things better* (Roschelle, Pea, Hoadley, Gordin, & Means, 2000). Although the purposes and intended audiences of each effort are quite different, each is focused on helping users both develop and model epistemic reasoning around science content.

In this response, I use these projects as springboards to discuss more general issues that I consider relevant to efforts that attempt to support science learning with technology systems that rely on engaged participation within a distributed cognition environment. One issue concerns whether and how authoritative expertise is introduced and supported within a distributed expertise environment. The second issue focuses on issues raised about the assessment and evaluation of participant learning in these environments.

What does it mean to *do things better?* One perspective (Working Group on Postsecondary Learning, 2013) describes substantive changes that can be supported with technology and that represent departures from traditional conceptions of learning and understanding:

- moving from thinking about something an expert *knows* and can articulate to a complex model of tacit (non-conscious) and conscious competencies;
- moving from knowledge and skills localized in a student's mind to distributed understandings and performances;

- moving from a focus on memorizing and applying facts, simple concepts, and straightforward procedures to "higher-level" conceptual and analytical capabilities deployed adaptively in diverse contexts; and
- recognizing how, beyond the conceptual and procedural aspects of learner competencies that are often described as "cognitive," complementary aspects of learner competencies, so-called "non-cognitive factors," are instrumental to successful postsecondary learning, work, and citizenship (pp. 2–3).

The efforts of Moon and Pigg et al. are both built on a foundation of distributed expertise in which learning occurs through shared participation where not only do individuals bring different understanding but the process of negotiating that understanding recapitulates many essential science practices. The goals of each project go well beyond the memorization of facts and certainly support the development of skills that extend beyond traditional academic targets.

The Next Generation Science Exemplar System (NGSX), described by Moon (this volume), is a professional development model built on a library of video exemplars and supporting content around critical science processes and practices. Modules currently being tested include those focused on models, argumentation, and reasoning from evidence. Additionally, the system encourages particular activity structures in which teachers take on different roles (e.g., rotating facilitators) to develop an understanding of the professional development content. Expertise derives both from the shared insights of the professional development group and from the authoritative information that is contained within the system's resources.

The *Science Buzz* blog, described by Pigg et al. (this volume), is a public website focused on *science in the news, emerging research,* and *seasonal science* and supported by a consortia of science museums. Pigg et al. focus on one extensive set of exchanges around chickens and egg fertilization. The initial question was raised by a museum educator and was picked up by other individuals who had interests and experiences relevant to the question at hand. In this informal science setting, participants were largely unknown to each other and established their own identities by sharing their expertise and experiences. Conversations are monitored and often facilitated by authorities (e.g., museum-affiliated experts). In this case, the distributed expertise among participants led to a change in understanding by the designated "authority."

The blog highlights the sophisticated argumentation processes in which a community of interested participants can engage. Conjectures are made, evidence is presented and evaluated, new information is injected into the discussion, and norms around scientific reasoning and communication are evident. The increased sophistication in understanding and reasoning within the community that occurs from this discussion is evident.

The two online systems of technology-supported communication vary in the explicitness of their goals, resulting in substantially different designs. The NGSX system is designed with a very clear goal in mind—to support the professional development of teachers in learning about and learning to teach well-defined critical aspects of scientific reasoning and communication. *Science Buzz* is much more open as discussions can arise from anywhere in the community, typically stemming from interest. There is no established curriculum as there is for NGSX.

The systems also vary in terms of motivation for participation. Teachers will participate in NGSX professional development for various reasons that may include, but certainly are not limited to, explicit interest in the content. For example, school districts might encourage their teachers to participate in such programs to improve and reform science teaching in the district. *Science Buzz* participants have little reason to participate other than because they have specific interest in the substance of particular discussion threads.

These two exemplars of technology-supported communication raise interesting issues about the nature and role of expertise in distributed learning systems and their implication for science learning. In the NGSX system, expertise and authority is generally not negotiable—participants are expected to refer to authoritative materials as they engage in professional learning activities. Teachers participating in professional development are not presumed to be expert in the content being discussed. *Science Buzz* is different in that many participants bring substantial expertise to discussion threads. Pigg et al. provide an illustration of a participant having certain understanding that was not previously available to the "authority" or to other participants. Expertise in such environments is more clearly shared from the outset.

Why might considerations of authority be so important in environments that rely on distributed cognition? For one, the absence of other sources of authority can lead to impoverished interactions among groups. Research on professional learning communities, even absent technology, identifies the importance of having external resources for participants to call on to inform and support interactions. In addition, groups most in need of such resources (e.g., schools with many new teachers) are those with the least access to them (e.g., through collaborations with universities) (Talbert, 2010).

In thinking about systems like NGSX, authoritative expertise is introduced to support specific knowledge (e.g., what is a model), epistemic reasoning (how to use evidence to support claims), and norms of effective discussion (how to facilitate discussions about particular professional development targets). *Science Buzz* discussions provoke questions of what can happen when the authority promotes fallacious reasoning. One need only turn to certain blogs dedicated to supporting intelligent design while discounting evolutionary theory (e.g., www.uncommondescent.com). These discussions share a great deal of communication characteristics with those on *Science Buzz*. Arguments are made, and evidence is cited to support intelligent design. In addition, evolutionary counterarguments are considered, evidence is cited to discount these positions, and evolutionary positions are rejected. The communication is often, but not always, thoughtful, well-written, and polite. The argument structure and communication strategies employed in these discussions are certainly consistent with the goals of science communication embodied in the NGSS (NGSS Lead States, 2013).

Of course, despite the form of these discussions, the science community would reject the substantive content of these discussions. The National Research Council (NRC) has made a persuasive case for the rejection of non-scientific approaches of creationism and/or intelligent design (NRC, 2008). Others (e.g., Sober, 2007) have rejected the approach as being non-scientific despite the veneer of scientific argumentation. Many problematic reasoning strategies that violate normative epistemic practice have been catalogued by Lindsay (2013).

Chinn and Buckland (2012) argue that conceptual change with respect to evolution requires changes in both theoretical concepts and epistemic practices, particularly with respect to how one considers evidence. They cite many problematic aspects of how evidence is considered by intelligent design proponents identifying epistemic practices that differentiate evolutionary scientists from the broad categorization of Young Earth Scientists:

- core aims of inquiry—accounting for empirical data vs. affirming Biblical truth;
- explanations—seeking detailed explanations vs. seeking explanations compatible with Biblical texts;
- methodological naturalism—seeking natural vs. supernatural explanations;
- methods—testing evolutionary theories through rigorous tests vs. avoiding testing with empirical data;
- empirical evidence—theory explaining diverse lines of evidence vs. inconsistent and often outdated evidence;
- disinterestedness—impartiality as a norm, even if against one's own interests vs. commitment to a particular view;
- human capacities—humans as capable of acquiring knowledge of the world vs. religious views skewing knowledge-forming practices; and
- debate tactics—argumentation guided by accepted inferential and discursive rules vs. focus on the scoring of points with audiences ignorant of the details of theory and evidence.

Arguments for intelligent design may be incorrect as science, but they are sophisticated nonetheless. The ability to call out fallacies in these arguments requires deep expertise of theory, evidence, and epistemic reasoning. Thus, particularly in contentious areas, it seems necessary to have expertise injected in discussions to respond to and correct scientifically invalid perspectives.

The two chapters also raise questions about the assessment and evaluation of participation and learning in distributed learning environments. The answers are conditional on the intended purpose(s) of the tools and/or interventions. I focus on three specific questions.

First, *what do we mean by engagement and participation*? In most formal educational settings, we gauge engagement and/or participation in terms of the proportion of individuals in the sample classroom or course who demonstrate evidence of engagement and participation and the extent to which the engagement and participation meet particular levels.

One could imagine such an approach being taken with NGSX. A school district deciding to adopt NGSX for professional development might inquire about the proportion of teachers participating and the extent to which teachers, individually and collectively, engaged in the various professional development activity structures.

Science Buzz presents a different set of issues. As a public forum, the sample is unknown and, in many ways, limitless. There is no denominator on which to consider participation rates. The goal of *Science Buzz* is not to contribute to the understanding of a specific group of individuals; as would be the case in a classroom or a district professional development effort. Rather, *Science Buzz* is intended to provide opportunities for self-selected individuals to participate in a community with shared interests. While a website might be deemed more successful if it had more active participants, it is extremely unlikely for an evaluator to determine the extent to which individuals exist who would have been strong participants under different circumstances. We can know how many people participated and the extent to which they show evidence of engagement, but do not know how many individuals are both interested in chicken fertilization and have access to the blog. Most importantly, the blog is not trying to reach all individuals with all discussion groups. Thus, for these two systems, we would define *participation* with very different metrics.

The second assessment and evaluation question concerns *the appropriate unit(s) of analysis and how different analyses can help address different questions*. In the NGSX system, it is possible that assessment and evaluation would be done at different levels of analysis. Specifically, the system is designed to help teachers learn specified concepts and skills, and the most direct way of understanding the system's impact is to assess each participating teacher's knowledge individually. These individual assessment scores could be pooled to make claims about overall effectiveness of the system.

However, evaluation at the community level consists of more than aggregating individual assessment scores. For example, we might be most interested in how NGSX changes the kind of science teaching that occurs in a school or district. Are different curricula and/or assessments designed or employed? Are there fundamental changes in the discourse and expectations in classrooms? Do teachers have different professional conversations among themselves outside of the NGSX curriculum and professional development time?

It is natural for *Science Buzz* to focus primarily on the community in its evaluation. Pigg et al.'s analysis of the talk that transpired around chicken fertilization is, in itself, an evaluation. Collectively, there is evidence that the community is gaining substantive knowledge and also engaging in productive epistemic reasoning processes. There are substantial and substantive interchanges that occur over time involving multiple individuals.

Yet, in such groups, we need to be concerned that engagement and participation is limited. Baskin and Anderson (2003) analyze the presence of technology in formal schooling and find that for a subset of students and teachers, there is evidence of profound engagement and transformation. However, others, for various reasons, never engage with technologies or the learning opportunities they afford as they serve yet another educational hurdle.

For systems like *Science Buzz*, it is important to understand the distribution and intensity of engagement and participation. If we queried each individual who made at least one blog contribution, would there be a proportion of individuals who voluntarily removed themselves from the discussion? If they did stop participating, what were their reasons for stopping? If we assessed individuals on their understanding of chicken fertilization, what would be the level of understanding across individuals? How many participants report changing their understanding as a result of participation?

The final question about assessment concerns *what should be assessed*. We have already suggested engagement and participation, but what about cognitive and epistemic facets of understanding? For NGSX, assessing the specific curricular goals of professional development seems reasonable. But one might also want to assess the extent to which teachers develop their broader views of science; how students learn science; and how teachers see themselves as colleagues within school and district science communities, as well as in a larger world. Specific targets of assessment are interwoven with decisions about selecting appropriate units of analysis.

What might one assess at the community level with respect to *Science Buzz*? One could examine the shared understanding of particular blog topics or, given the stated goals of the blog, evidence of science communication and processes as individuals engage in the blog. These kinds of issues, in addition to engagement and participation, seem consistent with the overall goals of the system.

At the individual level, the need for assessment is less obvious. Given the nature of the enterprise, individuals can opt in or out for a variety of reasons and, thus, are likely to be partial participants, not reading everything in the blog. It is doubtful that the current design is consistent with estimating an individual's understanding or, more specifically, the extent to which the system influenced that understanding. Nor is it clear how to interpret any aggregation of individual scores. Unless individuals are mandated to participate and use blogs in particular ways, a focus on individual scores may be misguided.

The Moon and Pigg et al. chapters, and the underlying work, are tremendously provocative. Each is built on comprehensive understandings of the learner, content, reasoning, and communication. Each raises critical questions for the field that extend far beyond their specific implementation. Careful design of systems along with insightful analyses and evaluations will surely have implications for technology-based STEM tools and their deployment.

Theme V References

Allen, C.D., & Penuel, W.R. (2014). Studying teachers' sensemaking to investigate teachers' responses to professional development focused on new standards. *Journal of Teacher Education, 66*(2), 136–49. http://doi.org/10.1177/0022487114560646

Alozie, N.M., Moje, E.B., & Krajcik, J.S. (2010). An analysis of the supports and constraints for scientific discussion in high school project-based science. *Science Education, 94*(3), 395–427. http://doi.org/10.1002/sce.20365

American Association for the Advancement of Science. (2011). *Vision and Change in Undergraduate Biology Education*. Washington, DC: AAAS.

Association of American Medical Colleges and Howard Hughes Medical Institute. (2009). *Scientific Foundations for Future Physicians* (Text). Washington, DC: AAMC. Retrieved from http://hhmi.org/news/aamc-hhmi-committee-defines-scientific-competencies-future-physicians

Banilower, E.R., Heck, D.J., & Weiss, I.R. (2007). Can professional development make the vision of the standards a reality? The impact of the national science foundation's local systemic change through teacher enhancement initiative. *Journal of Research in Science Teaching, 44*(3), 375–95. http://doi.org/10.1002/tea.20145

Barron, B. (2006). Interest and self-sustained learning as catalysts of development: A learning ecology perspective. *Human Development, 49*(4), 193–224. http://doi.org/10.1159/000094368

Baskin, C., & Anderson, N. (2003). The online classroom: A self-actualising theme park or a trial by multimedia? *Australian Educational Computing, 18*(1), 11–20.

Bauer, M.W., Allum, N., & Miller, S. (2007). What can we learn from 25 years of PUS survey research? Liberating and expanding the agenda. *Public Understanding of Science, 16*(1), 79–95. http://doi.org/10.1177/0963662506071287

Bereiter, C., & Scardamalia, M. (2012). *What will it mean to be an education person in the mid-21st century*. The Gordon Commission on the Future of Assessment in Education.

Bickerstaff, K., Lorenzoni, I., Jones, M., & Pidgeon, N. (2010). Locating scientific citizenship: The institutional contexts and cultures of public engagement. *Science, Technology & Human Values, 35*(4), 474–500. http://doi.org/10.1177/0162243909345835

Borko, H. (2004). Professional development and teacher learning: Mapping the terrain. *Educational Researcher, 33*(8), 3–15. http://doi.org/10.3102/0013189X033008003

Bricker, L.A., & Bell, P. (2008). Conceptualizations of argumentation from science studies and the learning sciences and their implications for the practices of science education. *Science Education, 92*(3), 473–98. http://doi.org/10.1002/sce.20278

Chinn, C.A., & Buckland, L. (2012). Model-based instruction: Fostering change in evolutionary conceptions and in epistemic practices. In K.S. Rosengren, S.K. Brem, E.M. Evans, & G.M. Sinatra (Eds.), *Evolution Challenges: Integrating Research and Practice in Teaching and Learning about Evolution* (1 edition, pp. 211–32). Oxford, UK: Oxford University Press.

Coburn, C.E. (2005). Shaping teacher sensemaking: School leaders and the enactment of reading policy. *Educational Policy, 19*(3), 476–509. http://doi.org/10.1177/0895904805276143

Darling-Hammond, L., Wei, R.C., Andree, A., Richardson, N., & Orphanos, S. (2010). *Professional development in the United States: Trends and challenges.* Dallas, TX: National Staff Development Council.

Dede, C., Ketelhut, D.J., Whitehouse, P., Breit, L., & McCloskey, E. (2009). A research agenda for online teacher professional development. *Journal of Teacher Education, 60*(1), 8–19. http://doi.org/10.1177/0022487108327554

Driver, R., Newton, P., & Osborne, J. (2000). Establishing the norms of scientific argumentation in classrooms. *Science Education, 84*(3), 287–312. http://doi.org/10.1002/(SICI)1098-237X(200005)84:3<287::AID-SCE1>3.0.CO;2-A

Duschl, R. (2008). Science education in three-part harmony: Balancing conceptual, epistemic, and social learning goals. *Review of Research in Education, 32*(1), 268–91. http://doi.org/10.3102/0091732X07309371

Duschl, R.A., & Osborne, J. (2002). Supporting and promoting argumentation discourse in science education. *Studies in Science Education, 38*(1), 39–72. http://doi.org/10.1080/03057260208560187

Duschl, R.A., & Ellenbogen, K. (2009). Argumentation and epistemic criteria: Investigating learners' reasons for reasons (pp. 111–118). Presented at the Educacion Quimica.

Fahnestock, J. (1986). Accommodating science the rhetorical life of scientific facts. *Written Communication, 3*(3), 275–96. http://doi.org/10.1177/0741088386003003001

Falk, J.H., Storksdieck, M., & Dierking, L.D. (2007). Investigating public science interest and understanding: Evidence for the importance of free-choice learning. *Public Understanding of Science, 16*(4), 455–69. http://doi.org/10.1177/0963662506064240

Fuller, E.J. (2015). *Distribution of teachers by state and geographic locale in 2012–13.* State College, PA: Center for Evaluation and Education Policy Analysis.

Gerber, B.L., Cavallo, A.M.L., & Marek, E.A. (2001). Relationships among informal learning environments, teaching procedures and scientific reasoning ability. *International Journal of Science Education, 23*(5), 535–49. http://doi.org/10.1080/09500690116971

Goldman, S., Duschl, R.A., Ellenbogen, K., Williams, S.M., & Tzou, C.T. (2003). Science inquiry in a digital age: Possibilities for making thinking visible. In H. van Oostendorp (Ed.), *Cognition in a Digital World* (pp. 253–84). Mahwah, NJ: Lawrence Erlbaum Associates.

Grabill, J.T., & Pigg, S. (2012). Messy rhetoric: Identity performance as rhetorical agency in online public forums. *Rhetoric Society Quarterly, 42*(2), 99–119. http://doi.org/10.1080/02773945.2012.660369

Gross, A.G. (1994). The roles of rhetoric in the public understanding of science. *Public Understanding of Science, 3*(1), 3–23. http://doi.org/10.1088/0963-6625/3/1/001

Irwin, A., & Michael, M. (2003). *Science, Social Theory and Public Knowledge.* Maidenhead, UK: Open University Press. Retrieved from http://site.ebrary.com/id/10409190

Jiménez-Aleixandre, M.P., Rodríguez, A.B., & Duschl, R.A. (2000). "Doing the lesson" or "doing science": Argument in high school genetics. *Science Education, 84*(6), 757–92. http://doi.org/10.1002/1098-237X(200011)84:6<757::AID-SCE5>3.0.CO;2-F

Johnstone, B., & Eisenhart, C. (2008). *Rhetoric in Detail Discourse Analyses of Rhetorical Talk and Text.* Philadelphia, PA: John Benjamins. Retrieved from http://public.eblib.com/choice/publicfullrecord.aspx?p=623063

Lampert, M. (2010). Learning teaching in, from, and for practice: What do we mean? *Journal of Teacher Education, 61*(1–2), 21–34. http://doi.org/10.1177/0022487109347321

Lave, J., & Wenger, E. (1991). *Situated Learning: Legitimate Peripheral Participation*. New York: Cambridge University Press.

Lemke, J.L. (1990). *Talking Science: Language, Learning, and Values*. Norwood, NJ: Ablex Publishing Corporation.

Lenhart, A., Purcell, K., Smith, A., & Zickuhr, K. (2010). Social media and young adults. Retrieved from http://pewinternet.org/2010/02/03/social-media-and-young-adults/

Lindsay, D. (2013). A list of fallacious arguments. Retrieved from http://don-lindsay-archive.org/skeptic/arguments.html#authority

Michaels, S. (2013, July). Personal communication.

Michaels, S., O'Connor, C., & Resnick, L.B. (2008). Deliberative discourse idealized and realized: Accountable talk in the classroom and in civic life. *Studies in Philosophy and Education*, *27*(4), 283–97. http://doi.org/10.1007/s11217-007-9071-1

Michaels, S., Sohmer, R., O'Connor, M., & Resnick, L.B. (2009). Guided construction of knowledge in the classroom: The troika of talk, tasks, and tools. In B. Schwarz, T. Dreyfus, & R. Hershkowitz (Eds.), *Transformation of Knowledge Through Classroom Interaction* (pp. 105–29). New York: Routledge.

Moon, J., Singer, S.R., & Reiser, B.J. (2012). Science standards require a teacher-learning rethink. *Education Week*. Retrieved from http://edweek.org/ew/articles/2012/11/30/13moon.h32.html

Moon, J., Passmore, C., Reiser, B.J., & Michaels, S. (2013). Beyond comparisons of online versus face-to-face PD commentary in response to fishman et al., "Comparing the Impact of Online and Face-to-Face Professional Development in the Context of Curriculum Implementation." *Journal of Teacher Education*, *65*(2), 172–76. http://doi.org/10.1177/0022487113511497

Nakamura, L. (2010). Race and identity in digital media. In J. Curran (Ed.), *Media and Society* (5th Revised edition, pp. 336–47). London: Bloomsbury Academic.

National Governors Association Center for Best Practices, & Council of Chief State School Officers. (2010). *Common Core State Standards for Mathematics*. Washington, DC: National Governors Association Center for Best Practices, Council of Chief State School Officers.

National Research Council. (2004). *The Engineer of 2020: Visions of Engineering in the New Century*. Washington, DC: National Academies Press. Retrieved from http://nap.edu/catalog/10999/the-engineer-of-2020-visions-of-engineering-in-the-new

National Research Council. (2007). *Taking Science to School: Learning and Teaching Science in Grades K-8*. Washington, DC: National Academies Press.

National Research Council. (2008). *Science, Evolution and Creationism*. Washington, DC: National Academies Press.

National Research Council. (2009). *Learning Science in Informal Environments People, Places, and Pursuits*. Washington, DC: National Academies Press. Retrieved from http://site.ebrary.com/id/10315878

National Research Council. (2012a). *A Framework for K-12 Science Education: Practices, Crosscutting Concepts, and Core Ideas*. Washington, DC: National Academies Press. Retrieved from http://nap.edu/catalog/13165/a-framework-for-k-12-science-education-practices-crosscutting-concepts

National Research Council. (2012b). *Education for Life and Work: Developing Transferable Knowledge and Skills in the 21st Century*. Washington, DC: National Academies Press.

National Research Council. (2015). *Guide to Implementing the Next Generation Science Standards*. Washington, DC: National Academies Press. Retrieved from http://nap.edu/catalog/18802/guide-to-implementing-the-next-generation-science-standards

Neumerski, C.M. (2013). Rethinking instructional leadership, a review what do we know about principal, teacher, and coach instructional leadership, and where should we go from here? *Educational Administration Quarterly*, *49*(2), 310–47. http://doi.org/10.1177/0013161X12456700

NGSS Lead States. (2013). *Next Generation Science Standards: For States, By States*. Retrieved from http://nap.edu/catalog/18290/next-generation-science-standards-for-states-by-states

Passmore, C.M., & Svoboda, J. (2012). Exploring opportunities for argumentation in modelling classrooms. *International Journal of Science Education*, *34*(10), 1535–54. http://doi.org/10.1080/09500693.2011.577842

Putnam, R.T., & Borko, H. (2000). What do new views of knowledge and thinking have to say about research on teacher learning? *Educational Researcher*, *29*(1), 4–15. http://doi.org/10.3102/0013189X029001004

Reiser, B.J. (2013). *What professional development strategies are needed for the successful implementation of the Next Generation Science Standards?*. Invitational Research Symposium on Science Assessment, Educational Testing Service.

Roschelle, J.M., Pea, R.D., Hoadley, C.M., Gordin, D.N., & Means, B. (2000). Changing how and what children learn in school with computer-based technologies. *Future of Children*, *10*(2), 76–101.

Sawyer, R.K. (2011). What makes good teachers great? The artful balance of structure and improvisation. In R.K. Sawyer (Ed.), *Structure and Improvisation in Creative Teaching* (pp. 1–24). Cambridge, UK: Cambridge University Press.

Singer, S.R. (2011). STEM education: Time for integration. *Peer Review*, *13*(3), 4–7.

Sober, E. (2007). What is wrong with intelligent design? *The Quarterly Review of Biology*, *82*(1), 3–8.

Sun, M., Penuel, W.R., Frank, K.A., Gallagher, H.A., & Youngs, P. (2013). Shaping professional development to promote the diffusion of instructional expertise among teachers. *Educational Evaluation and Policy Analysis*, *35*(3), 344–69. http://doi.org/10.3102/0162373713482763

Supovitz, J.A., & Turner, H.M. (2000). The effects of professional development on science teaching practices and classroom culture. *Journal of Research in Science Teaching*, *37*(9), 963–80. http://doi.org/10.1002/1098-2736(200011)37:9<963::AID-TEA6>3.0.CO;2–0

Talbert, J.E. (2010). Professional learning communities at the crossroads: How systems hinder or engender change. In A. Hargreaves, A. Lieberman, M. Fullan, & D. Hopkins (Eds.), *Second International Handbook of Educational Change* (pp. 555–71). New York: Springer.

Wellman, B., & Guilia, M. (1999). Net surfers don't ride alone: Virtual communities as communities. In B. Wellman (Ed.), *Networks in the Global Village: Life in Contemporary Communities*. Boulder, CO: Westview Press.

Wilson, S.M. (2013). Professional development for science teachers. *Science*, *340*(6130), 310–13. http://doi.org/10.1126/science.1230725

Windschitl, M., Thompson, J., & Braaten, M. (2008). Beyond the scientific method: Model-based inquiry as a new paradigm of preference for school science investigations. *Science Education*, *92*(5), 941–67. http://doi.org/10.1002/sce.20259

Windschitl, M., Thompson, J., Braaten, M., & Stroupe, D. (2012). Proposing a core set of instructional practices and tools for teachers of science. *Science Education*, *96*(5), 878–903. http://doi.org/10.1002/sce.21027

Working Group on Postsecondary Learning. (2013). *New technology-based models for postsecondary learning: Conceptual frameworks and research agendas*. Washington, DC: Computing Research Association. Retrieved from http://cra.org/uploads/documents/resources/rissues/Postsecondary_Learning_Report.pdf

Wynne, B. (1992). Misunderstood misunderstanding: Social identities and public uptake of science. *Public Understanding of Science, 1*(3), 281–304. http://doi.org/10.1088/0963-6625/1/3/004

Youngs, P., & King, M.B. (2002). Principal leadership for professional development to build school capacity. *Educational Administration Quarterly, 38*(5), 643–70. http://doi.org/10.1177/0013161X02239642

Yuen, A.H.K., Law, N., & Wong, K.C. (2003). ICT implementation and school leadership. *Journal of Educational Administration, 41*(2), 158–70. http://doi.org/10.1108/09578230310464666

Zappen, J.P. (2005). Digital rhetoric: Toward an integrated theory. *Technical Communication Quarterly, 14*(3), 319–25. http://doi.org/10.1207/s15427625tcq1403_10

Reflections and Summary

22

REFLECTIONS ON THE WATERBURY SUMMIT

STEAM and Systems Thinking

Stephanie E. Vasko

When I was initially asked to join the Waterbury Summit as a panelist for the "Systems Thinking" session, I had my definition in mind of what I meant by "systems" and "systems thinking." The way that I articulate these concepts in my teaching philosophy is that students should have an understanding of the social and global systems in which they live and be able to articulate the structures, stakeholders, and relationships present in the different parts of these systems. As I have not taught in a K-12 classroom, I was interested in how these concepts were (and are) applied in that space. I prepared for this panel by reading the papers for the Summit. From this, I walked away with a broad conceptualization of how systems and systems thinking are used with the K-16 audience. However, I also walked away with more questions on the roadblocks and strategies for the implementation of systems thinking in the classroom. After discussions with the participants at the Summit and after watching all of the presentations and panels, I decided to keep systems thinking in the background as an evolving concept as I continued to develop my own teaching philosophy.

The Waterbury Summit also inspired me to read more about standards, especially in light of the conception I had that often instructors and researchers see standards as restrictive. Both during and after the Waterbury Summit, I began to explore the Next Generation Science Standards (NGSS) (NGSS Lead States, 2013). This first meant understanding the difference between the Common Core Standards and the NGSS. Within the frequently asked question on the NGSS website "How was the development of the Next Generation Science Standards different than the development of the Common Core State Standards?," the NGSS are described as having taken into account "the importance of having the scientific and educational research communities identify core ideas in science and articulate them across grade bands" and creating a document (*A Framework for*

K-12 Science Education) that "presents and explains the interrelationships among practices, cross-disciplinary concepts and disciplinary core ideas" (NGSS Lead States FAQ, 2013 p. 3). I see these "cross-disciplinary" (and "cross-cutting," to use another term from the NGSS) concepts as the perfect space to integrate systems thinking into the classroom.

While digging deep into the NGSS, what most resonated with me in terms of systems, was part three of the "Recommendation for Core Ideas in K-12 Science Instruction," which states that a core idea should "relate to the interests and life experiences of students or be connected to societal or personal concerns that require scientific or technological knowledge" (NGSS Lead States FAQ, 2013). Without saying "systems thinking" explicitly, this recommendation illustrates that students should be presented with the scientific knowledge as contextualized in a larger global setting. In my interpretation of this recommendation, when we utilize "systems thinking" in the classroom, we need to engage students by presenting STEM (science, technology, engineering, mathematics) as lenses for understanding each other.

A conceptualization of STEAM, or STEM plus "the arts," has been articulated by many scholars and instructors. However, these same scholars and instructors disagree on how to define "the arts." Several of these definitions focus on the idea that innovation comes from the integration of art and STEM (Maeda, 2013; White, 2010), some conceptualizations add a "D" in for "design" or incorporate the "D" into the "A" (Britton, 2014), and others add an additional "M" for music (Picking up STEAM[M]) or change the "M" from "mathematics" to "manufacturing" (DeMio, 2015). Additionally, the concepts of "STEAM" and the maker movement can be intertwined or overlap, which also makes untangling what we mean when we say "the arts" even more complicated (Britton, 2014; Vasko, 2015a, 2015b, 2015c, 2015d, 2015e). Regardless of the current slew of acronyms that seek to integrate art and STEM, one thing is for sure; the interest in moving in this direction exists. I do not claim to present an exhaustive survey of STEAM in this piece, rather, just to raise awareness of what it could stand for, some of the challenges and opportunities that exist, and my plans for moving forward with STEAM in graduate education, professional collaborations, and personal identity formation (Vasko, 2015a, 2015b, 2015c, 2015d, 2015e). As discussed above, STEM/arts integration may "relate to the interests and life experiences of students" (NGSS Lead States FAQ, 2013). As one author stated, "many know that STEAM education isn't just the content, but the process of being scientists, mathematicians, engineers, artists and technological entrepreneurs" (Miller, 2014, p. 1).

That is not to say that there will not be challenges in moving from STEM to STEAM, especially in the face of standards and testing. As some scholars state, ". . . arts infused instruction in STEM disciplines is often times not the norm in the US educational system, and educational policies often make it difficult to teach this way without deviating from standards-driven curriculum

(Fusarelli, 2004)" (Henriksen, 2014, p. 1). One of these issues, as Ann Jolly (2014) puts it, is how do we "include the arts in STEM in an authentic way" (How Do We Solve the STEM vs. STEAM Conundrum? section, para. 1). The way that I interpret "authentic" is that we need to move towards designing best practices that do not use STEM and the arts in service of one another (Vasko, 2015a, 2015b, 2015c, 2015d, 2015e). An additional challenge facing moving towards a STEAM curriculum is integration. Susan Riley (2012), in "Use Arts Integration to Enhance Common Core," states

> Integration requires collaboration, research, intentional alignment and practical application on behalf of the teachers who take on this challenge. From the students, integration demands creativity, problem-solving, perseverance, collaboration, and the ability to work through the rigorous demands of multiple ideas and concepts woven together to create a final product. Integration is not simply combining two or more contents together" (2012, p. 1).

Integration requires time, effort, and support on the parts of districts, instructors, and communities. According to Moore, Tank, Glancey, and Kersten (2015), "even with the articulation of high-quality standards, schools, and teachers generally must develop their own structures for integration, as there are not many resources (Roehrig, Moore, Wang, & Park, 2012)" (p. 297). It is my firm belief, in light of the Waterbury Summit and my own experiences as an instructor and scientist, that STEAM can expand engagement opportunities for students in our classrooms, but that integration may be the hardest step. Additionally, perhaps training for integration in teacher education (for both K-12, undergraduate, and graduate classrooms) will ease the integration barrier. As Bybee notes "the changes implied by the Framework for K-12 Science Education (NRC, 2012a) and NGSS (NGSS Lead States, 2013) imply dramatic changes in teacher education programs" (Bybee, 2014, p. 217). While he outlines his own strategies for changes to teacher education, I urge that any changes consider an inclusion of the arts for the cross-disciplinary/crosscutting concepts aspects of the NGSS.

My reflections on STEAM, systems, and my own learning experiences and research, as well as my interpretation of the NGSS, and the interactions during the Waterbury Summit have led me to the position that we *must* create (or at least support) STEAM spaces in education (Vasko, 2015a, 2015b, 2015c, 2015d, 2015e). Some universities and schools are already taking the plunge with regard to spaces for this type of engagement, and as mentioned above, there has been some research on STEAM and implementing it in the classroom. In my own line of thinking on STEAM, I am not suggesting that students be directed to the STEM pipeline or a humanities pipeline, but rather, to allow students the option to be engaged in both simultaneously and to explore using each field to investigate the other. Additionally, one field should not be used in the service of the other; rather, they can (and should) be used synergistically (Vasko, 2015a, 2015b, 2015c, 2015d, 2015e). The Waterbury Summit inspired me to integrate systems thinking into my own teaching philosophy, but also to think about the nature of systems and what systems are in play for students in STEM classrooms. I believe there is still much to do in terms of public outreach, educational usage, and identity formation, and I look forward to pursuing and implementing these research areas in the future (Vasko, 2015a, 2015b, 2015c, 2015d, 2015e).

Work was made possible in part through funding from the Penn State Rock Ethics Institute and the National Science Foundation. The opinions and views expressed in this piece are those of the author and do not represent or reflect the opinion, views, or policy of the Pennsylvania State University, the Rock Ethics Institute, or the National Science Foundation.

23

SUMMARY

Driving Change Forward

Amber S. Bismack, Yann Shiou Ong, Armend Tahirsylaj, and Richard A. Duschl

As we move forward with the new agendas emphasized in many major national reports (e.g., National Research Council [NRC], 2012a, 2012b), including the *Framework for K-12 Science Education* (henceforth referred to as the *Framework*) (NRC, 2012a) and the *Common Core State Standards for Mathematics* (*CCSS-M*) (National Governors Association Center for Best Practices and Council of Chief State School Officers, 2010), we find ourselves entering new waters for what the agendas mean for PreK-16 STEM education. The *Framework* and *CCSS-M* call for the integration of content with practices. "The important shift is away from an emphasis on merely knowing to an emphasis on using knowledge" (Duschl, Bismack, Greeno, & Gitomer, this volume, p. 7). The science-as-practice shift represents a large change that is occurring in pedagogical, psychological, and philosophical characterizations of STEM disciplines. The contributions and deliberations from the Waterbury Summit and the ideas presented in this volume suggest that this shift will impact many stakeholders, in particular teachers, affect all education levels, and spawn new conceptualizations of research.

The Summit proceedings and chapters in this volume unfold a series of ideas, frameworks and perspectives regarding complex changes necessary for generating high quality STEM education systems. While the authors' contributions build from the assigned Summit themes and emphasize topical themes pertaining to STEM education, here in this summary contribution we identify five thematic strands that appear to span all of the chapters, as well as the panel talks and group discussions from the Waterbury Summit:

1. The emergence of new research agendas that take up equity, ethics, and epistemic issues necessary to support and guide change.
2. The coordination of preservice teacher education and teacher professional development (PD) programs that occur in tandem in order to enact the changes and transformations outlined in the *Framework*.

3. The use of technology and social media to facilitate, at a large scale, the teaching and learning of model-based reasoning and quantitative reasoning.
4. The importance for assessment tasks, observations, and methods to align with the science-as-practice curriculum/course frameworks and standards and with 'assessment for learning' teaching and learning practices.
5. The development and adoption of systemic approaches for policy development, PD, and curricula implementation of the *Framework's* three-dimensional view of knowledge.

Figure 23.1 provides a visual representation depicting the interactions between the five thematic strands and the overarching goals of STEM education. The thematic strands, we believe, represent salient domains that have the potential of impacting when and how the goals for STEM education are reached. We challenge and invite our readers to consider other thematic strands that emerged from the chapters.

The central box in Figure 23.1 presents the overarching STEM education goals (Duschl et al. this volume). The four surrounding boxes represent the thematic strands involved in achieving those goals with each strand situated within the larger educational system. Research within and across each thematic strand should be informed by a systemic approach to implementing and sustaining educational reforms. Due to the extensive impact of many educational reforms, it is

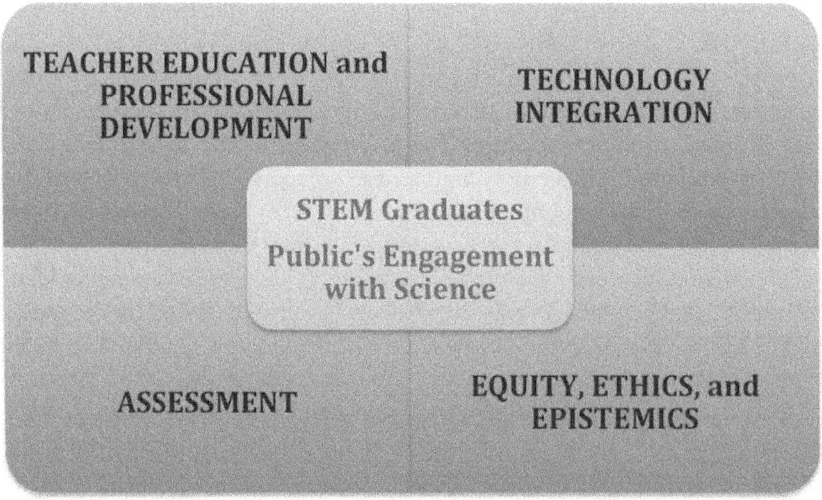

FIGURE 23.1 Systemic approaches to the implementation and sustainability of educational reforms

important to involve many different stakeholders, including teachers and administrators, in the initial and continued research surrounding the implementation and sustainability of these reforms (Penuel, this volume). If we truly intend to support the public's productive engagement with science and a growing demand for STEM graduates, then we need to be cognizant of the capabilities and resources necessary for supporting those who are impacted most by the reforms, particularly teachers and students (Coppola, this volume). We now examine in more depth each of the five thematic strands, beginning with the most prominent and impactful theme—Equity, Ethics, and Epistemics—as identified during the Summit, as well as throughout the chapters in this volume.

New Research Agendas for Equity, Ethics, and Epistemics (3Es) in STEM Learning and Decision Making

As we consider the integration of STEM disciplines and of the science-as-practice fusion of content with practices, it is important to consider the equity, ethics, and epistemic impacts that are part and parcel elements of systems thinking, model-based reasoning, and quantitative reasoning. The 3Es will impact (a) how students, and STEM professionals, engage with and potentially join the STEM fields, (b) how research is designed and conducted, and (c) how resources are distributed among research and development efforts. To engage in the challenging discussions and studies surrounding these topics, it is important to involve diverse groups of stakeholders (Brickhouse, this volume; Tuana, 2010). Tuana (2010) explains that science research should include philosophers of science and science researchers, in order "to identify and appreciate the full range of ethical issues relevant to scientific practice" (p. 473), and to "make important contributions to policy debates" (p. 486). The intersection of policy and science is often muddled and conflicted, which means greater stakeholder participation in research and development deliberations would help facilitate different perspectives and inform decision-making. This raises issues about how to develop equitable access for stakeholders and how to address the two overarching goals of PreK-16 STEM education: 1) Increase and sustain a STEM workforce, and 2) Enhance the public's understanding of and engagement with STEM knowledge and practices (Duschl et al., this volume).

When considering what it means to provide equitable access to STEM knowledge and practices, some questions arise (Farmer, session 4 panel and Q&A discussion).[1] Are STEM fields inclusive for all students and if not, why not? Who are not included in STEM disciplines? Why not? What needs to be done to improve the inclusiveness? One possibility is to rethink how students might "enter" STEM fields so there are multiple entry points (Schauble, session 4 panel and Q&A discussion). This can be a challenge, however, due to the increasing complexity of the STEM disciplines (Plummer, session 4 panel and Q&A discussion), which increases the difficulty for students to navigate different entry points. Another area of need to support further equity involves research around "productive failure." Many students leave STEM fields due to poor performance early in their studies. One way to counteract this migration out of STEM is to support research regarding "productive failure" and developing ways to reinvigorate students (Coppola, this volume; Duschl, keynote address; Penuel, session 4 panel and Q&A discussion).

To engage students in STEM knowledge and practices, there is a need for further research and PD to support teachers in engaging all students in the disciplines. Such teacher preparation involves recognizing situations when students are not given equitable access to STEM disciplines in order to support an inclusive learning environment (Farmer, session 4 panel and Q&A discussion). This will also involve PD for teachers in how local issues can be incorporated in curricula to support students' learning of how STEM concepts apply to their daily lives (Biddle & Schafft; this volume). As Biddle and Schafft (this volume) identify, there is also a need for policy decisions that support the inclusion of local issues in school settings.

As global societies continue to face difficult science and engineering based issues, the importance of ethical literacy and value judgments becomes more evident in STEM topics and research (Tuana, this volume). For example, Tuana (2010; this volume) highlights the importance of ethics in scientific decision-making regarding climate change for both researchers and policy-makers. Scientific issues continue to pervade social and political realms and will have further need for the inclusion of ethical literacy. This means science curricula should also include a focus on ethical literacy and value judgments (Biddle & Schafft, this volume) and ethics should be emphasized in teacher education and PD for future and current teachers, respectively.

Ethical literacy also has implications for STEM disciplinary research through decision-making within and from STEM research. For example, there is a need for (1) further attention to the ethical use of research practices and sharing of data and results (Responsible Conduct of Research), (2) the impacts of scientific outcomes on policy and society and vice versa (External factors), and (3) ethical concerns embedded within the production of scientific research (Internal factors) (Tuana, 2010). Those topics are the tip of the iceberg when considering the needed research around ways ethical literacy and science relate. Other research needs to include characterizing responsible science, identifying ways values contribute to science, characterizing the impact of values on science decisions (Tuana, this volume), and the inclusion of ethical questions in national science and mathematics curricula (Biddle & Schafft, this volume), just to name a few.

With a shift from siloed school subject areas to a greater focus on integrated STEM education, we need to also consider the epistemology of STEM disciplines. This shift focuses more on the "how and why" of science, engineering, and mathematics, that is how such knowledge comes to be established and why it counts as such knowledge, instead of the "what" (i.e., knowledge itself) (Osborne, this volume). This means a shift in curriculum, instruction, and assessment from addressing "what do we want students to *know* and what do we need to do to know it" to emphasizing, "what do we want students to *do* and what do we need to know to do it" (Duschl, keynote address). This has implications for research about scientific reasoning, particularly the need for greater, more detailed descriptions of procedural and epistemic knowledge. Though Osborne (this volume) briefly characterizes *procedural knowledge* and *epistemic knowledge*, further depth in our understanding of how these two types of knowledge play out in learning and teaching is needed. Once descriptions are determined, there needs to be research in how the types of scientific knowledges are taught and learned, which would have direct impact on curriculum, teacher education, PD, and assessment.

By including studies of equity, ethics, and epistemics within and across the STEM disciplines, researchers and students can work to expand our understandings of how disciplines relate to and are distinct from one another and how STEM explanatory models affect societal decision-making. As science continues to move beyond the academic world and into the social and policy realms, there is a greater need for understanding the integrated issues that emerge from STEM domains. This, in turn, has implications for how we are preparing our future generations to engage in STEM knowledge and practices. Today's students are tomorrow's decision-makers and we need to prepare *all* students to make ethical decisions for themselves, their communities, and the world.

Teacher Education and Professional Development

Due to the intentional integration of the three dimensions in the *Framework*, teaching will be quite different and will require further demands on teachers (Moon, this volume; Wilson, 2013) in order to prepare students to be able to *know* and *do* science and mathematics (Borko, this volume), as well as engineering. To support this ambitious science and mathematics teaching, teacher education and PD, as well as curriculum, will need to be carefully and intentionally designed in ways that support teachers' understanding of how to engage students in the *doing* of both science and mathematics (Bismack, Arias, Davis, & Palincsar, 2015; Borko, Jacobs, & Koellner, 2010; Davis et al., 2014; Davis & Krajcik, 2005; Schwarz & Gwekwerere, 2007). This has implications for how we use the structure, content, and research of teacher education (TE) programs and PD opportunities to support teachers, as both have a renewed emphasis to prepare teachers for teaching science "as a way to understand and make sense of the world around us" (Benson, this volume, p. 116).

Organizational Structures and Platforms for Productive PD

When considering PD programs necessary to support teachers learning how to engage their students in the doing of science and mathematics, the structure needs to be integrative, scalable, and sustainable (Borko, this volume). We need to reconceptualize PD programs by considering PD programs as systems in and of themselves and how they relate to and interact with K-12 schools and undergraduate programs, as well as informal environments. As Penuel, Borko, Moon, and Fuller (this volume) each emphasize, the attention needs to consider the systemic implications for many different people. The fields must address PD from a systemic approach by engaging participants "not just as objects to be changed but as co-learners and agents in design" (Penuel, this volume, p. 98).

In order to address PD from a systemic perspective, it is important to include many different stakeholders in the design, organization, and implementation of the PD programs. These should include PD designers, PD leaders, teachers, school leaders, and policy makers, among others. When engaging many different stakeholders, barriers do arise. For example, as Fuller (this volume) points out, school leaders can substantially influence the implementation and effectiveness of particular educational reforms and PD efforts.

Current research (Borko et al., 2010; Desimone, 2009) highlights certain characteristics of high-quality PD opportunities. Two important factors are sustainability (Borko, this volume) and scalability (Moon, this volume). This can be a challenge to accomplish due to the cost and locations of teachers, particularly those in small, rural schools, or urban settings (Fuller, this volume). One way to address this challenge can be the use of technology and social media implementation tools to reach many teachers for minimal costs (see Technology section, p. 315).

New Practices and Content for Mathematics and Science PD Opportunities

Various authors (e.g., Benson, this volume; Johnson, this volume; Litzinger, this volume) have emphasized the importance of three practices that span mathematics and science disciplines, as well as the ethical issues and value judgments that foreground scientific decision-making (Tuana, this volume). Due to the increased expectations of teachers to teach all three dimensions (NRC, 2012a), it is important to highlight the connections between both science and mathematics. This becomes challenging when the disciplines are often "siloed" during K-12 education (Bunt, session 3 panel and Q&A discussion; Malcom, keynote address). There needs to be "a shift from a linear, silo subject, mechanistic view [of the disciplines] to a recursive, interdisciplinary, systems view" (Mayes, this volume, p. 186). To shift the focus, teachers will need to be educated in how the disciplines relate and how to engage their students in the interdisciplinary nature of the subjects. This will require shifts in the content and format of TE and PD.

Some practices common to STEM disciplines are systems thinking, model-based reasoning, and quantitative reasoning. Systems thinking has been characterized as "being able to explain the system or components of a system under study and then predict the impacts the system will have on surrounding systems" (Duschl et al. this volume, p. 2). By teaching TE and PD with a systems thinking perspective, teachers will be able to consider components of curricula and performance assessment tasks as integrated problems, which require both scientific and mathematical knowledge—providing a way to de-silo the disciplines (Minstrell, session 3 panel and Q&A discussion).

Model-based reasoning and quantitative reasoning are emphasized in the *Framework* as scientific practices and in the CCSS-M as mathematics practices, thereby offering once again the opportunity to de-silo the two subjects. Borko (this volume) expands upon the importance of model-based reasoning, by providing a sample overview of a PD program focused on model-based reasoning. Due to the importance of these practices in both science and mathematics, particularly with their connections to systems thinking (Duschl et al., this volume), it is important to support teachers' understandings of the practices and how to engage their students in them.

Examples of Teacher Education Orienting around Systems Thinking and Model-Based Reasoning

Teacher education faces many challenges when attempting to prepare and empower pre-service teachers to address the challenges in K-12 classrooms across the country. It is presumptuous to assume that TE programs can prepare teachers for all of the content and learning they will need in order to be effective teachers. Instead, the goal is to prepare "well-started beginners who can continue to learn from practice" (Zembal-Saul, Session 4 Poster Presentation). By supporting pre-service teachers' teaching and learning around the relationships between the STEM disciplines, the hope is that they will build a firm foundation that continues to develop throughout their career. Though this is a high goal to achieve amidst the many expectations placed on TE programs, it is possible and is occurring in some programs. For example, at Penn State University, specialized content courses have been developed for pre-service teachers through collaborative work between science discipline faculty and science education researchers (Haefner, Friedrichsen, & Zembal-Saul, 2006; Ward & Zembal-Saul, 2007; Zembal-Saul, 2009). One goal of these co-designed courses is to incorporate the three dimensions (NRC, 2012a) around authentic science experiences. The pre-service teachers collect and analyze real data, identify patterns, construct, critique, and revise scientific explanations and arguments, consider children's science learning and misconceptions, and engage in teaching applications around the various science disciplines (see Posters 1–4, this volume; Haefner & Zembal-Saul, 2004).

Two examples of these specialized co-designed content courses were presented at the Summit: 'Fundamentals of Climate Science' and 'Fossil Mammals and Biotic Response to Climate Change' (see Posters 3 and 4, this volume). In these courses, preservice teachers work with data sets and phenomena in order to engage in science discourse about climate change. Climate science has been a useful context for supporting preservice teachers' understanding of systems thinking and how different components of climate change interact to inform the broader global climate system. When working with the data sets, the students engage in quantitative reasoning to create models that help inform their understanding of the climate system in which they are working and discussing. The course content has also provided the students a context to discuss ethical dilemmas related to climate change and their epistemology of science. These courses have supported pre-service teachers' attention to evidence and argumentation, which has helped them (1) shift away from an "activity" mindset of science teaching, (2) consider the content more thoughtfully, (3) view classroom 'talk' as essential to science learning, and (4) adopt a role of the teacher as assessing and monitoring student thinking (Merritt, Zembal-Saul, Larcom, Crane, & Furnam, 2011; Posters 3 and 4, this volume; Zembal-Saul, 2009). These courses represent examples of ways to support teachers' understanding and engagement with the cross-disciplinary practices outlined in this volume.

Due to the new agendas in STEM education, there are large redesigns necessary for the structures and content of effective PD and TE. Based on the ideas presented here and throughout the book (e.g., Borko, this volume; Moon, this volume), we recommend that PD should:

- take a systemic approach to the design and implementation of PD programs by involving many different stakeholders;
- be scalable in ways that support teachers in a variety of educational environments;
- be ongoing and sustainable by creating intentional communities of practice that are designed to continually support teachers;
- be embedded within teachers' practice in ways that build on students' ideas and engage teachers in inquiry about their own practice;
- support teachers' understanding of and attention to the equitable challenges facing the STEM disciplines, as well as the ethical implications surrounding research in the disciplines;
- integrate science and engineering practices and the practices of mathematics, with science and mathematics content; and
- emphasize systems thinking, model-based reasoning, and quantitative reasoning as practices that cross mathematics and science in order to emphasize the cross-disciplinary nature of the disciplines and problems facing the world today.

These goals will have extensive and far reaching implications for TE and PD supporting the ideas presented in the *Framework* and *NGSS*.

Technology Integration

Roschelle, Pea, Hoadley, Gordin, and Means (2000) advocate that technology be used as a tool to help students learn things better and learn better things. This notion of using technology to *learn better* applies to four potential areas where technology and social media could alter learners' engagement with practices in STEM disciplines. These potential areas of change include:

1. scaling-up learning,
2. integrating instruction and assessment,
3. learner motivation and interest, and
4. teacher PD.

The first potential area of change is scaling up the extent of reach to learners while providing customized support for learning. As noted by Fonseca (session 1 panel and Q&A discussion), the use of online education such as massive open online courses (MOOCs) allows learning scientists and faculty to examine individual's learning patterns while reaching hundreds or thousands of learners at once. What this means is while large numbers of individual learners are enrolled in the same online education course, it is possible to customize the learning content to support individual learners' needs. For example, customized and timely feedback on learner's performance can be provided using an embedded intelligent tutor. An individualized learning pathway in terms of the types of practices or additional courses a learner might need can thus be suggested based on each learner's performance (Anderson, 2008). Providing well-designed online courses targeted at model-based reasoning and quantitative reasoning gives many individual learners the opportunity to develop these essential practices in STEM. Borge (this volume) suggests educational environments that meet the needs of a range of learners can be created by "redefining the role of the teacher [and] using technology as a means to support and enhance cognition" (Table 4.1, p. 83). The purposeful use of technology and social media seems a promising approach to provide customized support for learners with different learning needs.

Duschl et al. (this volume) emphasize that "[b]oth model-based reasoning and quantitative reasoning involve an iterative process of analyzing, modeling, communicating, evaluating, and redesigning models to explain scientific phenomena or processes" (p. 3). Learning environments can thus be infused with technology (e.g., digital tools) to support individual learners as they engage in these processes. An example of technology use is described in Johnson's (this volume) utilization of software to create dynamic environments that help middle school students develop their continuous covariational reasoning, an aspect of quantitative reasoning. In addition to the development and testing of various technological tools for supporting students' development of STEM practices, future research agendas should include the customization of these tools according to learners' needs, as well as the impact of technology use on the teaching and learning of STEM practices.

The second potential area of change is using technology to integrate instruction and assessment, so as to support 'assessment for learning' where assessment is used to inform future learning. Through the use of technology, a large amount of measurement data can be captured, analyzed, and displayed in a relatively short amount of time. For example, dashboards can be used to visually display data related to learners' activities captured within a technological system such as an online learning platform (US Department of Education, 2012). Such measurements can be interpreted as indicators of student learning, and can be selectively displayed to stakeholders, including individual learners, all learners, teachers, parents, and/or school administrators, at various grain sizes and time scales (e.g., immediately, daily, monthly, or annually). By taking actions based on sound interpretation of the dashboard data, stakeholders could improve future teaching and learning, thus leading to continuous improvement (US Department of Education, 2012). As different stakeholders may require different types of data for decision-making, future research agendas should include evaluating the usability of different types of data visualizations for various stakeholders, based on the decisions and choices they may want to make (US Department of Education, 2012). In addition, there is a need for more research on designing and validating automated assessments of complex STEM practices such as systems thinking, model-based reasoning, and quantitative reasoning.

Technology can also be used to create a more integrated assessment and learning system that supports teachers and students in their decision-making to improve student learning. An example of a system for supporting teachers' decision making is Diagnoser by Minstrell, Anderson, and Li (this volume). Details of the 'assessment for learning' aspect of Diagnoser are highlighted in the *assessment* strand below. Diagnoser uses online support tools to facilitate teachers in their implementation of a blended approach to assessment, instruction, and learning by mediating teachers' interpretation of student thinking based on students' responses to questions. Future instructional moves are recommended to teachers based on the diagnosed facets of students' thinking. Reflective assessments support students in making decisions to improve their own learning. They promote learner control as students take on the role of assessing and reflecting on their own understanding and performance based on a given set of criteria (Borge, this volume). Technology-infused reflective assessments can provide automated prompts for learners to reflect on the best strategy or tool for improving their understanding and performance (e.g., Shimoda, White, Borge, & Frederiksen, 2013).

The third potential area of change is the promotion of learner motivation for engaging in STEM practices through the use of technology and social media. For example, learner motivation can be promoted through the gamification of learning and awarding of micro-credentials for achievement of learning, such as digital badges (Vasko, session 1 panel and Q&A discussion). Zimmerman and Malcom (session 3 panel and Q&A discussion) suggest the use of mobile technology to support students from less privileged backgrounds to collect data in out-of-school learning contexts, such as around their community. Such data may be of greater interest and relevance to the students than the data they typically work with in school. In addition, Pigg et al. (this volume) describe how social media such as *Science Buzz*, a science museum blog, can be utilized to create a space for shared conversation about scientific issues among scientists and the public participants. Such purposeful use of social media taps into the interests of the public participants to facilitate their engagement with scientific concepts and scientific ways of thinking and meaning making (Pigg et al. this volume). Further research could be done on the impact of technology and social media use on learner motivation and their learning of complex practices such as systems thinking, model-based reasoning, and quantitative reasoning.

The fourth potential area of change is PD for teachers. Technology can be used to bring about changes in teacher PD at a larger scale than previously possible. An example is the Teacher Learning Journeys, which incorporates individualized PD and a badging engine based on a flexible MOOC model (Peck, session 5 panel and Q&A discussion,). Not only does technology serve as a platform for such large-scale PD opportunities, it has also enabled the social networking of teachers who are geographically dispersed through the use of the Internet (Benson, session 2 panel and Q&A discussion).

A challenge identified with virtual PD is the lack of organized professional communities that provide strong and continued support systems for the teachers participating in the PD. Moon (this volume) and colleagues have been developing a PD program (Next Generation Science Exemplar Learning System) that addresses this challenge by researching how to virtually support teachers' learning and teaching of science practices in certain content areas, while also developing a community of learners within the structure of the PD. The program has the potential to bridge the scalability challenge due to its virtual nature, while also addressing the sustainability challenge facing PD opportunities through the creation of intentional communities of practice (e.g., Lave & Wenger, 1991).

Assessment

The fourth thematic strand is a need to align assessment tasks, methods and tools for measuring and interpreting students' "grip" of STEM practices (Duschl & Bybee, 2014) with the science-as-practice curriculum/course frameworks, and with 'assessment for learning' teaching and learning practices. An example of such alignment is Diagnoser (Minstrell et al., this volume), whereby assessment assumes a diagnostic nature and is blended with teaching and learning under the Building on Learner Thinking (BOLT) framework. The BOLT framework was conceptualized for designing and implementing learning environments with an emphasis on science practices such as developing and validating models through observation, argumentation, consensus building, and knowledge generation. As discussed in the *technology* strand above, Diagnoser utilizes technology to facilitate this necessary assessment alignment with the goal of supporting 'assessment for learning.'

In the NRC report *Developing Assessments for the Next Generation Science Standards* (2014), the report committee recognizes that existing science assessments do not capture the three dimensions of science learning intertwined in the NGSS (disciplinary core ideas, crosscutting concepts, and scientific and engineering practices). Hence, new approaches are required to develop such desired science assessments. The committee's main recommendations for developing assessments for the NGSS include the following points:

1. Tasks for assessing the NGSS performance expectations should:

    a. "include multiple components reflecting the connected use of various scientific practices in the context of disciplinary core ideas and crosscutting concepts,
    b. address the progressive nature of learning by providing information on where students land on a continuum between expected beginning and ending points in a given unit or grade, and
    c. include an interpretive system to evaluate students' artifacts to help teachers interpret the students' responses and provide tools for helping teachers determine the next steps in instruction" (p. 3).

2. Measurement of the NGSS performance expectations requires a system of assessments that assess and provide relevant information about student learning to various stakeholders (e.g., students, teachers, administrators, policy makers, and the public). Such a system should include:

    a. "assessments designed to support classroom instruction,
    b. assessments designed to monitor science learning on a broader scale, and
    c. a series of indicators to monitor that students are provided with adequate opportunity to learn science in the ways laid out in the K-12 Framework and the NGSS" (p.4).

3. A "bottom-up" rather than "top down" approach should be taken when developing the appropriate assessment system—start by designing assessments for the classroom, then move towards assessments for monitoring.

Various chapters in this volume act as extended conversations about the committee's recommendations for designing assessments aligned with science-as-practice curriculum/course frameworks. For example, design-based implementation research (DBIR), which can be used to "address and study problems of implementation" (Penuel, this volume, p. 100), provides an approach to designing the assessment system articulated by the committee. In addition to the committee's assessment tasks recommendations (point 1 above), assessment of practices should be grounded in content around core concepts (Duschl et al. this volume) and the disciplinary context (Coppola, this volume). This means the tasks for assessing a practice may differ across the content areas—assessment of systems thinking in engineering may be different than assessment of systems thinking in life science (Litzinger, this volume).

In relation to the recommendation for a subsystem of assessments to support classroom instruction (point 2a above) or 'assessments for learning,' given that knowing and doing in STEM disciplines involve the complexities of the *what, how,* and *why* of the disciplines, student populations from different socio-cultural-historical contexts would likely demonstrate different extent of knowing and doing in the STEM disciplines, even at the same grade level. If we take a human-centered design approach to redesigning learning environments (Borge, this volume), learners with different skills and experiences should receive different productive learning experience in STEM that best supports their future learning (Greeno, session 4 panel and Q&A discussion). In other words, the learning experience or scaffolding that students at the same grade level need in order to reach a higher level of understanding or engage in more sophisticated practices may differ. Hence, students may need to take different learning pathways to eventually achieve the same learning outcomes. In view of the call for assessments to be aligned with classroom teaching and learning practices (Minstrell et al. this volume), assessment ought to be localized and aligned with the students' learning experiences. For example, Coppola (this volume) explains how he and his colleagues re-designed an introductory freshmen chemistry course and its assessment in order to develop students' understanding and use of models with authentic, unfamiliar data, based on the students' high school chemistry experiences. If students are to take different learning pathways, then aligning assessments with students' learning experiences implies the notion of a single, standardized version of assessment for the purpose of supporting learning should be given up (Lehrer, session 4 panel and Q&A discussion; Duschl et al., this volume).

As for the recommendation for a subsystem of assessments for monitoring science learning at a broader scale (point 2b above), or 'assessment of learning,' the revised Advanced Placement (AP) biology examination (Benson, this volume) provides an example of alignment between large scale, standardized assessment tasks, and a science-as-practice curriculum/course framework. The AP biology examination was revised to align with the AP biology course curriculum that incorporates elements of systems thinking, model-based reasoning, and quantitative reasoning. At the international level, the Programme for International Student Assessment (PISA) 2015 science assessment framework (Organization for Economic Cooperation and Development [OECD], 2012) frames scientific knowledge as comprising content knowledge, procedural knowledge, and epistemic knowledge of science (Osborne, this volume). Assessment tasks targeted at evaluating students' scientific content, procedural, and epistemic knowledge will be implemented for the PISA 2015 assessment. The inclusion of science practices in a large scale assessment for monitoring science learning at the international level (based on the participating countries) prompts a research agenda for studying the impact of changing an international benchmark assessment, such as PISA, on curriculum frameworks and classroom practices in countries that participate in PISA testing.

For better or worse, PISA has become an influential instrument for benchmarking student performance worldwide. A recent study found that PISA results have pushed many participating countries and economies to initiate education policy reforms that introduce changes in ways countries deliver education, teach curriculum, and assess learning (Breakspear, 2012). Assuming that PISA's newly revised assessment framework that integrates content, procedural, and epistemic knowledge represent the 'state-of-the-art' among large-scale standardized science assessments, it may be hypothesized that such an assessment is worth adapting the national or state assessments to, and that it serves as a form of benchmark about aggregate national performance. However, there is one key difference between the PISA assessment and when individual countries adopt PISA-like assessment frameworks and items: when those items become part of the national assessment, they have high-stakes for the student as well as for teachers and schools, especially in countries with accountability models in place. And also considering that assessments often drive what teachers do, countries might expect widespread implications for teaching and learning. The thematic strands elaborated in this chapter and programs and practices discussed throughout this volume may serve as helpful and constructive references to address those implications and respond with appropriate changes. Ultimately, systemic approaches to developing and implementing any educational intervention will be required to enact sustainable change.

Systemic Approach to Teaching and Learning

There is a growing understanding and broadening consensus among the scholars that STEM education has reached a point where it has attracted quite some attention all the way from the federal, to the state, to the district level. Malcom (keynote address) reminded us of this change, where the field moved from no conversation at all a decade ago, to too many conversations during the current decade. These have led to disconnected conversations resulting in silos that miss or avoid what others in the same problem space are doing. The situation has reached an extreme, according to Malcom, where stakeholders, including STEM policy-makers, researchers, and practitioners, among others, have produced silos that do not serve the purpose of providing the best possible STEM education for students. As a result, the stakeholders have developed the habit of "talking past each other." The general plea is to consider the larger social and policy contexts where teaching and learning and teacher education takes place, with whom and under what conditions. This fifth thematic strand represents the larger context within which the other four thematic strands reside.

Today's global educational world is quite different from ten years ago (Fuller, this volume). In a world of interconnectedness and interdependence, no system can afford to operate under such disconnected scenarios. As reflected by the perspectives of the Summit participants, a new era of cooperative and collaborative systemic-based work and system-wide approaches with a broad range of STEM education stakeholders must become a reality. Otherwise, the field will not yield the best results from the current momentum and policy attention it has attracted from various policy levels—especially at the federal level in the US It will fail to deliver on the STEM workforce and publics' engagement goals of STEM education (Duschl et al., this volume). A review of past educational reforms will quickly point to more failures than successful efforts.

Systemic educational reforms have a poor track record. Educational historians have concluded that educational reforms have only been successful to make incremental changes to the existing school practices at best (Tyack & Cuban, 1995; Tyack & Tobin, 1994). Why should this time be different? Or, better put, what is different this time around? One possibility is the momentum that STEM fields have reached, especially with the direction given with the NGSS. For the US context, the NGSS and Common Core are unprecedented educational events. With a tradition of locally-rooted, fragmented, and often small district education systems that have resisted any idea of a "national curriculum," the US states that have opted in for adopting the Common Core and NGSS are moving forward on the same educational standards for all students. Concerted efforts at the federal level, as well as strong political support from state governors and financial support from the business community have supported the new momentum where all students in the system are held accountable for the same standards. Yet, there are serious questions and concerns raised by teachers and by parents about the new assessment systems that will restrain implementation. However, the Common Core and NGSS are beginning steps, and importantly they provide a roadmap for all stakeholders (Malcom, keynote address). In the world of education policy, building the bridge between policy and practice is the most difficult, but the most necessary move to implement a reform agenda and enact change.

A systems-thinking perspective can contribute to a systemic STEM education reform. Litzinger (this volume) reminded us that having students who are attentive and considerate to environmental, social, societal, and political implications of their work requires them to be systems thinkers. When each STEM stakeholder is attentive to implications that their actions have in a broader sense, we can begin to see a better alignment of policy intervention and implementation. Understanding that nothing exists in a bubble, and that interdependence is the new norm, will assist STEM stakeholders to make better decisions, policies, and interventions that are more effective and efficient from policy to classroom level.

Authors and the Summit participants emphasized the need to teach systems thinking in K-16, which will arguably lead toward more STEM graduates and a public engaged in science topics. However, integration of systems thinking into education programs cannot take place without changing educational systems. Penuel (this volume) promotes a bottom-up or mixed perspective on educational design using the DBIR approach. He rejects top-down approaches, which usually leaves practitioners voiceless. And still, design for changing the system is only the first step—broad partnerships need to be put in place for the implementation process—partnerships that need to do the implementation on the ground without ever losing sight of the macro-level perspective. However, it depends on the goals of the change as to what strategy is utilized to reach those goals. Systemic approaches might help when large-scale reforms are introduced, while improvement science could be a better strategy if the goal is only to improve the existing system.

Improvement science emphasizes the shift from *going fast and learning slow* to *learning fast and implementing well* (Bryk, Gomez, Grunow, & LeMahieu, 2015). The objective of improvement science is to improve rather than reform the systems. Three core improvement questions are asked while improvement in learning outcomes is sought, "What is the specific problem I am now trying to solve? What change I might introduce and why? And, how will I know whether the change is actually an improvement?" (Bryk et al., 2015, p. 9). Further, improvement science fully integrates systemic thinking into the improvement efforts. Bryk et al. (2015) highlight that improvement science never adopts a top-down approach, but always requires full participation of teachers, principals, and educational leaders who act as improvers of the system and collaborate with external researchers to generate evidence about how to achieve better learning outcomes. The entire process is a learning journey revolving around the concept of networked improvement communities (NICs) whose goal is to innovate and learn together toward educational improvement. In other words, improvement science adopts a bottom-up approach with wide participation of educational stakeholders in the improvement efforts. This strategy suggests that to make improvement science productive for system-wide reforms, a large number of big and small NICs spread out across the system might be required to bring about systemic changes.

However, to enact change in the system, the entire system must be the focus of design. Nothing exists independent of other things in the system (Penuel, this volume), and there are many components of the education system affecting and being affected by other components simultaneously (Fuller, this volume). Therefore, applying systems thinking at different levels of policy design and implementation (federal to state to district to school to classroom) feeds and strengthens partnerships at any level for more coherent actions and sustained progress with STEM reform. And then intervention is designed based on the level. This has led to seemingly opposing views on designing change among scholars. For example, Borge (this volume) suggests that when applying human-centric design, whenever education change occurs, students' needs should be considered first because if priorities of other stakeholders are addressed first, "such an approach would likely not be very productive or effective" (p. 71). Penuel (this volume), on the other hand, argues that scholars and other educational stakeholders need to design for systems, not for classrooms, thus taking a much more macro-level perspective than Borge. Still, the goal in either case is to improve teaching and learning environments.

While the consensus on the direction of STEM reform seems to be broadening and deepening, the road ahead is not easy. Summit contributors have pointed to issues that need special attention as more work is completed to implement intended educational reform. Malcom (keynote address) warns that STEM education stakeholders need to prepare for the worst before situations get better, and notes that funding issues and student changing demographics need to be seriously considered. Further, Malcom (keynote address), Penuel (this volume), and Fuller (this volume), among others, have emphasized learning from past and ongoing failures in the systemic reform process so that the reform agenda keeps progressing forward. They also point to issues of principal turnover, teacher evaluation, curriculum, instruction and assessment alignment, as well as PD and teacher education as major sources of tension for the change to be sustained. However, the positive momentum needs to be maintained and further supported so that STEM fields produce a much desired next generation of STEM graduates to further develop and transform the very field they will be in, and much aspired scientifically literate citizenship. Not easy targets to be achieved, but the momentum and support are in place to address the research and policy changes necessary for STEM reform.

Conclusion

As we seek to move forward with the directions outlined in this volume and toward the goals of developing a strong STEM workforce and scientifically engaged citizenry, there are a few challenging and critical research agendas to address. These agendas involve answering the following questions:

- What equitable challenges face the STEM disciplines and how do we engage *all* students in the disciplines?
- What are the ethical components and implications of the STEM disciplines?
- How can we support teachers and students in understanding the ethics surrounding the STEM disciplines?
- How do ethics and epistemics interact?
- How can we take a systemic approach toward teacher education and PD in ways that support the scalability and sustainability of reform-oriented programs?
- How can we use *systems thinking*, *model-based reasoning*, and *quantitative reasoning* to support an interdisciplinary approach toward PD and teacher education?

- How can PD and teacher education be embedded in teachers' practice with a focus on using students' ideas to inform instruction?
- How can we develop curricula that supports both the teachers' and students' understanding of the dimensions outlined in the *Framework*?
- How can we use technology to integrate curriculum, instruction, and assessment in creative ways that reaches and motivates many students?
- How can we support and assess students' *coherent* understanding of science and mathematics when the standards and expectations span such broad science and mathematics disciplines?
- How can we align assessments and standards in ways that are aligned with the dimensions of the *Framework*?
- What would assessments look like that use *systems thinking, model-based reasoning*, and *quantitative reasoning* to integrate the STEM disciplines?
- The *Framework* argues that the integrative nature of the dimensions and the greater depth will increase students' interests, dispositions, and conceptual orientations toward science, but is this really the case and how do we go about assessing these other measures?
- How can observation protocols be useful in "describing the characteristics of instruction that are associated with particular valued practices," while also giving "guidance to how the teaching of core ideas looks in [the] classroom" (Duschl et al., this volume, p. 24)?
- With all these challenges in mind, as well as the reform outlined in the *Framework*, what steps do we, as an education community, need to take to develop a "common understanding of what it means to teach and learn content, to engage in meaningful questioning and feedback, and to support students' reasoning and problem solving" (Duschl et al., this volume, p. 24)?

To engage in these many research agendas, a systemic approach will be necessary to collaboratively involve many stakeholders in ways that link research, policy, and practice. This will be challenging work and will require regular communication in order to support strong STEM education. Today's students are tomorrow's decision-makers and we need to prepare *all* students to make informed, ethical decisions regarding challenging scientific issues that currently face the world, as well as those that have yet to occur. We have set the stage to take on the complex change necessary for an education system intent on preparing a more informed citizenship. Now we call the education community to action.

Note

1. For further information about the Waterbury Summit, as well as video recordings of the various presentations, panel discussions, and question and answer sessions, see the Waterbury website at www.waterbury.psu.edu.

Reflections and Summary: References

Anderson, T. (Ed.). (2008). *The Theory and Practice of Online Learning: Second Edition* (2nd edition). Edmonton: University of Washington Press.

Bismack, A.S., Arias, A.M., Davis, E.A., & Palincsar, A.S. (2015). Examining student work for evidence of teacher uptake of educative curriculum materials. *Journal of Research in Science Teaching, 52*(6), 816–846.

Borko, H., Jacobs, J., & Koellner, K. (2010). Contemporary approaches to teacher professional development. In P.L. Peterson, E.L. Baker, & B. McGaw (Eds.), *International Encyclopedia of Education* (Vol. 7, pp. 548–56). Oxford, UK: Elsevier. Retrieved from http://public.eblib.com/choice/publicfullrecord.aspx?p=634860

Breakspear, S. (2012). *The Policy Impact of PISA: An exploration of the normative effects of international benchmarking in school system performance* (OECD Education Working Papers). Paris: Organisation for Economic Co-operation and Development. Retrieved from hwww.oecd-ilibrary.org/content/workingpaper/5k9fdfqffr28-en

Britton, L. (2014). STEM, DASTEM, and STEAM in making: Debating America's economic future in the 21st century. *University of Washington: Information School.* Retrieved from http://tascha.uw.edu/2014/09/stem-dastem-and-steam-in-making-debating-americas-economic-future-in-the-21st-century/

Bryk, A.S., Gomez, L.M., Grunow, A., & LeMahieu, P.G. (2015). *Learning to Improve: How America's Schools Can Get Better at Getting Better.* Cambridge, MA: Harvard University Press.

Bybee, R.W. (2014). NGSS and the next generation of science teachers. *Journal of Science Teacher Education, 25*(2), 211–21. http://doi.org/10.1007/s10972-014-9381-4

Davis, E.A., & Krajcik, J.S. (2005). Designing educative curriculum materials to promote teacher learning. *Educational Researcher, 34*(3), 3–14. http://doi.org/10.3102/0013189X034003003

Davis, E.A., Palincsar, A.S., Arias, A.M., Bismack, A.S., Marulis, L.M., & Iwashyna, S.K. (2014). Designing educative curriculum materials: A theoretically and empirically driven process. *Harvard Educational Review, 84*(1), 24–52.

DeMio, T. (2015, February 1). From the brainy minds of kids: NKY Makerspace. *Cincinnati.com.* Retrieved from http://cincinnati.com/story/news/2015/02/01/steam-makerspace-boone-county-schools-robotics-northern-kentucky/22587043/

Desimone, L.M. (2009). Improving impact studies of teachers' professional development: Toward better conceptualizations and measures. *Educational Researcher, 38*(3), 181–99. http://doi.org/10.3102/0013189X08331140

Duschl, R.A., & Bybee, R.W. (2014). Planning and carrying out investigations: An entry to learning and to teacher professional development around NGSS science and engineering practices. *International Journal of STEM Education, 1*(1), 1–9. http://doi.org/10.1186/s40594-014-0012-6

Fusarelli, L.D. (2004). The potential impact of the No Child Left Behind Act on equity and diversity in American education. *Educational Policy*, 18, 71–94.

Haefner, L.A., & Zembal-Saul, C. (2004). Learning by doing? Prospective elementary teachers' developing understandings of scientific inquiry and science teaching and learning. *International Journal of Science Education, 26*(13), 1653–74. http://doi.org/10.1080/0950069042000230709

Haefner, L.A., Friedrichsen, P.M., & Zembal-Saul, C. (2006). Teaching with insects: An applied life science course for supporting prospective elementary teachers' scientific inquiry. *The American Biology Teacher, 68*(4), 206–12. http://doi.org/10.1662/0002-7685(2006)68[206:TWIAAL]2.0.CO;2

Henriksen, D. (2014). Full STEAM ahead: Creativity in excellent STEM teaching practices. *The STEAM Journal, 1*(2), 1–7.

Jolly, A. (2014). STEM vs. STEAM: Do the arts belong? *Education Week*. Retrieved from http://www.edweek.org/tm/articles/2014/11/18/ctq-jolly-stem-vs-steam.html

Lave, J., & Wenger, E. (1991). *Situated Learning: Legitimate Peripheral Participation*. New York: Cambridge University Press.

Maeda, J. (2013). STEM +Art = STEAM. *The STEAM Journal, 1*(1). Retrieved from http://scholarship.claremont.edu/steam/vol1/iss1/34

Merritt, M., Zembal-Saul, C., Larcom, E., Crane, R., & Furnam, T. (2011). Activities grounded in evidence based explanations help pre-service elementary educators move towards deeper understanding of climate science. In Geological Society of America Annual Meeting and Exposition, 9–12 October 2011. Minneapolis, MN.

Miller, A. (2014). PBL and STEAM education: A natural fit. *Edutopia*. Retrieved March 29, 2015, from http://edutopia.org/blog/pbl-and-steam-natural-fit-andrew-miller

Moore, T.J., Tank, K.M., Glancy, A.W., & Kersten, J.A. (2015). NGSS and the landscape of engineering in K-12 state science standards. *Journal of Research in Science Teaching, 52*(3), 296–318. http://doi.org/10.1002/tea.21199

National Governors Association Center for Best Practices, & Council of Chief State School Officers. (2010). *Common Core State Standards for Mathematics*. Washington, DC: National Governors Association Center for Best Practices, Council of Chief State School Officers.

National Research Council. (2012a). *A Framework for K-12 Science Education: Practices, Crosscutting Concepts, and Core Ideas*. Washington, DC: National Academies Press. Retrieved from http://nap.edu/catalog/13165/a-framework-for-k-12-science-education-practices-crosscutting-concepts

National Research Council. (2012b). *Education for Life and Work: Developing Transferable Knowledge and skills in the 21st Century*. Washington, DC: National Academies Press.

National Research Council. (2014). *Developing Assessments for the Next Generation Science Standards*. Washington, DC: National Academies Press.

NGSS Lead States. (2013). *Next Generation Science Standards: For States, By States*. Retrieved from http://nap.edu/catalog/18290/next-generation-science-standards-for-states-by-states

NGSS Lead States FAQ. (2013). *Frequently asked questions*. Retrieved from http://nextgenscience.org/print/228

Organization for Economic Cooperation and Development (OECD). (2012). The PISA 2015 Assessment Framework: Key competencies in reading, mathematics and science. Retrieved from http://www.oecd.org/pisa/pisaproducts/pisa2015draftframeworks.htm

Riley, S. (2012). Use arts integration to enhance common core. *Edutopia*. Retrieved March 29, 2015, from http://edutopia.org/blog/core-practices-arts-integration-susan-riley

Roehrig, G.H., Moore, T.J., Wang, H. H. and Park, M.S. (2012). Is adding the E enough? Investigating the impact of K-12 engineering standards on the implementation of STEM integration. *School Science and Mathematics, 112*(1), 31–44.

Roschelle, J.M., Pea, R.D., Hoadley, C.M., Gordin, D.N., & Means, B. (2000). Changing how and what children learn in school with computer-based technologies. *Future of Children, 10*(2), 76–101.

Schwarz, C.V., & Gwekwerere, Y.N. (2007). Using a guided inquiry and modeling instructional framework (EIMA) to support preservice K–8 science teaching. *Science Education, 91*(1), 158–86. http://doi.org/10.1002/sce.20177

Shimoda, T., White, B.Y., Borge, M., & Frederiksen, J.R. (2013). Designing for science learning and collaborative discourse. In *Proceedings of the 12th International Conference on Interaction Design and Children* (pp. 247–256).

Tuana, N. (2010). Leading with ethics, aiming for policy: New opportunities for philosophy of science. *Synthese, 177*(3), 471–92. http://doi.org/10.1007/s11229-010-9793-4

Tyack, D., & Tobin, W. (1994). The "grammar" of schooling: Why has it been so hard to change? *American Educational Research Journal, 31*(3), 453–79. http://doi.org/10.3102/00028312031003453

Tyack, D.B., & Cuban, L. (1995). *Tinkering Toward Utopia: A Century of Public School Reform.* Cambridge, MA: Harvard University Press.

US Department of Education. (2012). *Enhancing teaching and learning through educational data mining and learning analytics: An issue brief.* US Department of Education. Retrieved from http://tech.ed.gov/learning-analytics/

Vasko, S.E. (2015a). *"Crafter" or "Maker"? STEM/STEAM, Education, and the Maker Movement.* Unpublished Abstract.

Vasko, S.E. (2015b). Crafts and STEM in Dialogue. In The Consortium for Socially Relevant Philosophy of/in Science and Engineering Annual Meeting. Detroit, MI.

Vasko, S.E. (2015c). *CREATE for STEM Science Seminar Series.* East Lansing, MI.

Vasko, S.E. (2015d). *Developing and Communicating Expertise between Crafting and Science/Technology/Engineering/Mathematics (STEM) Fields.* Unpublished Abstract.

Vasko, S.E. (2015e). *Creating craftier engineers. Slate.com.* http://slate.com/articles/technology/future_tense/2015/06/stem_classes_should_include_lessons_about_low_tech_crafts_like_sewing.html

Ward, A.R., & Zembal-Saul, C. (2007). Scientist-science teacher educator partnerships for developing and teaching science courses for preservice elementary teachers. *Pennsylvania Teacher Educator, 6*, 1–9.

White, H. (2010). STEAM not STEM–Whitepaper. Presented at the STEAM not STEM.

Wilson, S.M. (2013). Professional development for science teachers. *Science, 340*(6130), 310–13. http://doi.org/10.1126/science.1230725

Zembal-Saul, C. (2009). Learning to teach elementary school science as argument. *Science Education, 93*(4), 687–719. http://doi.org/10.1002/sce.20325

ABOUT THE AUTHORS

Ruth Anderson

Educational Researcher, FACET Innovations

Ruth's varied professional background includes K-12 classroom research, educational program evaluation in STEM learning, curriculum development, and second language instruction with diverse learner populations. Her ongoing research interests include formative assessment, blended learning environments, and effective instruction and literacy support for secondary students (L2 and native speakers) across disciplines.

Spencer A. Benson

Director, Centre for Teaching and Learning Enhancement, Professor in Faculty of Health Sciences at the University of Macau SAR, PRC.

In 2008–09, he was a Fulbright Fellow to Hong Kong working on the 3–3–4 reform and General Education. Before joining the University of Macau in 2013, he directed the Center for Teaching Excellence at the University of Maryland, MD, US. Research interests include the development of pedagogies for teaching science for all students, assessment of student learning, secondary science education, and the use of technologies for learner-centered 21st century education. He has published more than 50 articles on teaching and learning and has given talks and workshops on teaching and learning in more than a dozen countries around the world.

Catharine Biddle

Assistant Professor, University of Maine, Department of Higher Education, Educational Leadership and Early Childhood Education

Dr. Biddle received her Ph.D. in Educational Leadership from The Pennsylvania State University in 2015. She also had projects with Penn State's Center for Rural Education and Communities and the Center for Educational Evaluation and Policy Analysis. Catharine served as the Executive Director for the Nanubhai Education Foundation, a 501(c)3 organization working to improve teacher training in rural India and has been a middle and high school teacher. Her research interests include the impact of spatial differences on educational policy and practice in rural areas, as well as the complex relationships between schools and community-based organizations.

Amber S. Bismack

Ph.D. student, Department of Educational Studies (Science Education), The University of Michigan

Prior to pursuing her doctoral studies, Amber worked at The Pennsylvania State University as a Research Assistant and Program Manager supporting both the Waterbury Summit and the Childhood and Early Adolescent Education (CEAED) program. For the Waterbury Summit, Amber helped coordinate the development of academic papers, as well as the Summit logistics. For the CEAED program, Amber was involved with program development, curricular concerns, developing assessment reports for accreditation, and coordinating communication across Penn State's campus locations, faculty, and staff. Her research interests focus on elementary teachers' ambitious science teaching, as well as supporting preservice elementary teachers' understanding of and engagement with science content and practices.

Marcela Borge

Assistant Professor, Learning and Performance Systems, College of Education, The Pennsylvania State University

Dr. Borge's research interests are at the intersection of learning, cognition, and design. Her current research focuses on unpacking group cognition, engagement, and student needs in order to design new models of interaction and technological tools to enhance learning. Her work has been funded by the Office of Naval Research, The National Science Foundation, and the Center for Online Innovation in Learning.

Hilda Borko

Professor of Education, School of Education, Stanford University

Dr. Borko received her Ph.D. in educational psychology from the University of California, Los Angeles. Dr. Borko's research explores teacher cognition, the process of learning to teach, and the impact of teacher professional development programs on teachers and students. Her current program of research includes a design-based implementation research project to adapt the Problem-Solving Cycle professional development (PD) program for mathematics teachers and a study of the effectiveness of a PD program to improve discourse and inquiry in elementary science classrooms. She served as President of the American

Educational Research Association (2003–2004) and is a member of the National Academy of Education.

Nancy Brickhouse

Provost, Professor of Science Education, Saint Louis University

Dr. Brickhouse, Ph.D. Purdue University, had a distinguished career as a science education research and administrator at the University of Delaware. In 1992, she was a visiting Fellow at the University of Leeds. Dr. Brickhouse's research has focused on access and equity in science education. Dr. Brickhouse's research develops theories of science teaching and learning that link these practices to epistemologies of the sciences. Her recent work focuses on issues of science education policy, particularly as they relate to the implementation of standards and assessment and the provision of equitable access to science literacy. Dr. Brickhouse served a five-year term as editor of *Science Education.*

Brian P. Coppola

Arthur F. Thurnau Professor of Chemistry, Associate Chair of Educational Development and Practice, Department of Chemistry, College of Literature, Science, and the Arts, The University of Michigan

Dr. Coppola received his Ph.D. in organic chemistry from the University of Wisconsin-Madison in 1984. He was promoted to Full Professor of Chemistry in 2001. His publications range from mechanistic organic chemistry research in 1, 3-dipolar cycloaddition reactions to educational philosophy, practice and assessment. He is a member of the editorial boards of *The Chemical Educator, The International Journal of Science Education*, the *Journal of Science Education and Technology*, and the *Journal of Chemical Education.* He is an Associate Editor for *The Journal for Research in Science Teaching*, and he is the editor-in-chief of *The Hexagon*, the quarterly publication of *Alpha Chi Sigma,* the professional chemistry fraternity.

Richard A. Duschl

Kenneth B. Waterbury Chaired Professor in Secondary Education, Department of Curriculum and Instruction, College of Education, The Pennsylvania State University

Prior to joining Penn State in 2008, Dr. Duschl held the Chair of Science Education at King's College London. He recently served as Chair of the National Research Council research synthesis report *Taking Science to School: Learning and Teaching Science in Grades K-8.* His research focuses on establishing epistemic learning environments and on the role of students' inquiry and argumentation processes. Dr. Duschl has twice received the 'JRST Award' (1989; 2003) for the outstanding research article published in the *Journal of Research in Science Teaching.* He also served for more than a decade as editor of the research journal *Science Education.*

Kirsten Ellenbogen

President and CEO, Great Lakes Science Center

Dr. Ellenbogen's research focuses on understanding the role of museums in the broader learning ecology and facilitating digital environments to support STEM learning and scientific argumentation. She is co-principal investigator of the Center for Advancement of Informal Science Education (CAISE), and she was a member of the National Academy of Sciences' committee that produced the volume, *Learning Science in Informal Environments*. She serves on the board of trustees for the Rock and Roll Hall of Fame and Museum, the Cleveland Water Alliance, and the Association of Science-Technology Centers. She holds a Ph.D. in Science Education from Vanderbilt University and a B.A. from the University of Chicago.

Edward J. Fuller

Associate Professor, Department of Education Policy Studies, Executive Director Center for Evaluation and Education Policy Analysis, College of Education, The Pennsylvania State University

Dr. Fuller's research interests include the study of educators' quality, distribution, mobility, turnover, and career pathways; educator preparation; school improvement; evaluation; and charter schools. He is Director of the Penn State Center for Evaluation and Education Policy Analysis (PCEEPA) and Associate Director of Policy, University Council for Educational Administration. Prior to joining Penn State, Dr. Fuller was a special research associate and adjunct faculty at the University of Texas at Austin. He was the Director of Research for the Texas State Board for Educator Certification and Program Director for Evaluation, Charles A. Dana Center—The University of Texas at Austin.

Drew H. Gitomer

Rose and Nicholas DeMarzo Chair in Education, Graduate School of Education, Rutgers, The State University of New Jersey

As a member of the Learning and Teaching Department, Dr. Gitomer conducts research that centers on the assessment and evaluation of teaching and related policy issues in teaching and teacher education. His current work focuses on a range of constructs that are related to teaching quality, including the quality of classroom interactions, teacher knowledge, teacher beliefs, and student achievement. As part of this work, Gitomer and his colleagues are carrying out validity studies of a variety of measures, including classroom observation protocols, classroom assignment protocols, and new measures of teacher knowledge. He is Co-editor of the AERA *Handbook of Research on Teaching*, 5th edition.

Jeff Grabill

Professor and Chair, Department of Writing, Rhetoric, and American Cultures, Michigan State University, Senior Fellow, University Outreach & Engagement

Dr. Grabill is a senior researcher with WIDE Research (Writing in Digital Environments) and also a co-founder of Drawbridge Incorporated, an educational technology company. He studies how digital writing is associated with citizenship and learning. He has published two books on community literacy and articles in journals like *College Composition and Communication*, *Technical Communication Quarterly*, *Computers and Composition*, and *English Education*.

James Greeno

Adjunct Professor of Education, School of Education, University of Pittsburgh

Dr. Greeno is a learning scientist who studies understanding and learning with conceptual understanding. His recent research has focused on understanding is achieved in interaction, especially discourse. He is an Emeritus Professor of Education at Stanford University. He is a retired member of the National Academy of Education and a member of the Society of Experimental Psychologists.

William Hart-Davidson

Associate Dean for Graduate Studies, College of Arts and Letters, Michigan State University

Dr. Hart-Davidson earned his Ph.D. in 1999 in rhetoric and composition from Purdue University. He is a Senior Researcher at Writing in Digital Environments Research at Matrix. In 2014, began a three-year appointment as Associate Dean of Graduate Studies in the College of Arts and Letters. He is a co-inventor of Eli Review, a software service that supports writing instruction. Eli is a system based on research and pedagogy developed with his colleagues at WIDE.

Heather Lynn Johnson

Assistant Professor of Mathematics Education, Mathematics & Science Program, School of Education and Human Development, The University of Colorado, Denver

After teaching high school mathematics for 13 years, Dr. Johnson completed graduate work in mathematics education, earning a Ph.D. from The Pennsylvania State University. As a mathematics educator, Dr. Johnson's research investigates middle and high school students' quantitative and covariational reasoning related to rate of change. At the University of Colorado Denver, Dr. Johnson has been committed to connecting research and practice, receiving the NCTM's Linking Research and Practice Outstanding Publication Award in 2014.

Min Li

Associate Professor, College of Education, University of Washington

Dr. Li is an assessment expert with a deep interest in understanding how student learning can be accurately and adequately assessed both in large-scale testing and classroom settings. Her work reflects a combination of cognitive science and psychometric approaches in various projects, including examining the cognitive demands of science items, measurement issues in constructing instructionally sensitive test items, issues of testing linguistic minority students in mathematics and science, analyzing teachers' classroom assessment practices, development of instruments to evaluate teachers' assessment practices, and use of science notebooks as assessment tools.

Thomas A. Litzinger

Professor of Mechanical Engineering, Director of the Leonhard Center for the Enhancement of Engineering Education, Department of Mechanical and Nuclear Engineering, College of Engineering, The Pennsylvania State University

Dr. Litzinger joined Penn State in 1985, and in 1997 became Director of the Leonhard Center. From 1992 until his appointment as Director, he served as the Penn State Principal Investigator of the Engineering Coalition of Schools for Excellence in Engineering and Leadership (ECSEL). His work in engineering education involves teaching and learning innovations, curricular change, faculty development, and assessment. He also teaches and conducts research in combustion and thermal sciences. He is a Fellow of the American Society for Engineering Education (ASEE) and the American Society of Mechanical Engineers (ASME), and he serves as an Associate Editor of *Advances in Engineering Education.*

Robert Mayes

Professor, Department of Teaching and Learning, College of Education; Director, Institute for Interdisciplinary STEM Education, Office of Research and Economic Development, Georgia Southern University

Dr. Mayes has taught mathematics on the high school, community college, and university levels. His first professorship was at West Virginia University. He later joined the University of Northern Colorado and served as the Director of the Mathematics and Science Teaching Center (MAST). Dr. Mayes then returned to the WVU Mathematics Department to direct the Institute for Mathematics Learning (IML). He also served as the Vice-President for College on the West Virginia Council of Teachers of Mathematics Executive Committee. Subsequently, he joined the University of Wyoming, assuming the directorship of the Science and Mathematics Teaching Center (SMTC). His research focuses on the teaching and learning of problem solving, mathematical modeling, and quantitative reasoning in the sciences and social sciences.

Jim Minstrell

Senior Research Scientist and Co-Founder, FACET Innovations

Jim spent 30 years teaching mathematics, physics, and integrated science and mathematics at the high school level. About ten years into his teaching career, Jim became aware of "misconceptions" research. He began studying the learning and teaching of his students and those of his teacher colleagues. The negative connotation of the term misconceptions and the realization that there were productive resources, as well as problematic thinking in students' understandings prompted Jim and colleagues to coin the term "Facets of student thinking." He has been a principal investigator or co-principal investigator of projects related to assessment, classroom practice, professional development, and the development of www.Diagnoser.com, a set of diagnostic formative assessment and instruction tools. Jim's interest in formative assessment and the promotion of deep conceptual understanding and reasoning in learners informs all of his work.

Jean Moon

Founder and President, Tidemark Institute

The Institute designs, implements, and participates in projects focused on K-16 STEM education especially in the areas of policy, teacher professional development, and innovations in learning environments particularly cyber-enabled learning environments. Dr. Moon is also a Visiting Scholar at Clark University and was the Director, Board of Science Education at the National Research Council. She consults widely on a variety of education issues. Currently, Dr. Moon is engaged in the design and implementation of a unique web-based professional development system (Next Generation Science Exemplar Learning System or NGSX) for science teachers, higher education faculty, science specialists, and educators working from informal science learning venues. As part of NGSX, Moon and her colleagues have designed a framework for peer-lead professional development facilitation to support the enactment of NGSX that is now being piloted in multiple states.

Yann Shiou Ong

Ph.D. Candidate, Department of Curriculum and Instruction (Science Education) The Pennsylvania State University

Prior to pursuing her doctoral studies, Yann Shiou taught physical sciences, physics, and science research at the middle and high school levels in Singapore. She was also part of the school leadership team. Yann Shiou's research interests can be characterized by the following key words: scientific practices, including argumentation, modeling, and systems thinking, productive disciplinary engagement, computer-supported collaborative learning (CSCL), classroom learning environment, and social network analysis (SNA). Her current research focuses on improving students' scientific argumentation through peer critique.

Jonathan Osborne

The Kamalachari Professor of Science Education, Graduate School of Education, Stanford University

Dr. Osborne was a co-editor of the report *Beyond 2000: Science Education for the Future*, and an advisor to the UK House of Commons Science and Technology Committee for their report on Science Education. He was President of the NARST (from 2006 to 2007) and has twice received the JRST Award for the best research publication in the *Journal of Research in Science Teaching* (2003 and 2004). He was a member of the U.S. National Academies Panel that produced the framework for the Next Generation Science Standards. Currently, he is Chair of the expert group that is producing the framework for the 2015 Organization for Economic Cooperation and Development (OECD) Program for International Student Assessment (PISA) science assessment and is a member of the expert group for designing the accompanying questionnaires. His research interests are in the role of argumentation in science and improving the teaching of literacy in science.

William R. Penuel

Professor, Educational Psychology and Learning Sciences Program, School of Education, University of Colorado at Boulder

Dr. Penuel's research focuses on teacher learning and organizational processes that shape the implementation of educational policies, school curricula, and afterschool programs. He examines learning and development from sociocultural, social capital, and complex social systems perspectives. One strand of his research focuses on designs for teacher professional development in Earth science education. A second strand examines the role of research-practice partnerships in designing supports for teacher learning in school districts. A third strand examines how children's interest in science develops over time and across different kinds of settings. This third strand includes a focus on young children's learning through digital media, including public television programs. He is currently on the editorial board for *Teachers College Record, American Journal of Evaluation,* and *Cognition and Instruction.*

Anthony J. Petrosino

Associate Professor, STEM Education and the Learning Sciences, Department of Curriculum and Instruction, College of Education, University of Texas at Austin

Petrosino's research focuses on children and teachers' scientific and mathematical reasoning, with an emphasis on activities and tools for developing thought. There are two major strands to this program. The first focuses on the creation and study of learning environments that foster the development and growth of expertise and inquiry in K-16 classrooms. The second focuses on investigating the opportunities for model-based reasoning that occur in typical STEM classrooms as students move conceptually from intuitive everyday understanding to more formalized scientific understanding. He has a strong interest in how computational technologies can scaffold learning in STEM fields, especially in supporting learner's access to and analysis of data. Petrosino is a founder of the nationally recognized UTeach pre-service teacher program.

Stacey Pigg

Assistant Professor, Department of Writing and Rhetoric, University of Central Florida

Dr. Pigg studies the impact of networked and mobile technologies on how we work, think, and learn. Her research has appeared in journals such as *Community Literacy Journal*, *Computers and Composition*, *Rhetoric Society Quarterly*, *Technical Communication Quarterly*, and *Written Communication*.

Kai A. Schafft

Associate Professor, Education and Rural Sociology, Department of Education Policy Studies, Children, Youth, and Families Consortium, College of Education, The Pennsylvania State University

Trained as a rural sociologist, Dr. Schafft's work broadly examines the relationship between the well-being of rural schools and the communities they serve. He is Director of the Center on Rural Education and Communities at Penn State. His current work focuses on the impacts of Marcellus Shale development in Pennsylvania and the ways in which schools may serve as a lens for assessing and understanding community change. Dr. Schafft's work has been published in *Social Problems*, *Rural Sociology*, *Teachers College Record*, *Adult Education Quarterly*, the *American Journal of Education*, *Agriculture and Human Values*, and *Society & Natural Resources*. His latest book, co-authored with David Brown, *Rural People and Communities in the Twenty-First Century*, was published in 2011. Dr. Schafft also edits the *Journal of Research in Rural Education*.

Armend Tahirsylaj

Dr. Tahirsylaj earned his PhD in educational theory and policy from The Pennsylvania State University in 2015. His research interests include international comparative education, K-12 national education policies, curriculum, and instruction, and developing students' creativity and innovation skills through education. Prior to joining Penn State, Dr. Tahirsylaj completed a master's degree in curriculum, teaching, and learning from the University of Calgary, Canada, taught at the University of Prishtina, Kosovo, and worked at Kosovo's Ministry of Education and numerous international educational organizations in Kosovo.

Nancy Tuana

DuPont/Class of 1949 Professor of Philosophy, Directorship of the Rock Ethics Institute, Department of Philosophy, College of the Liberal Arts, The Pennsylvania State University

Dr. Tuana's research and teaching specialties include philosophy of science, feminist science studies, moral literacy, and approaches to intersectionality. She is particularly interested in ethical issues in science and is doing work on coupled epistemic-ethical issues in climate science. She is the ethics lead for a new

National Science Foundation (NSF) research network, Sustainable Climate Risk Management Strategies (http://scrimhub.org), which focuses on addressing the question: "What are sustainable, scientifically sound, technologically feasible, economically efficient, and ethically defensible climate risk management strategies?"

Stephanie E. Vasko

Post-Doctoral Research Fellow, Department of Philosophy and AgBioResearch, Michigan State University

Dr. Vasko received her B.A. in chemistry from Carleton College and her Ph.D. in Chemistry and Nanotechnology from the University of Washington (2012). Her research interests include nanotechnology, STEM education, and the intersection and interplay between STEM, crafting, and the arts.

Annmarie R. Ward

Assistant Professor, Department of Curriculum and Instruction, Director of the Center for Science and the Schools (CSATS), College of Education, The Pennsylvania State University

As Director of CSATS, Dr. Ward works with Penn State scientist and engineering faculty, as well as school districts across Pennsylvania to build capacity for K-12 teachers to teach science using methods that incorporate the practices of science and engineering. She also engages other Penn State STEM outreach groups in activities designed to promote collaboration and enhance understanding of current STEM education issues. Her research interests include 1) developing strategies for enhancing K-12 teachers', and STEM undergraduate and graduate students' understanding of how scientists and engineers do research, and how to translate that knowledge into their teaching and 2) optimizing universities' STEM outreach efforts, allowing them to more strategically and meaningfully address the science education needs of students, K-12 teachers, higher education, and the community.

Rose Mary Zbiek

Professor and Chair, Department of Curriculum and Instruction, College of Education, The Pennsylvania State University

Complementing her work in mathematical modeling for the learning of curricular mathematics, Dr. Zbiek's current research blends a focus on the mathematical understandings of secondary mathematics teachers with a concentration in classroom use of mathematics technology to study relationships among mathematical processes, technological expertise, and classroom practice of beginning and experienced secondary school mathematics teachers. An immediate past-associate editor of the *Journal for Research in Mathematics Education*, Dr. Zbiek serves as a member of the writing team for *Guidelines for Assessment and Instruction in Mathematical Modeling Education*, a document co-sponsored by the Society for Industrial and Applied Mathematics and the Consortium for Mathematics and Its Applications. She is series editor of Essential Understanding, a 16-book series on mathematical and statistical content for teachers of mathematics in prekindergarten–grade 12 published by the National Council of Teachers of Mathematics. She currently serves as a member of the Board of Directors of the National Council of Teachers of Mathematics.

WATERBURY SUMMIT PARTICIPANTS

Welcome Address

Rodney Erickson, President, The Pennsylvania State University

Keynote Speakers

Shirley Malcom, Head of the Directorate for Education and Human Resources Programs, The American Association for the Advancement of Science
Richard Duschl, Kenneth B. Waterbury Chaired Professor in Secondary Education, Department of Curriculum and Instruction, College of Education, The Pennsylvania State University

Session 1: Systems Thinking

Papers

Thomas Litzinger, Professor of Mechanical Engineering, Director of the Leonhard Center for the Enhancement of Engineering Education, Department of Mechanical and Nuclear Engineering, College of Engineering, The Pennsylvania State University
James Minstrell, Director of Operations, Technology Project Manager, and Co-Founder, FACET Innovations

Respondents

Marcela Borge, Assistant Professor of Education, Department of Learning and Performance Systems, College of Education, The Pennsylvania State University

William Penuel, Professor of Education, Educational Psychology and Learning Sciences Program, School of Education, The University of Colorado, Boulder

Poster

Robert Crane, Professor of Geography, Director of the Alliance for Education, Science, Engineering and Development in Africa, Department of Geography, College of Earth and Mineral Sciences, The Pennsylvania State University
Carla Zembal-Saul, Professor of Education, Kahn Endowed Professorship, Department of Curriculum and Instruction, College of Education, The Pennsylvania State University
Mark Merritt, Instructor of Science Education, Department of Curriculum and Instruction, College of Education, The Pennsylvania State University

Panelists

Fred Fonseca, Associate Professor of Information Sciences and Technology, Co-Director of the Center for Online Innovation in Learning, College of Information Sciences and Technology, The Pennsylvania State University
Denice Wardrop, Senior Research Associate, Director of the Sustainability Institute, Associate Director of Riparia, Department of Geography, College of Earth and Mineral Sciences, The Pennsylvania State University
Stephanie Vasko, Senior Research Assistant, Rock Ethics Institute, College of Liberal Arts, The Pennsylvania State University

Session 2: Model-Based Reasoning

Papers

Spencer Benson, Director of the Center for Teaching and Learning Enhancement, Professor in the Faculty of Health Sciences, The University of Macau, SAR, PRC

Respondents

Hilda Borko, Professor of Education, School of Education, Stanford University
Brian Coppola, Arthur F. Thurnau Professor of Chemistry, Associate Chair of Educational Development and Practice, Department of Chemistry, College of Literature, Science, and the Arts, The University of Michigan

Poster

William Carlson, Professor of Education, Former Director of the Center for Science and the Schools, College of Education, The Pennsylvania State University

Panelists

Nancy Bunt, Former Program Director, Southwestern Pennsylvania's Math & Science Collaborative, Allegheny Intermediate Unit, Pittsburgh, PA
William Carlsen, Professor of Education, Department of Curriculum and Instruction, College of Education, The Pennsylvania State University
Mary Beth Rosson, Professor of Information Sciences and Technology, Interim Dean for College of Information Sciences and Technology, Co-Director of the Computer-Supported Collaboration and Learning Lab, The Pennsylvania State University

Session 3: Quantitative Reasoning

Papers

Heather Johnson, Assistant Professor of Mathematics Education, Mathematics & Science Program, School of Education and Human Development, The University of Colorado, Denver
Richard Lehrer, Frank W. Mayborn Professor of Education, Department of Teaching and Learning, The Vanderbilt University
Anthony Petrosino, Associate Professor of Science and Mathematics Education, Elizabeth G. Gibb Endowed Fellow, Department of Curriculum and Instruction, College of Education, The University of Texas at Austin

Respondents

Robert Mayes, Professor of Education, Department of Teaching and Learning, College of Education, Director of the Institute for Interdisciplinary STEM Education, Office of Research and Economic Development, Georgia Southern University
Rose Zbiek, Professor of Education, Mathematics Education, Department Head, Department of Curriculum and Instruction, College of Education, The Pennsylvania State University

Poster

Stephanie Knight, Professor of Education, Department of Educational Psychology, Counseling, and Special Education, Associate Dean for Undergraduate and Graduate Studies, College of Education, The Pennsylvania State University
X. Ben Wu, Professor of Ecology and holder of the John Kincaid Professorship for Undergraduate Teaching Excellence, Department of Ecosystem Science and Management, Texas A&M University

Panelists

Jonna Kulikowich, Professor of Education, Department of Educational Psychology, Counseling, and Special Education, College of Education, The Pennsylvania State University
Christopher Lazzaro, Director of Science Education, The College Board, New York, NY
Heather Toomey Zimmerman, Associate Professor of Education, Department of Learning and Performance Systems, College of Education, The Pennsylvania State University

Summary Comments and Open Discussion

Leona Schauble, Professor of Education, Department of Teaching and Learning, Vanderbilt University Peabody College

Session 4: Equity, Epistemic, and Ethical Outcomes

Papers

Kai Schafft, Associate Professor of Education and Rural Sociology, Director of the Center on Rural Education and Communities, Department of Education Policy Studies, Children, Youth, and Families Consortium, College of Education, The Pennsylvania State University
Catharine Biddle, Assistant Professor, University of Maine, Department of Higher Education, Educational Leadership and Early Childhood Education
Jonathan Osborne, Professor of Education, Kamalachari Professorship in Science Education, School of Education, Stanford University

Respondents

Nancy Brickhouse, Provost, Professor of Science Education, Saint Louis University
Nancy Tuana, DuPont/Class of 1949 Professor of Ethics, The Nancy Tuana Directorship of the Rock Ethics Institute, Department of Philosophy, College of the Liberal Arts, The Pennsylvania State University

Poster

Russell Graham, Professor of Geosciences, Director of the Earth and Mineral Sciences Museum & Art Gallery, Department of Geosciences, College of Earth and Mineral Sciences, The Pennsylvania State University

Carla Zembal-Saul, Professor of Education, Kahn Endowed Professorship, Department of Curriculum and Instruction, College of Education, The Pennsylvania State University
Elizabeth Hufnagel, Assistant Professor, Department of Instruction and Learning, School of Education, University of Pittsburgh
Mark Merritt, Instructor of Science Education, Department of Curriculum and Instruction, College of Education, The Pennsylvania State University

Panelists

Barbara Farmer, Senior Lecturer, Director of the Office of Multicultural Affairs, College of Information Sciences and Technology, The Pennsylvania State University
James Greeno, Adjunct Professor of Education, School of Education, University of Pittsburgh
Julia Plummer, Associate Professor of Education, Department of Curriculum and Instruction, College of Education, The Pennsylvania State University

Session 5: STEM Communication & Policy Outreach

Papers

Kirsten Ellenbogen, President and CEO, Great Lakes Science Center
Jean Moon, Founder and Principal, Tidemark Institute

Respondents

Edward Fuller, Associate Professor of Education, Department of Education Policy Studies, College of Education, The Pennsylvania State University
Drew Gitomer, Professor of Education, Rose and Nicholas DeMarzo Chair in Education, Graduate School of Education, The Rutgers University

Real Scientists Video

Tom Keiter, Partner, Storyshop, LLC
Joseph Myers, Partner, Storyshop, LLC

Panelists

David Bauman, State Science Education Advisor, Pennsylvania Department of Education
Mark McLaughlin, Director of Shaver's Creek Environmental Center, The Pennsylvania State University

Kyle Peck, Professor of Education, Co-Director of the Center for Online Innovation in Learning, Department of Learning and Performance Systems, College of Education, The Pennsylvania State University

Summary Comments and Open Discussion

Paul LeMahieu, Senior Vice President for Programs and Administration, The Carnegie Foundation for the Advancement of Teaching and Graduate Faculty, University of Hawai'i—Manoa

INDEX

Page references for tables are bold; page references for figures are italic.

CPSIA information can be obtained
at www.ICGtesting.com
Printed in the USA
BVHW09s2105080718
521007BV00008B/116/P

9 781138 901049